The Hydrogen Atom

The Hydrogen Atom

Proceedings of the Symposium,
Held in Pisa, Italy, June 30 – July 2, 1988

Editors: G. F. Bassani, M. Inguscio,
and T. W. Hänsch

With 157 Figures

Springer-Verlag Berlin Heidelberg New York
London Paris Tokyo

Professor G. Franco Bassani
Scuola Normale Superiore, Piazza dei Cavalieri, I-56100 Pisa, Italy

Professor Massimo Inguscio
Università di Napoli, Dipartimento di Scienze Fisiche,
Pad. 20 Mestra d'Oltremare, I-80125 Napoli, Italy

Professor Dr. Theodor W. Hänsch
Max-Planck-Institut für Quantenoptik, Ludwig-Prandtl-Straße 10,
Postfach 1513, D-8046 Garching, Fed. Rep. of Germany

ISBN 3-540-50579-2 Springer-Verlag Berlin Heidelberg New York
ISBN 0-387-50579-2 Springer-Verlag New York Berlin Heidelberg

Printed in the United States of America

2156/3150 - 543210 - Printed on acid-free paper

Preface

The idea of having a symposium on the hydrogen atom arose from our many discussions of the fact that the simplicity and generality of physics are often obscured by the endless sophistication of the theoretical analyses and the enormous variety of experimental techniques.

Nowadays we have periodic international conferences on atomic physics, on laser spectroscopy, and on quantum electronics, where all aspects of the theory of interaction of radiation with matter are discussed and new techniques and experimental results are presented. However, none of these conferences brings together and shows the connections between the major physical ideas on which the different branches of modern physics are based. It occurred to us that a good way to achieve this goal would be to organize a symposium around the simplest bound quantum system, the system which stimulated the birth of modern physics, the system in which effects associated with electro-magnetic interactions, quantum physics, and relativity can be measured with the highest accuracy, thus making possible the most rigorous tests of the theories of these effects. That system is, of course, the hydrogen atom.

The fact that the solutions of the Schrödinger and of the Dirac equations for hydrogen are given in textbooks as well as treatments of hyperfine splittings and quantum electrodynamic corrections, and the long history of these subjects, could make people think that this is an old-fashioned topic, not appropriate for a stimulating symposium. But as we went on to review the progress recently made in studies of the hydrogen atom, we found the opposite to be true. The exploration of hydrogen has led to the development of powerful new techniques and experimental tools, greatly advancing the frontiers of spectroscopy. The possibility of producing new hydrogenic atoms with all types of particles and antiparticles has opened up completely new fields, further extending the role of the hydrogen atom as an ideal testing ground for the theory of elementary particles and their interactions, and for the quantum theory of matter and radiation.

After that we invited people from all over the world, belonging to different scientific communities, to present the latest results of their investigations of hydrogenic systems. We were gratified by their enthusiastic response and by their participation. The symposium was held at the Scuola Normale in Pisa at the end of June, and we believe the participants became more convinced than ever of the unity of physics and of the key role that the hydrogen atom is playing and will continue to play in the future. We wish to thank all the participants, and in particular F.T. Arecchi, R. Barbieri, B. Cagnac, L. Foà, J. Gay, A. Gozzini,

H. Layer, G. zu Putlitz, L.A. Radicati, and F. Strumia, who agreed to chair the various sessions. We are also grateful to the sponsoring institutions listed before this preface.

In organizing the material for these proceedings, we decided to subdivide it into four chapters, according to the most recent developments in the exploration of hydrogenic systems.

In the first chapter we have included a number of subjects which originate from the enormous improvements in spectroscopic techniques, such as those involving two-photon spectroscopy, atom trapping and Rydberg state transitions. Values of universal constants of extreme accuracy, as well as of hyperfine splittings and of radiative corrections have been obtained by these methods.

In the second chapter we consider more exotic two-particle systems: the hydrogen-like ions with large relativistic effects, positronium and muonium with their large hyperfine splittings, antihydrogen, and a new atom formed by a proton and an antiproton – with excitation spectrum in the X-ray region.

In the third chapter we present the contributions dealing with quantum electrodynamics corrections, including the weak interaction effects which lead to parity non-conserving transitions. An original presentation of the field by C. Cohen-Tannoudji has not been reported here, because it appears in the new book on the interaction of photons and atoms by him, J. Dupont-Roc, and G. Grynberg.

Finally, in the last chapter, we include all papers dealing with transitions in strong fields to discrete and continuum states, where chaotic behavior becomes relevant.

Though many of the above described phenomena are treated in a number of books, and some of them in the new book on the spectrum of atomic hydrogen edited by G.W. Series, we hope that our effort to collect in one book information on as many as possible of the fascinating aspects of the physics of the hydrogen atom will be found useful.

Pisa, Italy
January 1989

F. Bassani
T.W. Hänsch
M. Inguscio

Contents

Part I Precision Spectroscopy of Hydrogen: Fundamental Physics and Universal Constants

Part II Positronium, Muonium, and Other Hydrogen-Like Systems

Part III Quantum Electrodynamics and Beyond

Part IV Hydrogen in Strong Fields and Chaos

Part I

Precision Spectroscopy of Hydrogen: Fundamental Physics and Universal Constants

The Hydrogen Atom (An Historical Account of Studies of Its Spectrum)

G.W. Series

Clarendon Laboratory, Parks Road, Oxford, OX1 3PU, U.K.

1. Introduction

I was honoured and pleased by the invitation from Professor Bassani to give this introductory talk. But I must confess that as I pondered upon the task I had undertaken, my heart began to sink. I have lived through the last forty years of the story, but the active history of the subject takes us back another hundred and forty years, so I went to that great source-book, KAYSER's Handbuch der Spectroscopie [1] and that is where most of my early references come from.

Kayser's Handbuch lists over 700 publications on the spectrum up to 1927 - sixty years ago, just the time when quantum mechanics had gripped the imagination of the rising generation of mathematically-minded physicists. And since most of you will be more or less familiar with the happenings of those famous days, I shall mention, but shall not dwell on them. I shall suppose, moreover, that you are very well informed about the Lamb shift and QED, so my story will end, not abruptly, but roughly, where Willis Lamb came in. Even so, my task is to summarise, shall we say, 1500 - 2000 papers in forty minutes, so you must excuse me if I do less than justice to the work of some of our predecessors.

2. That the Spectrum exists and betokens Hydrogen

1672. To each colour belongs a definite refrangibility. That had first to be proved, and it was NEWTON who proved it [2]. Until FRAUNHOFER, a hundred and fifty years later [3], made a diffraction grating, what spectroscopy there was rested upon the dispersion of light by glass prisms.

Newton's spectrum of the sun showed the colours spread out in a continuous band. 1802 WOLLASTON [4] introduced a slit about 1 mm broad and viewed the slit through a prism. He saw the solar spectrum crossed by a number (a small number) of dark bands. It is difficult to be certain what bands these were, but it seems likely that one of them - in the yellow - might be attributed today to sodium. This, then, would have been the first time that the sodium D lines had claimed the attention of an observer.

1815. Over ten years later Joseph von Fraunhofer, still under 30, introduced a lens from a theodolite to follow a prism and saw, for the first time, those dark lines

The Hydrogen Atom Editors: G.F. Bassani · M. Inguscio · T.W. Hänsch

crossing the solar spectrum which, in perpetuity, will bear his name. Fraunhofer had been studying *emission* spectra from a variety of light sources and lamps. He found a bright yellow line coming from everything he tried – what spectroscopist doesn't? His resolution, of course, was far better than Wollaston's had been. And then 1817 he turned his attention to sunlight, wondering whether this also would produce bright lines. He was taken completely by surprise to find here *dark* lines embedded in colour. He counted over 700 of them [5]. Their relative positions remained unchanged as he turned the prism or opened the slit. The line C (which we now call Balmer-α) was of considerable strength, he noted. F (that is, Balmer-β) was fairly strong. The D lines consisted of two dark lines separated by one bright line. He studied the light from heavenly bodies: Venus, in whose spectra the D and F lines were recognised, and Sirius, which displayed three bright lines, having no similarity to sunlight.

These observations of the C and F lines are surely the true beginning of our story. But to label spectral lines, and even to recognise them when you see them again falls far short of associating them with some particular atom or molecule; and we must not forget that a dark line on a bright background was not, at that time, necessarily to be associated with a bright line on a dark background seen in the same place in the spectrum. Key points in what follows are to see how our predecessors were led to the emission/absorption relationship, and to bring hydrogen into the story.

1821. In this year FRAUNHOFER invented the diffraction grating [6]. He was a superb engraver and was able to divide a circle with the utmost delicacy and accuracy. That is why, when he wound wire on fine screw threads and produced spectra whose angular deviations were typically of the order of one degree of arc, he was able to obtain values of wavelength consistent with one another to a few parts in a thousand. For example, he quotes the wavelengths of the C and F lines as 2.422×10^{-5} and 1.794×10^{-5} 'Pariser Zoll', respectively. The ratio is 1.350. The ratios of 6.563×10^{-5}, 4.861×10^{-5} cm, accepted these days as the wavelengths of Balmer-α and Balmer-β, is 1.350. (My conversion of Zoll to centimetres yields the factor 2.7070; a 'Zoll' was rather larger than an inch.)

And so we begin the era of measurement in spectroscopy. Ångstrom, a few years later, worked in units of 10^{-7} mm. In referring to Fraunhofer's measurements he used the term 'pouce' where Fraunhofer used 'Zoll'. I mention these points to recall to your attention the great importance of having an agreed system of units of measurement. The history of measurement of characteristic features of the hydrogen spectrum has been, and still is, intimately connected with the establishment and maintenance of a system of base units, as many of you well know.

So, then, Fraunhofer observed and labelled characteristic and recognisable dark lines in the solar spectrum and gave us the wavelengths of many of them. But what were these lines and why were they there?

J.F.W. HERSCHEL, in 1831, published a book Vom Licht [7] in which he describes studies of the absorption spectra of dyes and the emission spectra of flames and found that many substances give a characteristic spectrum, but he did not go so far as to say that a particular spectrum was characteristic of a particular substance.

D. BREWSTER [8] in 1836 gave the true explanation of the Fraunhofer lines, the absorption of light by gases in the outer layers of the sun, superimposed on a continuous spectrum emitted by the material in inner regions.

In 1840 the first photograph of the solar spectrum was published by HERSCHEL [9] - and in 1842 E. BECQUEREL [10] published also a Daguerreplate of the solar spectrum showing beautifully the Fraunhofer lines, but it was the eighteen seventies before the gelatin-silver bromide dry plates became available.

Again, in this period, it was controversial whether the Fraunhofer lines were a property of solar light or whether they were produced by the apparatus; and further, if not by the apparatus, then by absorption in the sun or in the earth's atmosphere? Indeed, some of the lines are so produced, and were identified as such.

1851 is a big date in our time-chart, for it is the date of a publication by A. MASSON [11] when a red line, observed by him in emission from a spark in air using electrodes of different metals, was observed to be present for most of the electrodes he used. And when the air was replaced by hydrogen gas, there was the same red line, 'très prononcée'. He even labelled it α. *So, for the first time, that line which surely is the best-known and most studied line in the spectra of the elements, was associated with hydrogen.* By 1855, MASSON had attributed three more lines to the presence of hydrogen [12].

A.J. ÅNGSTROM in 1855 reports a study of the spark spectra of many elements, including hydrogen [13]. 'Remarkable, in the case of hydrogen', he says, 'are the strongly luminous and wide lines at the red end of the spectrum . . .'. And earlier he has written, *'When the solar spectrum is compared with the electric one it is found that some of the lines, such as C, D, E, G (C is Balmer-α) have thin corresponding lines in the solar spectrum'*. In the same paper he clarifies the relationship between *absorption* and *scattering*. He makes the connection between the oscillation of particles in molecules and the oscillations of the aether for absorption lines, and he invokes a principle of Euler to declare that 'the rays which a body absorbs are precisely those it can emit when luminous'. This comes very near to the mathematical connection between 'powers of emission' and 'powers of absorption' which KIRCHHOFF [14] was to formulate some years later, 1859/60, and which, says Kirchhoff, provides us with an 'unexpected' explanation of the origin of the Fraunhofer lines.

ÅNGSTROM finally drives home the nail in 1862 [15]. He recapitulates the results of his earlier works, in particular, that they had demonstrated that the spectra of mixtures or compounds presented the same lines as the constituent substances. He then makes a correspondence between lines in the emission spectra of a number of elements taken one at a time with particular groups of Fraunhofer lines identified by their distinguishing letters, remarking that the Fraunhofer lines are the 'inversion' of the bright lines seen in electrical spectra, and eventually he turns to hydrogen:

'assume', he says, 'that the C–line belongs to hydrogen . . . '
and he continues, speaking of the relationship between absorption and emission and adducing evidence concerning the occurrence of two other lines in the electrical spectrum of hydrogen, to the conclusion that *the solar C and F lines do indeed belong to hydrogen.*

So now we are home and dry. Hydrogen manifests a line spectrum. The wavelengths of the first four lines have been measured to a few parts in 10^3. Lines attributable to hydrogen have been identified in absorption in the solar spectrum.

A few years later ÅNGSTROM published an atlas of the solar spectrum [16] and gave accurate values of the wavelengths of the lines: these stood for many years as standards for the determination of wavelengths. ROWLAND's famous atlas [17] followed in 1882.

We now, 1876, 1880, turn briefly to the investigations of HUGGINS [18] [19]. In the first paper he published a photographic spectrum of the star Vega (α–Lyrae) exhibiting a beautiful set of Balmer lines, with a solar spectrum for comparison, showing coincidences. In the second paper is a list of the wavelengths of 14 lines, of which 12 are recorded as 'very strong'. 'There is a high probability that these lines present the spectrum of hydrogen'. But only the first three of these lines can be identified confidently with lines in the solar spectrum.

While this study is important in its own right in the context of astrophysics, nevertheless its greater importance for us is that these were the measurements upon which Balmer, a few years later, was able to try out a formula he had deduced on the basis of the wavelengths of the first four lines only. In passing we notice also a paper by H.W. VOGEL [20] whose determinations of wavelengths of the 5th, 6th, 7th and 8th members of the series agree closely with those of Huggins. And furthermore, at a later stage in his work, Huggins 'was astonished to find' that, in the spectrum of Sirius, the hydrogen lines were all jerked a little to the side of the place they occupied in the spectra of his laboratory sources . . . !

3. The Centenary

I thought you might be amused to hear what was happening a hundred years ago. A paper was written in 1888, published in 1889 [21], by J. TROWBRIDGE and W.C. SABINE, reporting their work from the Jefferson Physical Laboratory, Cambridge, U.S.A. 'On the use of steam in spectrum analysis'. Light from an arc between carbon poles, they point out, gives spectra full of impurities from the carbons. The spark from a Ruhmkorff coil gives clean spectra, but the light is very feeble and needs long photographic exposure times. They had the idea of directing a jet of steam upon the spark, and were rewarded by a spectacular improvement in the brilliance of the spectra. The moral for us is clear: if your laser needs tweaking, try steam. But take a hint from Trowbridge and Sabine. Shoot the steam into the spot the light is coming from, not into the power supplies.

4. Playing with Numbers

It is more than a hundred years since JOHANN JAKOB BALMER, a schoolteacher and privatdocent in Basle, a man coming up to 60 years old, gave us, in 1885, the first clue to the understanding of the hydrogen spectrum [22].

But that wasn't the first published report of the mathematical relationships between numerical values of the wavelengths. Reported in some text books is STONEY's result [23] that the first, second and fourth lines of hydrogen can be represented as the 20th, 27th and 32nd harmonics of a fundamental vibration having a wavelength of 131,274.14 Å. A further attempt by Stoney is reported in a footnote to Huggins' paper [19]. This search for harmonics is evidence of the belief that was held that the properties of the light were direct reflections of anharmonic motions of particles in the emitting bodies. Ten years later Stoney's result was shown to be no more than coincidental by SCHUSTER [24]. A lesser known formula is that of D'ARREST: $\lambda_\gamma^3 = \lambda_\beta . \lambda_\delta^2$ [25], which we now understand to derive from the numerical result $27(21)^3 \approx (500)^2$.

How, then, did Balmer arrive at the formula we know so well? He first observes that the wavelengths of the first four lines bear the relations 9/5; 4/3; 25/21; 9/8 to one another, and that the common factor $h = 3645.6$ mm/10^7 converts these numbers to the wavelengths. Multiplying then the second and fourth fractions by 4, one arrives at 9/5, 16/12, 25/21, 36/32, of which the numerators are 3^2, 4^2, 5^2, 6^2. Representing these numbers by m^2, the fractions are given by $m^2/(m^2-n^2)$, where n, for this set, is 2.

Balmer's derivation of his formula is made on the basis of these four lines only. Professor Hagenbach, who has encouraged him in the project, now introduces him to Huggins' determinations of wavelengths from white stars. With ten more lines which fit his formula, the matter is clinched. (True, there are more or less progressive differences between observed and calculated wavelengths as one proceeds along the series, but these can throw no serious doubt on the underlying validity of the formula.) Of Stoney's attempt to interpret the numbers as harmonics of a fundamental, Balmer concludes politely, 'interesting as this attempt is, I am doubtful whether it really demonstrates an inner connection between the phenomena'.

Let experimentalists and empiricists gain strength from this work: harmonic oscillations and their overtones were surely at the heart of theoretical physics in the nineteenth century. We should not lose sight of the courage, and of the confidence in the accuracy of measurement, which were required of an elderly schoolmaster at that time, to fly in the face of received wisdom.

5. Fine Structure Observed

How well known is it, I wonder, that only two years after Balmer's publication, and twenty six years before Bohr's interpretation of it, MICHELSON and MORLEY, in 1887, showed that Balmer-α consisted of two lines [26]. In this research the prime objective was to investigate the possibility of making the wavelength of sodium light the actual standard of length. They first used an etalon of the Fabry-Perot type, then the interferometer. After the sodium yellow lines they studied the red hydrogen line. 'The fringes disappeared at about 15,000 wavelengths and again at about 45,000 wavelengths, so that the red hydrogen line must be a double line with the components about 1/60th as distant as the sodium lines.' The recognition of Balmer-α, then, as twinned, was one of the first fruits of Fourier transform spectroscopy!

In 1904 J. BARNES at Johns Hopkins finds three components in Balmer-α [27]. But the centre line was much stronger than the outer lines! No, this was not the fine structure we know today. Experienced spectroscopists will already have guessed the interpretation that was to come later: self-reversal in the light source. So now we begin to realise the need for restrained excitation. From now on, self-absorption, self-reversal and Stark broadening emerge as diagnostics for physical conditions in light sources. To study hydrogen itself we shall use Geissler tubes at low pressures, and we shall begin the struggle against the Doppler enemy.

6. Systematics and Spectral Series

Time is getting on and there is much more to tell, but there are one or two landmarks I must mention before theory begins to assert itself. So far, you will notice, the story has been of experiment guided by curiosity and accident. A great achievement which cannot be ignored is RYDBERG's great work of 1889 [28] a monumental assessment of series spectra, the recognition that reciprocals of wavelengths, wavenumbers, make more sense than wavelengths; that wavenumbers of lines may best be represented by differences of terms; the building of the ordinal number of a line in a sequence into a formula for the term values; the existence of a universal constant; and finally, the recognition that Balmer's formula was a very special case of his own. And then, closely related to Rydberg's work, came RITZ's Combination Principle in 1908 [29].

On the experimental side came the discovery of other spectral series in hydrogen which beautifully fit Balmer's formula. 1908 - Paschen's series ($m = 3$) in the infrared, followed by Brackett's ($m = 4$) in 1922, Pfund's ($m = 5$) in 1924 - these are mentioned in all the text-books, but you may supplement them by Humphrey's series ($m = 6$) in 1953. Further into the infrared, in more recent times GRIEM *et al.* in 1973 [30] observed 11α, 12α, 13α at about 100 μ, and in the microwave region the line 109α at 5 GHz has made its appearance as a maser in the sky [31]. The series in the ultraviolet discovered by LYMAN in 1916 ($m = 1$) [32] has a longer history in astrophysics and is no less important in the laboratory. Many of you have already devoted years of your lives to enhancing the spectral purity of Lyman-α. And other series of lines were found, before the end of the last century, lines which appeared in astrophysical but not in laboratory sources. PICKERING, in 1896, discovered these first in stellar spectra [33]. Later, FOWLER found a related line at 4686 Å in the sun's chromosphere [34]. The common feature of all these lines was that they fitted Balmer's formula if *half-integral* quantum numbers were permitted. They were at first attributed to hydrogen, but that was wrong. BOHR, in 1913, [35] gave the correct interpretation, that they belonged to an element having $Z = 2$. Glancing ahead to Bohr's formula $E = -R_H Z^2/n^2$, we write this as $-R_H/(n/Z)^2$ and identify the $\frac{1}{2}$ as $1/Z$. This interpretation was at first doubted, but two further observations settled the matter. One was the refinement of his theory by Bohr in admitting finite nuclear mass M: the replacement of electron mass m_0 by reduced mass $m_0M/(m_0+M)$ accounted precisely for small differences in wavelength - which had been noted experimentally - between the hydrogen lines and the Pickering lines; the other was the observation [36] of the line 4686 Å in a discharge which showed no trace of the hydrogen lines.

7. At last, theory: the Old Quantum Theory

Not always emphasised in discussions of Bohr's theory (1913) is its success in accounting for the *stability* of atoms. We should be in trouble, should we not, if there were no stationary states. Right at the beginning of his famous paper [37] BOHR discusses this question of stability: Thomson's model (which is to be discarded) is actually superior to Rutherford's (until Bohr quantizes it) because it allows certain configurations and motions of electrons for which the system is in stable equilibrium.

And observe that Bohr introduces h, not by going to angular momentum, but by way of Planck's quanta of energy and the relation between the harmonic vibrations of an atom and the frequency of the light it emits. The binding energy of the electron is emitted as n quanta of radiation of frequency ν (this is a preliminary postulate, later he talks of single quanta at the harmonic frequency), which frequency is set equal to the average of the orbital frequencies of the electron at infinity, zero, and when bound in a stationary state, ω. So $W = nh\nu = \frac{1}{2}nh\omega$. Now use classical dynamics to relate ω to W and you arrive at the famous equation $W = -\text{const}/n^2$, with angular momentum $= h/2\pi$ as an incidental (though profoundly significant) result. The constant is that combination of e, h, m_0, c which we now call the Rydberg constant.

8. Fine Structure Explained

But no fine structure - yet - until in 1915 Bohr considered the effect of relativistic variation of mass with velocity in elliptical orbits under the inverse square law of binding, and pointed out that the consequential precessional motion of the ellipses would introduce new periodicities into the motion of the electron, whose consequences would be satellite lines in the spectra. The details of the dynamics were worked out independently by SOMMERFELD [38] and WILSON [39] in 1915/16 based on a generalisation of Bohr's quantization, namely, the quantization of action: the values of the phase integrals $J_i = \oint p_i.dq_i$ of classical mechanics should be constrained to assume only integral multiples of h.

To each degree of freedom we now have an independent quantum number, of which one, the space quantum number, has no effect on the energies of the system in free space. The relativistic energy is now

$$E_{rel} = m_o c^2 \left\{ 1 + \frac{\alpha^2 Z^2}{\left[n' + \sqrt{k^2 - \alpha^2 Z^2}\right]^2} \right\}^{-\frac{1}{2}}$$

with $n' = 0, 1 \ldots, k = 1, 2 \ldots$..

The energy levels generated by this formula are those you are all entirely familiar with. They are the Dirac energy levels. I need hardly say that α is the fine structure constant, now written as $e^2/\hbar c[\mu_0 c^2/4\pi]$, that you will recognise k as $(j+\frac{1}{2})$ and discover that $(k+n')$ has the same values as our present integer n, which is Bohr's n.

We all know that Sommerfeld's action integrals are no longer taken as the basis of quantization, and we may recall evidence against them from specific heats and from molecular spectra, but, you may say, if they lead to Dirac's energy levels they have a good claim on our attention, haven't they? Did we *have* to give them up?

Yes, we did. They didn't offer a satisfactory basis for predicting the relative *intensities* of spectral lines, and insofar as intensities could be predicted, the predictions were not supported by experiment.

But experiments to resolve the fine structure of the Balmer lines were difficult: as you all know, resolution was impeded by the Doppler broadening of components. So ionized helium comes into the picture, because, as Sommerfeld's formula predicted, fine structure intervals are a function of $(\alpha Z)^2$, so in helium they are of order four times as wide as in hydrogen and one has more chance of resolving the Doppler-broadened lines. So PASCHEN [40], in 1916, undertook an extensive study of the He^+ lines and in particular, 4686 Å ($n = 4 \rightarrow 3$). Fine structure, indeed, was found and matched against Sommerfeld's formula. The measurements were used to determine a value of α. But the structure did not really match the theory in that the quantum numbers bore no imprint of electron spin, so even the orbital properties - which dominated the intensity rules based on a correspondence with classical radiation theory - were wrongly associated with components, and the value of α derived from this first study was later abandoned.

It may surprise you to know that, right up into the nineteen fifties, experimental determinations of the fine structure constant were based upon measurements of spin doublet intervals in X-ray spectra, that is to say, effectively the $2P_{1/2} - 2P_{3/2}$ interval belonging to the hole in the L-shell. But the interpretation at this time rested upon the new form of quantum mechanics, and perhaps more important than the new mathematics, a piece of physics which had first to be discovered: *the spin and magnetic properties of the electron.*

9. The New Quantum Theory

The story is too well known to call for much elaboration here. We simply recall some names and important results.

1925/6 Schrödinger, Heisenberg. Non-relativistic quantum mechanics. Wave equation yields Bohr energy levels.

1928 Goudsmit and Uhlenbeck: electron spin. Thomas: spin–orbit energy. Heisenberg and Jordan: relativistic correction to energy. Net result: recovery of Sommerfeld energy levels, different quantum numbers.

1928. Dirac theory applied to hydrogen. Closed form of solution for energy levels first obtained by TEMPLE [41], 1930.

It is now appropriate to look at an actual term diagram: we choose Balmer-α, figure 1.

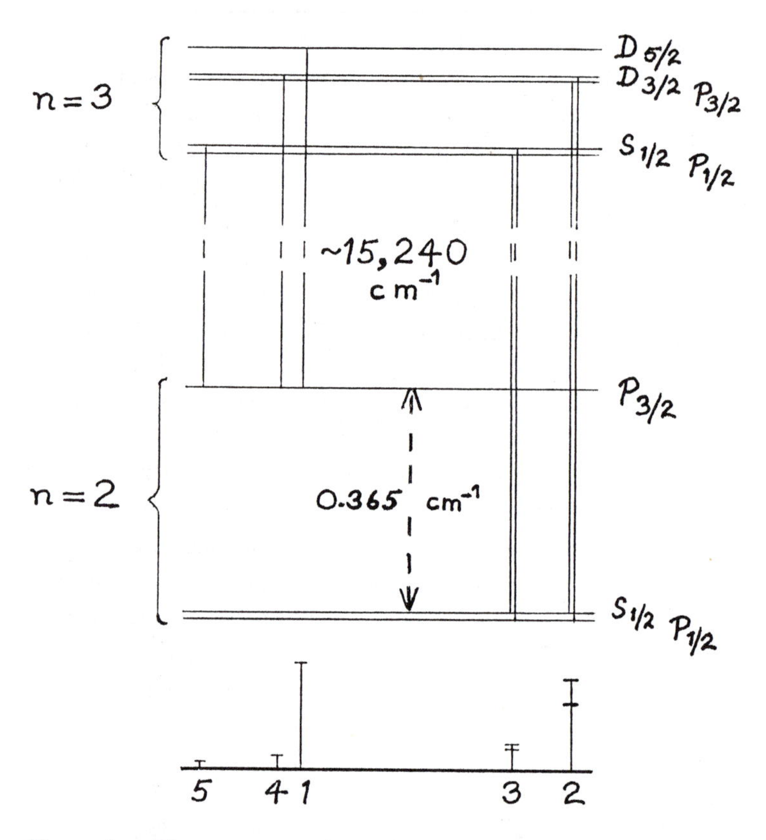

Figure 1 Fine structure of Balmer-α according to the Dirac theory.

We notice the following points:

(i) the interval 0.365 cm^{-1} in $n = 2$ is the spin–orbit interval $P_{1/2} - P_{3/2}$. This is essentially the doublet interval first observed and measured by Michelson, and is characteristic of all the Balmer lines.

(ii) levels of the same j but different l (e.g. $S_{1/2}$, $P_{1/2}$) are degenerate. This is certainly the prediction of theory in that the energy eigenvalues are functions of

(n,j), not of l. But the implications of this and the evidence from experiment was the occasion of profound and protracted controversy in the nineteen thirties: a controversy decidedly settled by the famous experiment of Lamb and Retherford published in 1947.

(iii) the fine structure is predicted to consist of seven components of which two pairs are degenerate.

10. Experimental Studies of Fine Structure

With improvements in spectroscopic techniques and particularly with the recognition that Doppler broadening was the obstacle to resolution of the fine structure, efforts were concentrated on improving light sources: spectroscopic resolving power was no problem. Resolution was improved by running discharges at low current, by cooling the discharge tubes, and above all by the use of deuterium instead of hydrogen: a factor 1.4 in the reduction of Doppler broadening was thereby gained painlessly. Results of wavelength determinations were used (i) to determine the Rydberg constant (with reliance on the validity of Dirac's energy levels), (ii) to check Dirac's theory (by concentrating on the fine structure), and (iii) to obtain a value for e/m_o by combining the Faraday constant with values of m_o/M_p derived from measurements of the mass-dependent isotopic shift between the lines of hydrogen and deuterium (about 2 Å for Balmer-α).

Experimental recordings of Balmer-α showed three components: Michelson's doublet - (5,4,1) of fig. 1 appeared as one strong line, 2 as a somewhat weaker line, and 3, weak and barely resolved even in the best work, as the third. Attention focussed on the position of 3. In 1938 the most distinguished schools of spectroscopists were divided as to whether it was, or was not, in the position predicted by Dirac's theory. R.C. WILLIAMS [42] asserted that it was not so, and PASTERNACK [43] pointed out that William's observations could be explained on the assumption of an upward shift of the $2S_{1/2}$ level by 0.03 cm^{-1}. And that is how it is.

11. Interaction with Radiation

It is remarkable how enormously successful were the theories of Bohr, Sommerfeld, Schrödinger and Dirac, of an atom in an empty universe. Agreement with experiment is secured before the admission of QED at a level better than 1 part in 10^6 of the total energy. This is an accuracy surpassing all ordinary needs of measurement and statistics. What millionaire knows his fortune to within one dollar, what psephologist can interpret a swing in votes, even of 1%? But physicists know that danger and divergences are hidden in the seemingly-innocuous empty universe, and the vacuum asserted itself powerfully in DIRAC's theory of radiation [44]. The interaction energy between a charged particle and even the zero-point radiation field, which is zero in first order of a perturbation expansion, diverges in second and higher orders. This is not the occasion to introduce QED; I wish merely to emphasise that theorists had been worried about divergences for twenty years before BETHE, in 1947, produced the first quantitative resolution of the problem [45]. An excellent account of that period has been given by WEISSKOPF [46].

I take it that the experimental work of those days is known to you: the epoch-making determination by LAMB and RETHERFORD, in 1947, of the finite, non-zero value of the $2S_{1/2} - 2P_{1/2}$ interval in hydrogen [47] and the subsequent determination of this interval, of order 1000 MHz, in hydrogen and in deuterium, to an accuracy of 0.1 MHz. Notice how this experiment marks a complete break with classical, optical spectroscopy. Atomic beams rather than gas discharges; stimulated resonances as against the spontaneous decay of excited atoms; the determination of fine structure intervals by direct measurement rather than as the difference between two gross structure lines of enormously greater frequency. Notice also, since Doppler broadening is proportional to frequency, that the microwave spectra are effectively free of it - even when, as some studies have contrived, the microwave resonances are elicited from hydrogen atoms moving randomly in a gas discharge.

Before I turn from fine to hyperfine structure, you will not deny me, I trust, the pleasure of describing my own contribution to the story. In 1947 my thesis supervisor, Heini Kuhn, invited me to 'have another go' at Balmer-α by classical, interferometric spectroscopy, hoping to gain improved resolution and firm confirmation of the Lamb shift by the use of liquid hydrogen as coolant for a gas discharge. The experiment was successful - though the accuracy of our measurement fell short of Lamb's by two orders of magnitude. However, I can claim, in addition to having secured good resolution of component 3, the first observation of component 5, from which it was possible to measure for the first time the displacement of $3S_{1/2}$ from the Dirac position. I went on from this to study the 4686 Å line of He^+, and here I was able to resolve a pair of components ($3S_{1/2}$, $P_{1/2}$ - $4P_{1/2}$, $S_{1/2}$) which showed the Lamb shift, not as the *displacement* of a component, but as a *splitting* into two distinguishable components of a pair which, on the basis of Dirac's theory, would have been strictly superimposed, one on top of the other.

This work, in its turn, was superseded in the late nineteen fifties and sixties before tunable lasers changed the whole complexion of optical spectroscopy. An account of this period is given in the recently published text: The Spectrum of Atomic Hydrogen: Advances [48].

12. Hyperfine Structure and the g-Factor of the Electron

Barely had spectroscopists absorbed news of the Lamb shift when another bombshell burst: the g-factor of the electron was not exactly 2. You, who (for the most part) have grown up knowing this will find it difficult to realise how shattering it was to see that beautiful symmetry destroyed. The news came also from Columbia in 1947: NAFE, NELSON and RABI [49] discovered that the hfs of the ground state of hydrogen was 1 part in 10^3 *larger* than its value evaluated from the Fermi formula using $g_e = 2$. There followed a succession of experiments (also described in [48]) whereby the numerical value of this important interval was progressively sharpened by new techniques in microwave spectroscopy until we arrived at the hydrogen maser - a story in its own right.

To the *theoretical* interpretation of the 1420 MHz value belong also puzzles and controversies centering round the value of α, the fine structure constant. The value obtained from hyperfine structure differed quite significantly from the value obtained from fine structure. That particular controversy is now dead: it was settled in favour of the hyperfine value by determinations of $2e/h$ based on the Josephson effect, and of the gradual building up of confidence in those values. Buttressing these values nowadays are determinations of e^2/h from the quantized Hall effect, and most especially, values of α itself obtained by comparing highly accurate determinations of $g-2$ for the free electron with exceedingly far-reaching evaluations of QED formulae.

That brings us to the nineteen eighties, our feet now firmly in the laboratory. But we began with the sky, and let us finish there. In 1945 VAN DE HULST [50] predicted that hydrogen in celestial bodies would make itself felt through radiation at a wavelength of 21 cm, 1420 MHz, corresponding to the ground state hyperfine interval. It was observed in 1957 by EWAN and PURCELL [51] and independently by MULLER and OORT [52]. You will be well aware of the importance of the 21 cm line, the signature of atomic hydrogen, in mapping the sky.

Which of Balmer-α, the 21 cm line, or - nowadays, when we can get above the atmosphere if we have money enough - Lyman-α is the more important line for astrophysicists and cosmologists it would be unprofitable to speculate. But certain it is that our study, the spectrum of atomic hydrogen, takes us deeply into theory and far into the universe, and gives us an inexhaustible mine for employment of the sharpest tools which experimental physics can provide.

References

1. H. Kayser: Handbuch der Spectroscopie, Vols 1-7 (Leipzig, 1900-1930)
2. I. Newton: Phil. Trans. Roy. Soc. Lond. 6, 3075 (1672)
3. J. von Fraunhofer: Gesammelte Schriften (König. Bayer. Akad. Wiss., München, 1888)
4. W.H. Wollaston: Phil. Trans. Roy. Soc. Lond. 92, 365 (1802)
5. J. von Fraunhofer: Gilbert's Ann. (Ann. der Physik) 56, 264 (1817)
6. J. von Fraunhofer: Denkschr. d. k. Akad. d. Wissensch. zu München 8, 1 (1821); Gilbert's Ann. 74, 337 (1823)
7. J.W.F. Herschel: Vom Licht, Engl. trans. J.C.E. Schmidt (Stuttgart, 1831)

8. D. Brewster: Phil. Mag. 8, 384; Pogg. Ann. 38, 50 (1836)
9. J.F.W. Herschel: Phil. Trans. Roy. Soc. Lond. 1, 1 (1840)
10. E. Becquerel: Biblioth. Univ. de Genève 40, 341 (1842)
11. A. Masson: Ann. Chim. et Phys. 31, 295 (1851)
12. A. Masson: Ann. Chim. et Phys. 45, 385 (1855)
13. A.J. Ångstrom: Pogg. Ann. 94, 141 (1855)
14. G. Kirchhoff: Monatsber. Berl. Akad. Wiss., 622 (1859); Pogg. Ann. 109, 148, 275 (1860); Ann. Chim. et Phys. 58, 254 (1860); 59, 124 (1860)
15. A.J. Ångstrom: Pogg. Ann. Phys. u. Chem. 117, 290 (1882)
16. A.J. Ångstrom: Recherches sur le Spectra Solaire (Uppsala, 1868)
17. H.A. Rowland: Atlas of the Solar Spectrum, Johns Hopkins Univ. Circ. 17 (1882); Phil. Mag. 13, 469 (1882); Nature 26, 211 (1882).
18. W. Huggins: Proc. Roy. Soc. 25, 445 (1876)
19. W. Huggins: Phil. Trans. Roy. Soc. Lond. 171, 669 (1880)
20. H.W. Vogel: Monatsber. d. K. Acad. der Wiss. zu Berlin, July 10 (1879)
21. J. Trowbridge and W.C. Sabine: Phil. Mag. 27, 139 (1889)
22. J.J. Balmer: Pogg. Ann. d. Phys. u. Chem. 25, 80 (1885)
23. G.J. Stoney: Phil. Mag. 41, 291 (1871)
24. A. Schuster: Proc. Roy. Soc. 31, 337 (1881)
25. H.L. d'Arrest: (1871) see article by K. Lundmark in Proc. Rydberg Cent. Conf. Lund 1955; Lunds Univ. Årsskrift, N.F. Avd. 2, 50, 103 (1955)
26. A.A. Michelson and E.W. Morley: Phil. Mag. (5) 24, 463 (1887)
27. J. Barnes: Phil. Mag. (6) 7, 485 (1904)
28. J.R. Rydberg: K. svenska Vetensk. Akad. Handl. 23, No. 11 (1889)
29. W. Ritz: Gesammelte Werke, 162 (Schweiz. Physikal. Gesellschaft 1908)
30. H. Griem *et al.*: Phys. Rev. Lett. 30, 944 (1973)
31. P.G. Mezger: in Phys. of the One- and Two-Electron Atoms, eds. F. Bopp and H. Kleinpoppen, 801 (North-Holland, 1969)
32. T. Lyman: Phys. Rev. 3, 504 (1914)
33. E.C. Pickering: Astrophys. J. 4, 369 (1896)
34. A. Fowler: Phil. Trans. Roy. Soc. Lond. 197, 202 (1901)
35. N. Bohr: Phil. Mag. 26, 476 (1913)
36. E.J. Evans: Nature 92, 5 (1913)
37. N. Bohr: Phil. Mag. 26, 1 (1913)
38. A. Sommerfeld: Ann. Phys. Lpz. 51, 1 (1916)
39. W. Wilson: Phil. Mag. 29, 795 (1915); 31, 156 (1916)
40. F. Paschen: Ann. Phys. Lpz. 50, 901 (1916)
41. G. Temple: Proc. Roy. Soc. 128A, 487 (1930)
42. R.C. Williams: Phys. Rev. 54, 558 (1938)
43. S. Pasternack: Phys. Rev. 54, 1113 (1938)
44. P.A.M. Dirac: Proc. Roy. Soc. A114, 243, 710 (1927)

45. H.A. Bethe: Phys. Rev. 72, 339 (1947)
46. V.F. Weisskopf: Rev. Mod. Phys. 21, 305 (1949)
47. W.E. Lamb and R.C. Retherford: Phys. Rev. 72, 241 (1947)
48. The Spectrum of Atomic Hydrogen: Advances, ed. G.W. Series, World Scientific (1988)
49. J.E. Nafe, E.B. Nelson and I.I. Rabi: Phys. Rev. 71, 914 (1947)
50. H.C. van de Hulst: Nederl. Tij. Natuurkunde 11, 201 (1945)
51. H.I. Ewen and E.M. Purcell: Phys. Rev. 83, 881A (1951), Nature 168, 356 (1951)
52. C.A. Muller and J.N. Oort: Nature 168, 357 (1951)

Interference of the Hydrogen Atom States (n=2)

Yu.L. Sokolov

I.V. Kurchatov Institute of Atomic Energy, SU-123182 Moscow, USSR

The phenomenon of atomic state interference has been investigated. The frequency of the ($2s_{1/2}$, F=0) - ($2p_{1/2}$, F=1) transition in the hydrogen atom has been measured using atomic interferometry. The Lamb shift was found to be δ = 1057. 8514 ± 0.0019 MHz, where the uncertainty is the statistical standard deviation of a single measurement.

1. Introduction

The develoment of a quantum mechanical approach to nature has given rise to a number of key problems which have as yet no comprehensive solution. Thus, further study of wave properties in microobjects and the comparison of obtained results and theory is necessary. However, one should note that the number of modern experiments demonstrating the phenomena of particle-wave duality is rather limited and they do not allow one to begin to solve vaguely formulated problems. It is probably fair to say, therefore, that some small-scale effects, which nevertheless play an important role, are neglected in many experimental techniques. If one seeks new approaches, it makes sense to study the interference of atomic states, since the interference pattern is extremely sensitive to the characteristics of its components which can manifest themselves in some new, previously unknown ways.

Reference to atomic interference is of interest for other reasons, too. One can study the properties of elementary particles by observing and precisely measuring various fine effects in the bound states. From such measurements, in principle, one can obtain such details as the behaviour of interactions at short distances, which otherwise are manifested only at very high energies. Optical measurements are probably among the most precise. If one adopts this form of measurement it is natural to ask whether it is possible to consider some phenomena of atomic physics within spec-

The Hydrogen Atom Editors: G.F. Bassani · M. Inguscio · T.W. Hänsch

troscopy, not in the sense of observing radiation, but, roughly speaking, by considering the atoms as waves and observing the interference of various atomic states. Then we have the following scheme: we can determine the properties of bound states by interference, and thereby determine the characteristics of elementary particles.

2. The Atomic Interferometer

A few years ago I proposed a method (the "atomic interferometer" method) which allows one to observe a stationary interference pattern for a long time while being able to arbitrarily change the phase shift, thus noticeably improving the accuracy of measurement. The interference of atomic states can be observed, in principle, with the aid of a device similar in main details to a standard two beam optical interferometer (e.g. Michelson's interferometer).

Imagine a beam of metastable $2s_{1/2}$ atoms of hydrogen passing in succession through two spatially separated zones I and II (Fig. 1). Inside these zones the atoms are subject to a perturbation that enables them to go to other states, say $2p_{1/2}$ and $2p_{3/2}$. A perturbing factor of this kind might be an electric field E that varies non-adiabatically within each of the zones. The criterion of non-adiabaticity is that the transit frequency $\omega = v/d$ be larger than, or of the order of, the Lamb frequency (for the $2s_{1/2}$ - $2p_{1/2}$ transition) or of the fine structure frequency (for $2s_{1/2}$ - $2p_{3/2}$); here v is the velocity of the atom and d is the breadth of the zones I and II, i.e. the distance over which the field changes abruptly. To simplify things, we shall continue our analysis with the two-level $2s_{1/2}$ - $2p_{1/2}$ system as an example. This is justified for fields that are not too strong; the effect of the $2p_{3/2}$ level can be taken into account by making small corrections.

It follows from the foregoing arguments that in the simplest version of the interferometer the role of zones I and II can be played by the boundaries of the field E, localized in a specified region. Then, when they cross boundary I, the atoms of the beam experience the perturbing influence of a growing field and go into a superposition of eigenstates ϕ_1 and ϕ_2 with energies ϵ_1 and ϵ_2 determined by the value of the field intensity E. At boundary II, where the field decreases to zero, beam components representing both the state $2s_{1/2}$ and the state $2p_{1/2}$ are produced, and each of

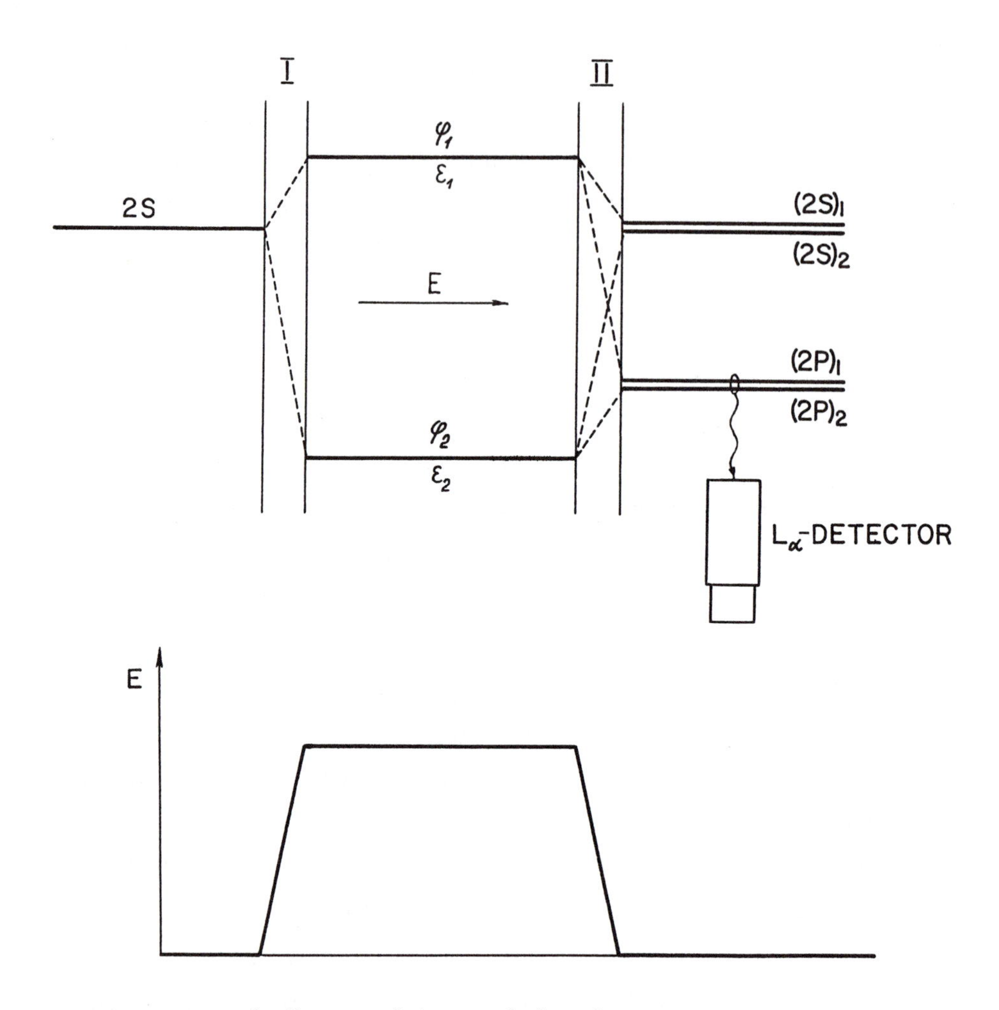

Fig. 1 Schematic diagram of the atomic interferometer

terms ϕ_1 and ϕ_2 initiates a pair of such states: ϕ_1: $(2s)_1 + (2p)_1$, and ϕ_2: $(2s)_2 + (2p)_2$. Thus, in the field-free region adjacent to boundary II, the state of the atom is described by a superposition of the four components: $(2s)_1$, $(2s)_2$, $(2p)_1$, $(2p)_2$.

Outside the field the amplitudes of the $2s_{1/2}$ and $2p_{1/2}$ eigenstates are defined by the transition amplitudes of, and the phase difference between, the components of each pair $(2s)_1$ - $(2s)_2$ and $(2p)_1$ - $(2p)_2$, which depend on the time of flight in the field and on the transition frequency between the ϕ_1 and ϕ_2 terms split by the electric field. The magnitude of such a splitting is entirely determined by the strength of the field E. Thus, when the field is continuously varied, periodic intensity oscillations of the H_{2s} and H_{2p} atom fluxes (occurring in counterphase) will be observed due to the interference of the $(2s)_1$ - $(2s)_2$ and $(2p)_1$ - $(2p)_2$ waves arising on boundary II. A similar effect is seen when the time of flight T, i.e. the distance between the field boundaries, is changed.

The interference pattern of the $(2p)_1$ - $(2p)_2$ components can be recorded by measuring the intensity of the short lived 2p part of the beam once it has passed through the interferometer. Thus, the detector placed behind zone II (i.e. in the field-free region) must count quanta corresponding to the single-photon transition 2p - 1s, i.e. the resonant line of the Lyman series (λ = 1216 Å). One can also observe the interference of the $(2s)_1$ - $(2s)_2$ components occurring in counterphase, for which purpose the beam should be passed through the additional field, either rf or constant.

To simplify the analysis, it is reasonable to divide the field strength range into "normal" and "strong". Normal should be taken to mean those fields for which the condition $x = \langle d \rangle E/\pi\hbar\delta \propto 1$ is satisfied, i.e. the Stark shift of the $2s_{1/2}$ and $2p_{1/2}$ levels, caused by the fields, proves to be of the same order as the Lamb shift (here $\langle d \rangle$ is the matrix element of the $2s_{1/2}$ - $2p_{1/2}$ transition, E is the field strength and δ is the Lamb shift).

In the two level case and assuming the field terminates abruptly at the boundaries, the yield of the H_{2p} - atoms is proportional to the quantiy I such that /1/

$$I = \sum_i c_i \frac{x_i^2}{1+x_i^2} \left[\cosh \frac{s_i T}{2\tau \sqrt{p_i + x_i^2}} - \cos 2\pi T\delta \frac{x}{x_i} \sqrt{1+x_i^2} \right] e^{-\frac{T}{2\tau}(1+\kappa x^2)} \quad (1)$$

where τ is the lifetime of the H_{2p} atom, T is the time of flight in the field E, and δ is the Lamb shift; c_i, s_i and p_i are constants; $\kappa = \frac{\delta}{2\nu_1}$; ν_1 is the $2p_{3/2}$ - $2p_{1/2}$ splitting frequency. The factor κx^2 is the $2p_{3/2}$ - level effect correction.

In the $2s_{1/2}$ - $2p_{1/2}$ system there are transitions in the electric field between the s and p hyperfine structure sublevels with total angular momentum projections 1, 0

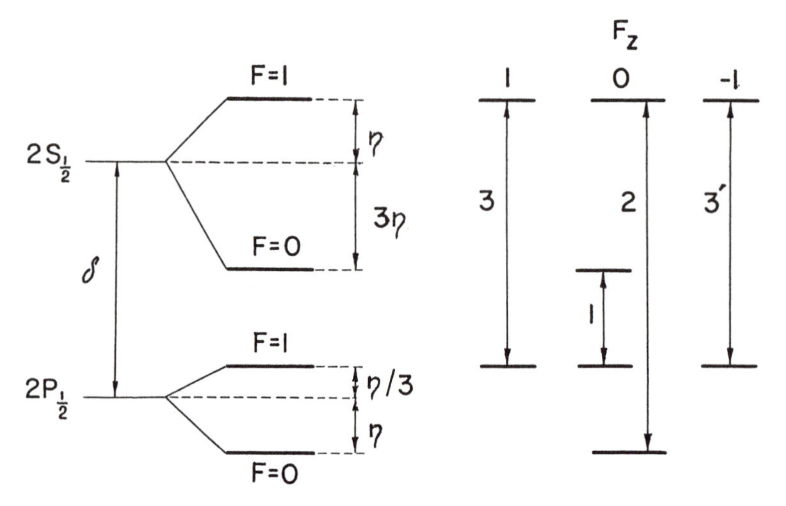

Fig. 2 Hyperfine structure of the 2s levels of the hydrogen atom

and -1 (Fig. 2). The energy differences of 3 and 3′ transitions coincide, so that the summation in (1) is carried out for three components of the hyperfine splitting with the following values of x_i:

$$x_1 = x \Big/ \left(1 + \frac{2}{3}\frac{\eta}{\delta}\right);\ x_2 = x \Big/ \left(1 - \frac{10}{3}\frac{\eta}{\delta}\right);\ x_3 = x \Big/ \left(1 + 2\frac{\eta}{\delta}\right) \qquad (2)$$

where η is the hyperfine splitting frequency.

It follows from (1) that the dependence of the H_{2p} atom yield on E and T found experimentally allows one, in principle, to determine the values of η and δ. It should be noted, however, that the method can not be realised fully in the case of the simple interferometer described. For example the determination of δ to an accuracy of 1 - 2 ppm, entails many practically insurmountable difficulties.

At the same time, a two electrode interferometer, which is convenient in operation, and which can be adjusted relatively easily with respect to a strictly collimated beam of hydrogen atoms, permits the study of many specific features of atomic interference.

Figure 3 shows the two electrode interferometer.

Fig. 3 Interferometer with two electrodes (longitudinal field)

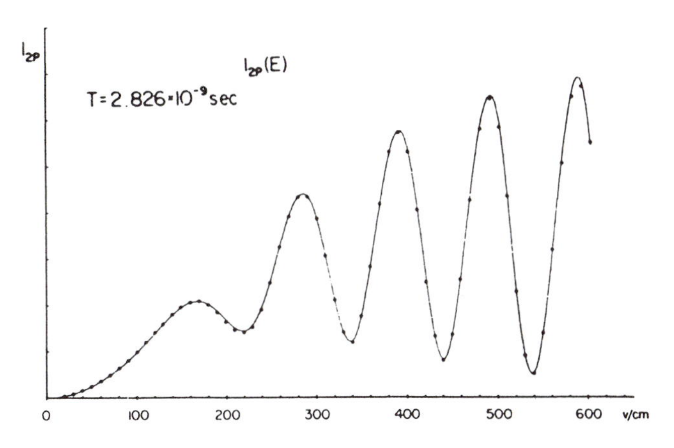

Fig. 4 Plot of $I_{2p}(E)$ $\ell = 0.5$ cm

Figures 4 and 5 show typical plots of $I_{2p}(E)$ and $I_{2p}(\ell)$, where ℓ is the flight distance. Both curves reveal a distinct interference pattern which is the optical analogue of the effect predicted by Pais and Piccioni for the system of K^0 and $\overline{K}^0$ mesons.

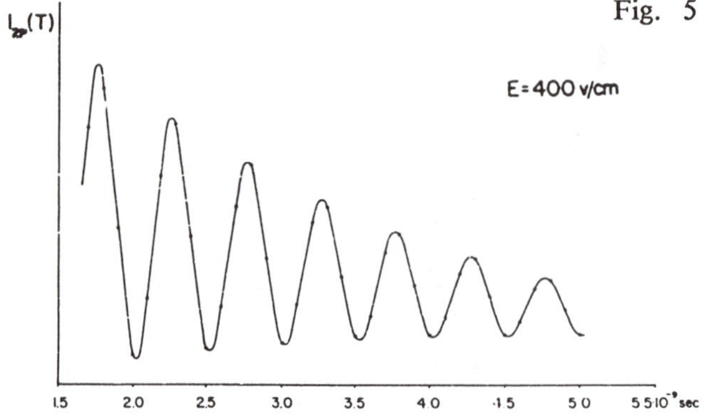

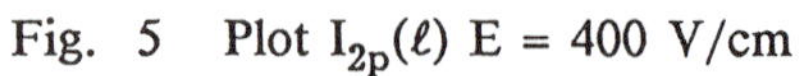

Fig. 5 Plot $I_{2p}(\ell)$ E = 400 V/cm

Fig. 6 Interferometer for fields up to 8000 V/cm

Figure 6 shows the interferometer designed for strong fields.

Figure 7 shows the typical interference curve at strong fields.

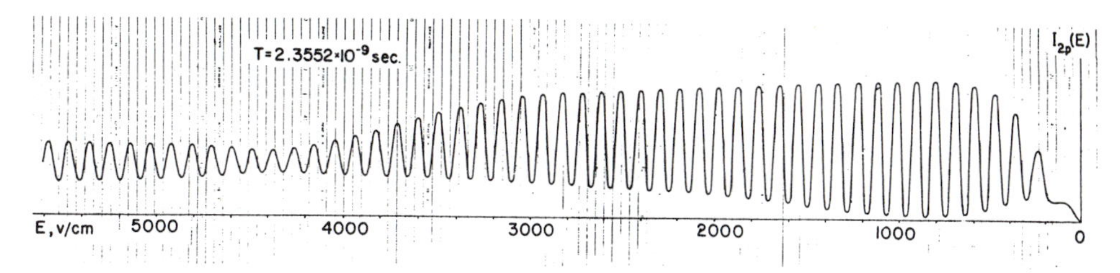

Fig. 7 Interference of $2p_{1/2}$ - state of hydrogen atom

In the experiments described above, it has been shown that when the two phase-shifted components of the 2p (or 2s) atomic hydrogen state interfere, some net curve - the superposition of separate curves corresponding to transitions between the components of the hyperfine 2s and 2p level structure - is registered. Further study of atomic interference has shown that hyperfine splitting can also be obtained in other ways.

The experimental scheme is shown in figure 8. The H_{2s} beam 1 with an energy of about 20 keV passes successively through rf fields A and B, which have frequencies of 1147 and 1088 MHz corresponding to transitions 2 and 3 (see Fig. 2), and then via interferometer 4 with Lyman α detector 5. In such a case only hyperfine structure component I with frequency ν = 909 MHz passes through the interferometer.

The electric field mixes states of opposite parity. Therefore, if the atom entering the interferometer is in a state with definite parity (e.g. in the 2s state), the probability of it emerging in the 2s or 2p state does not depend on the sign of the field.

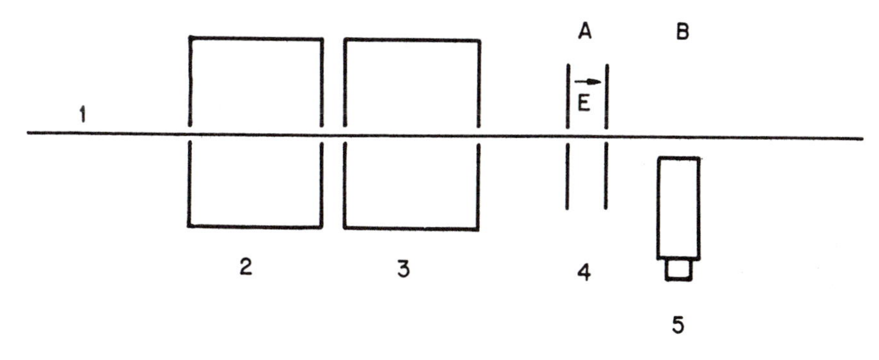

Fig. 8 Experiment on the interferometer field direction

On the other hand, if the initial wave function is a superposition of states of opposite parity (2s and 2p), the emergence probabilities for opposite field directions differ by an amount proportional to the product of the amplitudes of the atomic states 2s and 2p in the initial wave function.

The interferometric curves obtained at the rf fields 2, 3 switched "on", i.e. for the transition 1 at 909 MHz (since the 2s state component with F=1 is absent in this case), are shown in figure 9.

One can see that the yield of 2p atoms really does not depend on the field direction in the interferometer: the curves for +E and -E are in sufficiently good agreement (error bars are shown; experimental points are designated with crosses and circles).

However, when the rf fields 2, 3 are "off", i.e. when the ($2s_{1/2}$, F=1) component also passes through the interferometer, the situation is abruptly changed: the yield of 2p atoms essentially depends on the field sign (Fig. 10).

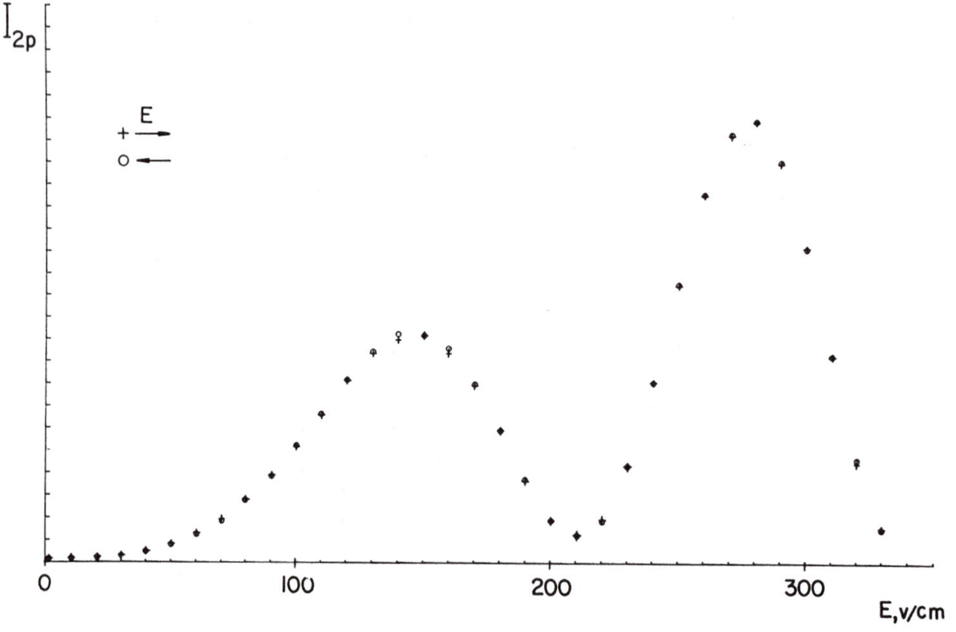

Fig. 9 Interference curves for opposing field directions (for transition 1)

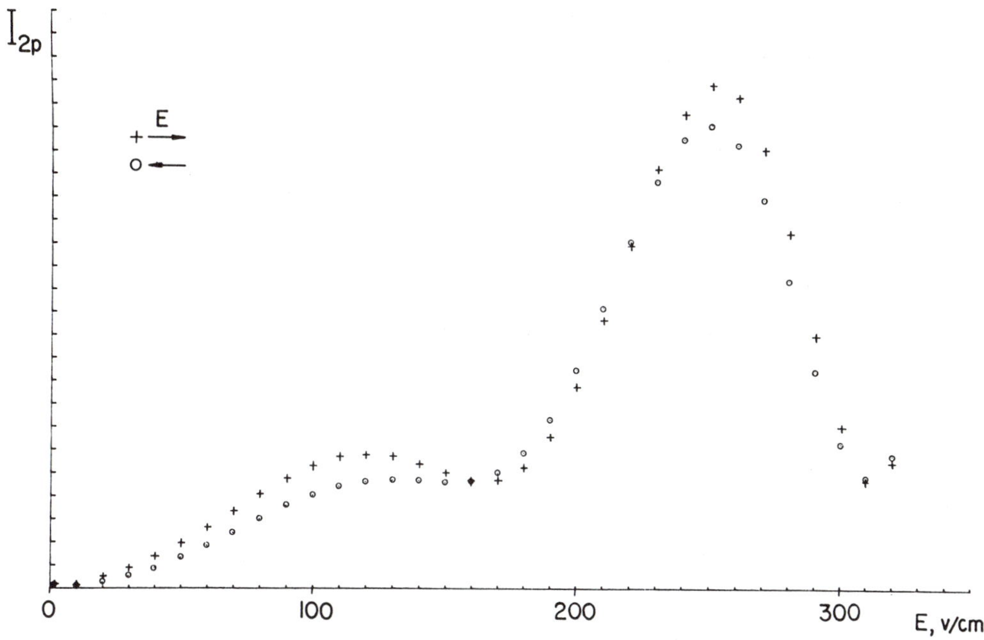

Fig. 10 Interference curves for opposing field directions (for transitions 1, 2, 3)

Although it is distinctly observed, we have not yet managed to describe the nature of the previously mentioned large-scale effect on the basis of quantum mechanical ideas concerning the behaviour of the hydrogen atom in an electric field.

3. The Lamb Shift Measurement

An interesting problem is the precise calculation and measurement of the Lamb shift δ which we describe here, commenting on the main points of interest. First, there is a disparity - not yet accounted for - both between the at present most precisely known theoretical values of δ, as well as between experiment and theory. Another important point is the opportunity provided to obtain information on the structure and properties of corrections which are not given directly by QED. In contrast to the anomalous magnetic moment, the Lamb shift characterizes the properties of bound electrons, i.e. it takes account of not only the QED effects but the effects arising from the nuclear structure. If the corrections independent of QED are far beyond the error limits of measurements for an anomalous magnetic moment, the corrections related to a finite size of proton for the Lamb shift will turn out to be within the range of modern experiments.

To determine the Lamb shift to within an error of the order of several ppm, the accuracy given by (1) is obviously insufficient. A computer calculation likewise cannot ensure the required accuracy, mainly because of the complicated behaviour of the atom in the interferometer and the uncertainty in the field characteristics at the boundaries, i.e. near the entrance and exit openings in the electrodes.

These difficulties can be eliminated by using an interferometer consisting of two independent systems I and II (Fig. 11), separated by a variable gap ℓ. An atom travelling at a velocity v through such a double interferometer is acted upon by nonadiabatic fields in each system, and thus a mixing of the states 2s and 2p results.

In the gap between the systems, i.e. in the region where there is no field, 2s and 2p are eigenstates, and their evolution can be determined accurately. It follows that an exact expression containing several parameters determined by the action of the fields E_1 and E_2 on the systems can be written for the probability $W(\ell)_{E_1E_2}$ of the

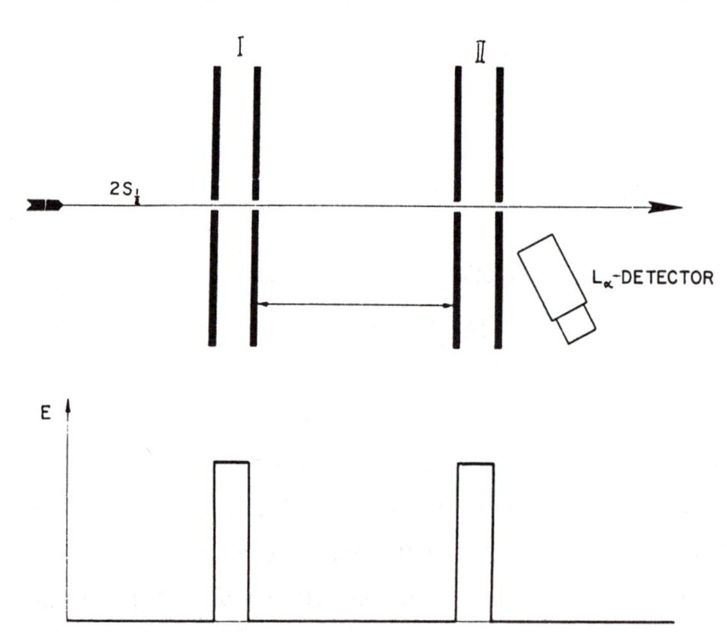

Fig. 11 Diagram of the "dual" atomic interferometer

emission of 2p atoms following the flight through the double interferometer, as a function of the length ℓ (or the time of flight $T=\ell/v$). If the conditions in the systems are kept constant while the length ℓ is varied, these parameters are fixed and need not be calculated (as shown below, the number of these parameters can be decreased to one by suitable reduction of the $W(\ell)$ curve).

It is important that when the function $W(\ell)$ is determined experimentally the variable ℓ can be chosen to be not the absolute value of the flight length, but its increment $\Delta\ell$ reckoned from a certain arbitrary null point.

As a result of the reduction of the experimental curve $W(\ell)$ we obtain an expression that does not contain unknown phases in the arguments of the cosines:

$$\Phi(\Delta\ell) = \cos\left[\frac{\omega}{v}(1-v^2/c^2)^{1/2}\,\Delta\ell\right] + \ell\cos\left[\frac{\omega_1}{v}(1-v^2/c^2)\,\Delta\ell\right] \qquad (3)$$

The described experimental data reduction procedure was developed by Dr. V.P. Yakovlev and Dr. V.G. Palchikov /2/, /3/.

The values of ω/v and ℓ in (3) were obtained using a least-squares program. In the case where the velocity remains constant, reduction of the initial data yields the values of ω/v and ℓ with an accuracy of not less than 5 ppm. It follows from (3) that to determine the Lamb frequency ν it is necessary to measure the velocity of the 2s atoms, using an independent method.

In all experiments the velocity was measured by observing the decay of the 2p atoms produced from the 2s atoms under the action of a nonadiabatic field. To determine the velocity from the experimentally obtained decay length $\ell_o = v\tau$, we must know τ, i.e. the lifetime of the 2p state. The value of τ was calculated: the error was estimated to be of the order of 1 ppm, which is acceptable for the determination of the Lamb shift with approximately the same accuracy.

It followed from the results obtained, that in 42 cases the velocity of 2s atoms could be regarded as constant. The values of ν obtained by reducing the data corresponding to the constant velocity selection criterion constitute a set of 42 values determined accurately to within several ppm. The equalisation of the accuracy is due to the fact that all the elements of the considered investigation were so constructed that the initial experimental and theoretical data, from which ν was calculated (i.e. ω/v and γ), were determined with practically the same relative error.

Figure 12 shows a histogram of the values of ν. They form a compact group with a mean value

$$\nu = 909.8934 \pm 0.0019 \text{ MHz}$$

(the error is assumed to be equal to one standard deviation). The corresponding value of the Lamb shift is

$$\delta(\text{H},n=2) = 1057.8514 \pm 0.0019 \text{ MHz}$$

If a sufficiently accurate theoretical value for the Lamb shift is available, then, as has been mentioned above, comparison with the highly precise measurements might be used to obtain information on the nuclear structure.

Theoretical values of δ were calculated by Mohr (δ = 1057.864(14) /4/ and by Erickson (δ = 1057.910(10)) /5/. Experimental values of the proton radius were obta-

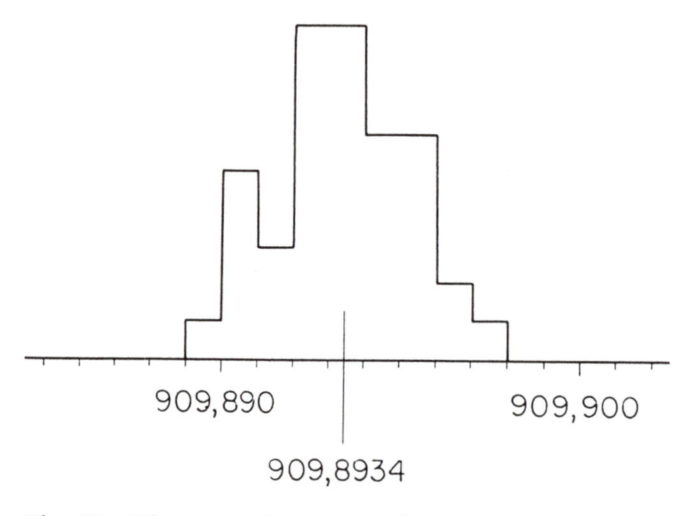

Fig. 12 Histogram of the ν - values

ined by Simon, Schmitt, Borkowsky and Walter ($\langle r^2\rangle^{1/2}$ = 0.862(12) fm) /6/ and by Hand, Miller and Wilson ($\langle r^2\rangle^{1/2}$ = 0.805(11) fm) /7/. Comparison of experimental values of the proton radius with theoretical values of δ were performed by Bhatt and Grotch /8/ and by Kinoshita and Sapirstein /9/. A set of new theoretical corrections introduced by these authors results in

$$\delta = \begin{cases} 1057.852(11), & \langle r^2\rangle^{1/2} = 0.805(11) \\ 1057.870(11), & \langle r^2\rangle^{1/2} = 0.862(12) \end{cases}$$

$$\delta = \begin{cases} 1057.857(14), & \langle r^2\rangle^{1/2} = 0.805(11) \\ 1057.875(13), & \langle r^2\rangle^{1/2} = 0.862(12) \end{cases}$$

We should emphasize the fact that the progress made by us in measuring the Lamb shift to higher precision allows one to determine the radius of the proton within the error limits 0.007 fm from the data obtained. Thus one can conclude that precise atomic spectroscopy is quite competitive in the study of interaction dynamics between electrons and protons. The advantages of such an approach are the opportunity of observing atomic states for a longer period of time and also that the corresponding experimental facilities both in size and cost are considerably more attractive than modern accelerators.

References

1. Yu.L. Sokolov: In: Atomic Physics 6 ed. by R. Damburg (Plenum Press, New York 1978), p. 207.
2. Yu.L. Sokolov and V.P. Yakovlev: Zh.Exp.Teor.Fiz., 83, 15 (1982).
3. V.G. Palchikov, Yu.L. Sokolov and V.P. Yakovlev: Metrologia 21, 99 (1985).
4. P.J. Mohr: Phys.Rev.Lett., 34, 1050 (1975).
5. G.V. Erickson: J.Phys.Chem.Ref.Data, 6, 831 (1977).
6. G.G. Simon, Ch. Schmitt, F. Borkowski, V.W. Walter: Nucl.Phys. A333, 381-391 (1980).
7. L.M. Hand, D.D. Miller, R. Wilson: Rev.Mod.Phys., 35, 335 (1963).
8. G.C. Bhatt, H. Grotch: Annals of Phys., 178, 1 (1987).
9. T. Kinoshita, I. Sapirstein: In: Atomic Physics 9, University of Washington, Seattle, 1984.

Separated Oscillatory Field Measurement of the Lamb Shift in H, n=2*

F.M. Pipkin

Lyman Laboratory of Physics, Harvard University,
Cambridge, MA 02138, USA

This paper reports a precision measurement of the Lamb shift in the n=2 state of hydrogen using separated oscillatory fields in conjunction with a fast atomic beam. Atoms in the $2^2S_{1/2}$ metastable state are produced through charge capture by protons in a gas target. The atoms pass thrugh two separated oscillatory fields whose relative phase is switched between 0 and π. A rf quenching field and a solar blind photomultiplier tube provide a monitor of the number of atoms in the $2^2S_{1/2}$ state that pass through the spectroscopy fields. The procedure used to take and correct the data is described and the major sources of residual error are discussed. The result of the measurement is compared with the current theoretical value. An experiment for further improvement in the measurement of the Lamb shift is outlined.

1. Introduction

The difference in the energy of the $2^2S_{1/2}$ and $2^2P_{1/2}$ levels in hydrogenic atoms is a purely electrodynamic effect due to the interaction of the bound electron with the quantized electromagnetic field. The measurement of this splitting was a major stimulus for the development of renormalization theory and still provides an important test of Quantum Electrodynamics. The precise measurement of this splitting is difficult because of the short radiative lifetime of the $2^2P_{1/2}$ state. The ratio of the linewidth to the transition frequency is roughly 1/10. Thus a high precision measurement requires a detailed understanding of the line shape and good control of the variables that shift and distort the resonance line.

The fast beam separated oscillatory field technique (SOF) provides a method through which one can obtain a series of lines whose widths are less than the natural line width with a good understanding of the factors which determine the line shape and the line center. This paper summarizes a separated oscillatory field measurement of the Lamb shift in hydrogen.[1]

*This work was carried out in collaboration with S. R. Lundeen and many of the innovative ideas which lead to the success of the experiment are due to him. His present address is Department of PHysics, University of Notre Dame, Notre Dame, IN 46556, USA.

The Hydrogen Atom Editors: G.F. Bassani · M. Inguscio · T.W. Hänsch

2. Description of Method

Figure 1 shows schematically the origin of the narrowed signal. An atom in the $2^2S_{1/2}$ state enters the first rf region where, in general, the rf field produces a superposition state in which the wavefunction for the atom has components in both the $2^2S_{1/2}$ and $2^2P_{1/2}$ states. The atom then passes through a free field region into a second rf region where the superposition state is changed further. The probability for the atom to emerge from the second rf field in the $2^2S_{1/2}$ state contains an interference term of the form shown in Fig. 1 whose frequency width depends on the separation in time T of the two rf regions rather than the natural linewidth. By changing the relative phase of the two rf regions from 0^0 to 180^0, one can isolate the interference signal.

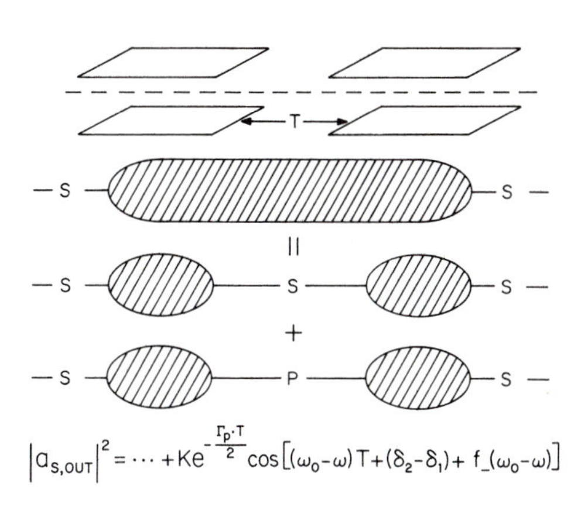

Fig. 1 Schematic diagram showing origin of the interference signal

Since the interference term depends on the amplitude for an atom to make a transition to the $2^2P_{1/2}$ state in the first rf region and back to the $2^2S_{1/2}$ state in the second rf region, it essentially selects atoms in the $^2P_{1/2}$ state that live for a time T. As the time separation between the two SOF regions is increased, the frequency width of the interference signal decreases as 1/T and the size of the interference signal decreases exponentially with half the decay constant for the $2^2P_{1/2}$ state. One can trade signal size for a decrease in line width and still retain a high signal to noise ratio for the narrowed signal.

To carry out this scheme, a fast atomic beam ($v/c \simeq 0.01$) is used to translate the required nanosecond time intervals into convenient laboratory distances. To avoid complications due to motional electric fields, the entire experiment is performed in zero magnetic field and the resonance is tuned through directly by changing the frequency of the applied rf field. Other rf fields are used to select one hyperfine state so as to simplify the line shape.

3. Description of Measurement

Figure 2 shows a schematic diagram of the fast beam apparatus. The fast beam is obtained by charge capture from a 50-100 keV proton beam produced by a commercial 150 keV accelerator. The separated rf fields are produced by pairs of plates each of which is a section of a 50Ω transmission line. The two plates of each pair are driven 180^0 out of phase to insure that the midplane containing the beam axis remains at ground potential and that the rf electric field seen by an atom traveling along the axis is uniformly polarized transverse to its velocity.

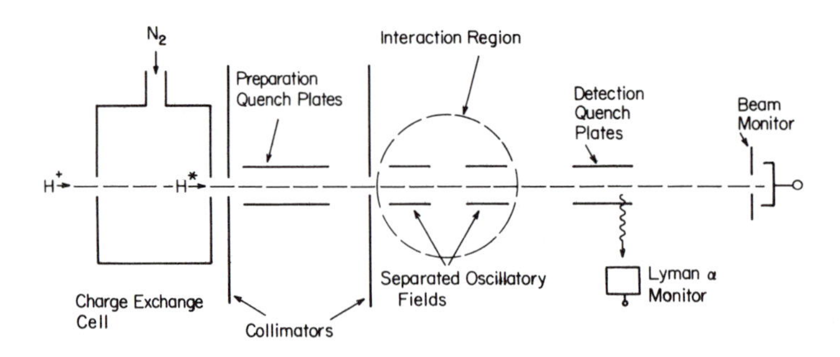

Fig. 2 A schematic diagram of the fast-atomic beam apparatus used in this measurement

To select one hyperfine state the preparation quench plates are driven at 1110 MHz and the detection quenching plates are driven at 910 MHz. Figure 3 shows the microwave system used to produce the two separated oscillatory fields. A high precision coaxial magic Tee drives the two rf regions so that they have relative phases of 0^0 or 180^0. Figure 4 shows the hyperfine state selection.

The signals are defined as the fractional decrease in the Lyman-α photocurrent N produced by a fixed amplitude rf electric field in the SOF region through the equations

$$S^0 = (N_{off} - N^0_{on})/N_{off},$$

$$S^\pi = (N_{off} - N^\pi_{on})/N_{off},$$

where N_{off}, N^0_{on}, N^π_{on} are the photocurrents when the rf is off, on with 0^0 relative phase, and on with 180^0 relative phase, respectively. In terms of these primary signals, the "interference" and "average quench" signals (I and $\bar{Q}$) are defined as

$$I = S^0 - S^\pi,$$

$$\bar{Q} = (S^0 + S^\pi)/2$$

Figure 5 shows the average quench and interference signals obtained with a 100 keV beam and increasing separation of the rf regions. The envelope of the interference signal is determined by the spatial distribution of the rf field in the two separated oscillatory field regions.

To eliminate the residual first order Doppler shift due to the failure of the direction of propagation of the rf field to be precisely perpendicular to the fast beam, measurements were taken with the rf drive on both the right and left sides of the beam. To eliminate the frequency shift due to phase errors in the rf drive system, measurements were made with the entire rf system, including the spectroscopy region, rotated 180^0 about an axis passing through the midpoint between the two

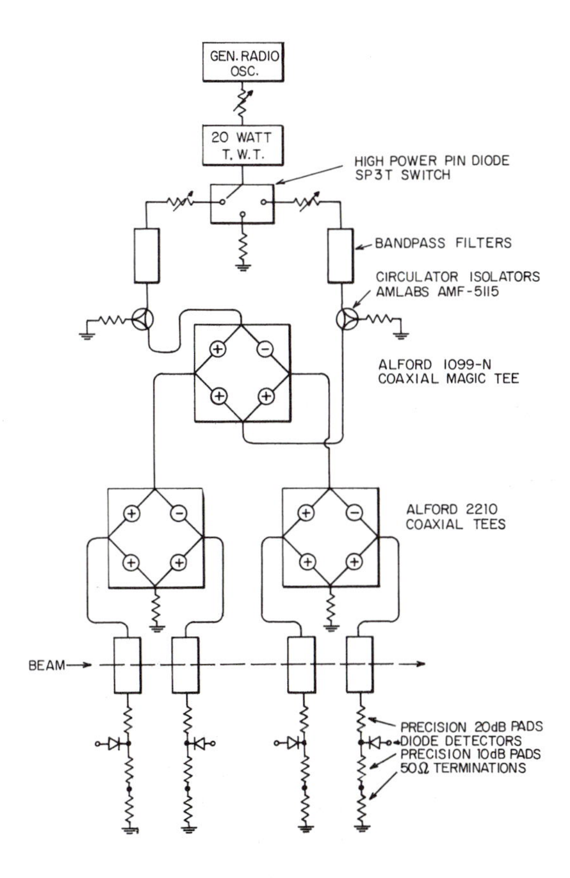

Fig. 3 The system used to power the separated oscillatory fields

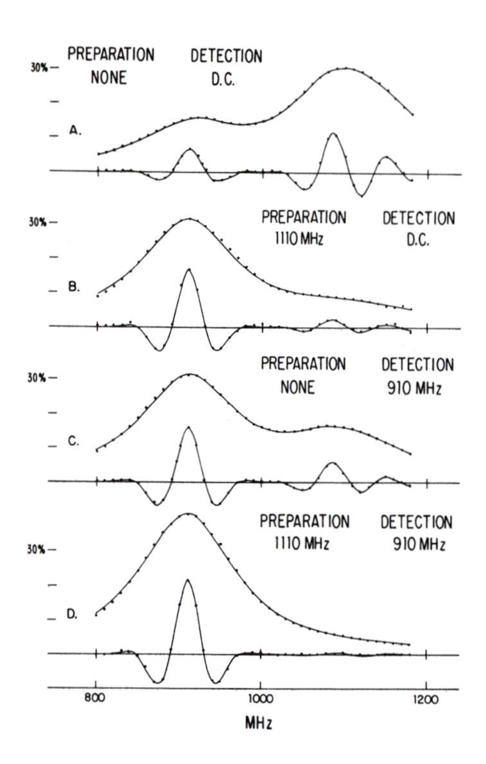

Fig. 4 A series of plots of the average quench and interference signals showing the hyperfine state selection due to the continuous rf fields in the state selection region and in the quench region.

SOF regions and parallel to the direction of propagation of the rf field. This exchange reverses the order in which an atom encounters the two rf regions and thus cancels phase errors.

Power monitoring diodes were used to set the rf power so the rf electric field did not vary as a function of frequency. The diodes were calibrated using a Hewlett Packard 432A Power Meter with a 8478B thermistor power sensor and calibrated attenuators. The power measurement was used in conjunction with slotted line studies of the rf interaction regions to determine the rf electric field at each frequency.

The method of symmetric points was used to determine the center of the interference curve. Extensive calculations showed that the line profile should be symmetric about the center frequency. The line center was then corrected for the second order Doppler shift, The Bloch-Siegert and rf Stark shifts, coupling between the rf plates, the residual F=1 hyperfine component, and distortion due to off axis electric fields. A small residual asymmetry in the average quench curve was attributed to a residual variation of the rf electric field across the line and corrected for on the assumption this was the correct explanation. Table 1 shows the measured interval and the corrections for one of the 8 data sets used to determine the final result.

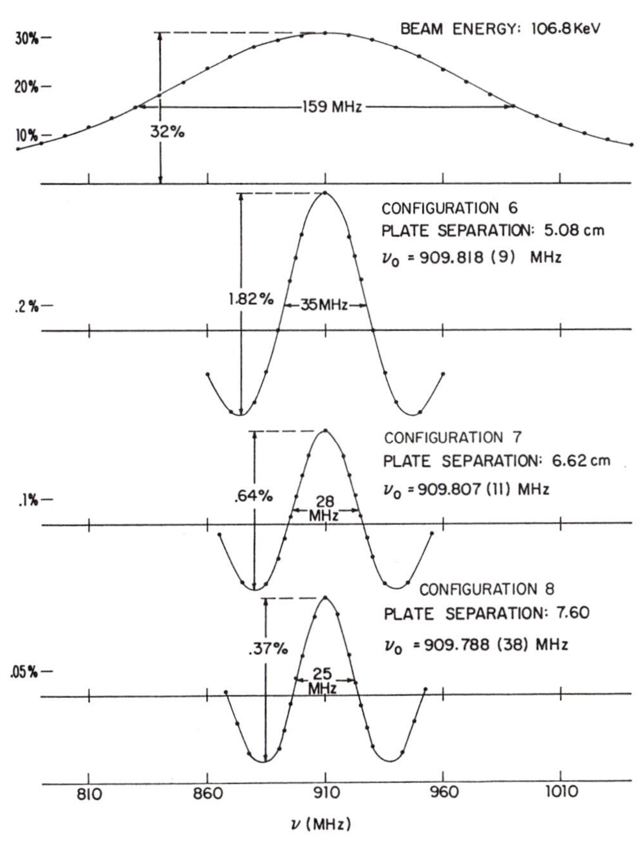

Fig. 5 Samples of the average quench and interference signals obtained in this experiment with increasing separation of the two rf regions

Table 1 The measured value and corrections for data point 6. The energy was 100 keV, the field separation 5.06(5) cm and the full width at half maximum 40.1 MHz. All the entries are in MHz.

Raw Center	909.818(9)
Corrections	
1) Time dilation	+0.104(4)
2) Bloch-Siegert & rf Stark shift	-0.027(2)
3) Plate coupling	0.000(0)
4) Residual F=1 component	0.001(0)
5) Incomplete $\bar{Q}$ subtraction	0.000(4)
6) Off-axis distortion	-0.003(2)
7) Variation of rf field	-0.001(1)
8) Additional rf field variation	-0.005(2)
$\nu[2^3S_{1/2}\ (F=0) \leftrightarrow 2^2P_{1/2}\ (f=1)]$	909.887(11)
Hyperfine structure	147.958
S(n=2)	1057.845(11)

4. Results

Figure 6 shows in graphic form the values obtained for the eight points at which data were taken. The final value for the Lamb shift is

$$S(H,n=2) = 1057.845(9)\ \text{MHz}$$

Figure 7 shows a plot of all the reported direct measurements of the Lamb shift in the n=2 state of hydrogen.

The principal uncertainty in the theoretical value for the Lamb shift is due to the radius of the proton. The radius determined by the early measurements by workers centered around Stanford [1] is

$$\langle r_p^2 \rangle^{1/2} = 0.805(11)\ \text{fm}.$$

The radius determined more recently by the German scientists [1] is

$$\langle r_p^2 \rangle^{1/2} = 0.862(12)\ \text{fm}.$$

If one includes the recently calculated recoil correction of BHATT and GROTCH [2] the theoretical value for the Lamb shift is

$$1057.852(11)\ \text{MHz if}\ \langle r_p^2 \rangle^{1/2} = 0.805(11)\ \text{fm},$$

$$1057.870(11)\ \text{MHz if}\ \langle r_p^2 \rangle^{1/2} = 0.862(12)\ \text{fm}.$$

The experiment and theory are in excellent agreement if the old value for the proton radius is used; the agreement is poor if the new value of the proton radius is used. Because of the discrepant values for the proton radius, one cannot say how well experiment and theory agree. There are also uncalculated theoretical contributions which could be as large as 10 kHz.[3]

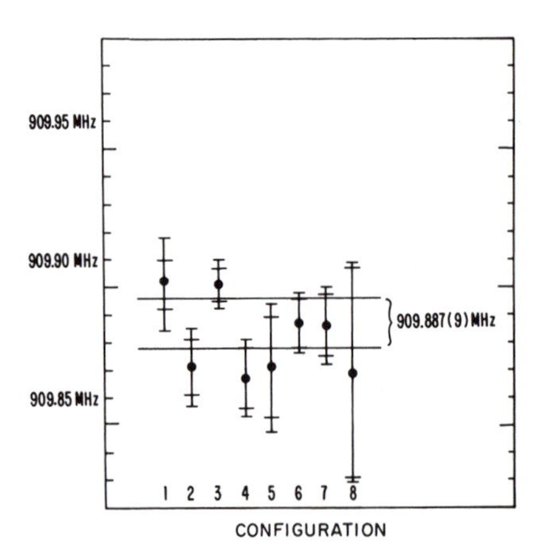

Fig. 6 A plot of the final line centers for the 8 configurations in which data were taken. The smaller error bars are the statistical uncertainties. The outer error bars include the systematic uncertainties.

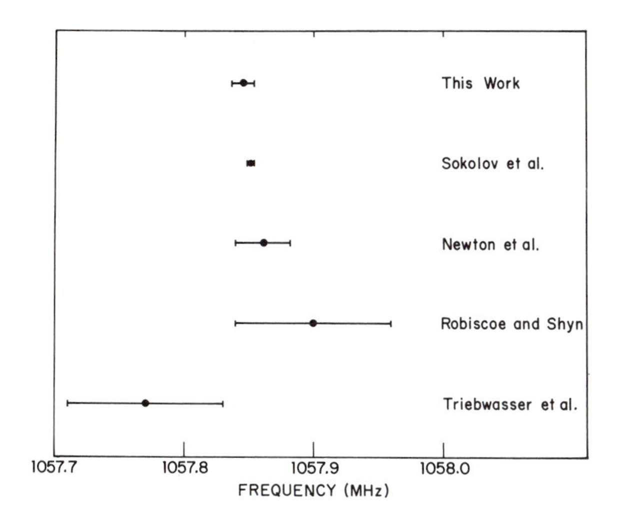

Fig. 7. A plot of all the direct precision measurements of the Lamb shift in the n=2 state of hydrogen

5. Prospects for Higher Precision

The principal limitations in the precent measurement are due to the line width and the uncertainty in the rf field as a function of frequency. A large solid angle detector such as the Lyman alpha detector [4] that has been used for measurements on He^+(n=4) could be used to provide an increase in signal by a factor of a hundred and thus make feasible measurements at wider separation with a line width of 20 MHz. The uncertainties in the rf field strength could be reduced by measuring the $2^2S_{1/2}$ - $2^2P_{3/2}$ transition near 10,000 MHz. At this frequency, one could use waveguides for which the microwave properties are more readily measurable and simpler to understand. At this frequency the necessary bandwidth required to study the transition is also a smaller fraction of the transition frequency. It is estimated that using this transition one could determine the Lamb shift to 1 in 10^6.

The uncertainty due to the nuclear radius could be reduced by making measurements on both hydrogen and deuterium. The ratio of the two radii is known better than the individual radii and the two measurements could be combined to reduce the uncertainty due to the nuclear radius.[5]

6. Conclusions

The fast beam separated oscillatory field method has been used to measure with high precision the Lamb shift in the n=2 state of hydrogen. The agreement between the measured value and the theoretical value is obscured by the discrepant values for the nuclear radius. This technique could be used to improve further the precision and to reduce the uncertainty due to the nuclear radius.

7. Acknowledgements

This research was made possible by a series of grants from the National Science Foundation. The current grant is PHY87-04527. It has also benefited from many individuals at Harvard and elsewhere.

8. References

1. This paper is an abbreviated version of a longer article which provides a complete description of this measurement. S. R. Lundeen and F. M. Pipkin, Metrologia 22, 9 (1986). The references here will be primarily to work reported since the earlier article.
2. G. C. Bhatt and H. Grotch, Ann. Phys. (N.Y.) 178, 1 (1987).
3. H. Grotch, paper presented at Adriatico Conference on "Vacuum in non-relativistic matter radiation systems" (14-17 July 1987), TRIESTE.
4. J. J. Bollinger and F. M. Pipkin, Rev. Sci. Instrum. 52, 936 (1981).
5. F. A. Bumiller, F. R. Buskirk, J. W. Stewart, and E. B. Dally, Phys. Rev. Lett. 25, 1774 (1970).

Hydrogen Spectroscopy and Fundamental Physics

W. Lichten

Physics Department, Yale University, P.O. Box 6666,
New Haven, CT 06511, USA

"Dann muss man doch bescheiden sein."
Albert EINSTEIN [1]

"One must be modest" is an appropriate quotation to describe laser spectroscopy of the hydrogen atom. This is a field of modest scientific accomplishments, despite its awesome technical machinery. To illustrate, the first demonstration of laser spectroscopy in hydrogen was made by HÄNSCH, SHAHIN, and SCHAWLOW in 1972. [2] Now, sixteen years later, the most precise laser measurement of the Lamb shift in the ground state of hydrogen is by the Oxford group and is known to one part in ten thousand. [3] This is just equal to that achieved by LAMB and coworkers thirty-five years ago. [4] In the meanwhile, radio frequency and other techniques have pushed the measurement of the H(2S) Lamb shift an order of magnitude beyond the accomplishments of laser spectroscopists. [5]

This talk will concentrate on the relation of the hydrogen atom to fundamental physics. It will start with laser spectroscopy of hydrogen and determination of R_∞. It will then discuss other systems, in particular the electron-positron system, as possible new tests of fundamental physics.

1. Laser Spectroscopy of Hydrogen and the Measurement of R_∞

The Rydberg constant is the scale factor that connects all theoretical calculations and experimental measurements of energy levels in any system involving electrons. This includes all atoms, molecules and condensed matter. In simple systems, such as hydrogen, positronium, muonium, and possibly helium, the theoretical accuracy is comparable to that of experiments. In this case, experimenters can be said to measure the Rydberg constant, if not to test theory. Laser spectroscopy, at the moment, is the method par excellance to measure R. Measurement of the Rydberg constant R is a simple matter. One measures the wavelength or frequency (the velocity of light is defined to be 299 792 458 m/sec) in a system, such as hydrogen, where theoretical calculations are expected to be accurate to within the experimental error. One then compares this measurement (in Hz or cm^{-1}) with the theoretical calculation (in atomic units), thereby finding the atomic unit in Hz or cm^{-1}. Half of the atomic unit is the Rydberg

The Hydrogen Atom Editors: G.F. Bassani · M. Inguscio · T.W. Hänsch

constant. At present, the most precise measurements of R are in the H,D system in a variety of states. All current measurements agree within experimental errors (see Table I). Within the near future, one might expect improvement in both measurements and calculations in the helium atom to provide an alternative determination of the Rydberg constant.

Table I. Comparison of measurements of Rf_r=R-10973 731 m^{-1}, errors and variances[a]

Group	Transition	R_{fr} (error)	Variances (parts in 10^{20})				
			Standard	Relative Wavelength (frequency) measurement	Statis-tical	Other	Total
Yale-NBS	2S - 3P	0.569(7)	3	10	25	5	42
Stanford	1S - 2S	0.571(7)	16	18	3	---	38
Paris	2S-8D,10D	0.571(2)	3	---	---	---	3
Yale-NBS	2S - 4P	0.573(3)	3	---	2	2	7
Oxford	1S - 2S	0.573(3)	7	1	---	1	9

a. see references 3,9 and other talks in this session

2. Where Are We Now? Where Are We Heading?

Table I breaks down the variance V = σ^2 of the different measurements of R. Figure 1 shows this graphically. This gives insight into the current status of the field and could be used to address several questions about the future.

One could ask, "What would be the contribution of using special techniques, such as two-photon spectroscopy, to take advantage of the potentially high precision of ultra-narrow lines, such as the 1S-2S transition in H (width = one Hertz)? The statistical part of the variance in Figure 1 reflects the "Q" (ratio of frequency of line width) of the lines used by the various groups. Narrowing the line width does lower the variance, as is shown by the improvement in the results of the Yale-N.B.S. group in going from Balmer-α to Balmer-β with its threefold higher Q. But there are limits. These are shown most graphically in the case of the Paris group, which exploited the narrowness of the 2S-8D, 10D, 12D lines to the fullest. The rest of their variance is negligible. Their result has reached the best possible accuracy achievable by laser spectroscopy, less than two parts in 10^{10}. Further improvement is not possible without a better frequency standard in the optical region.

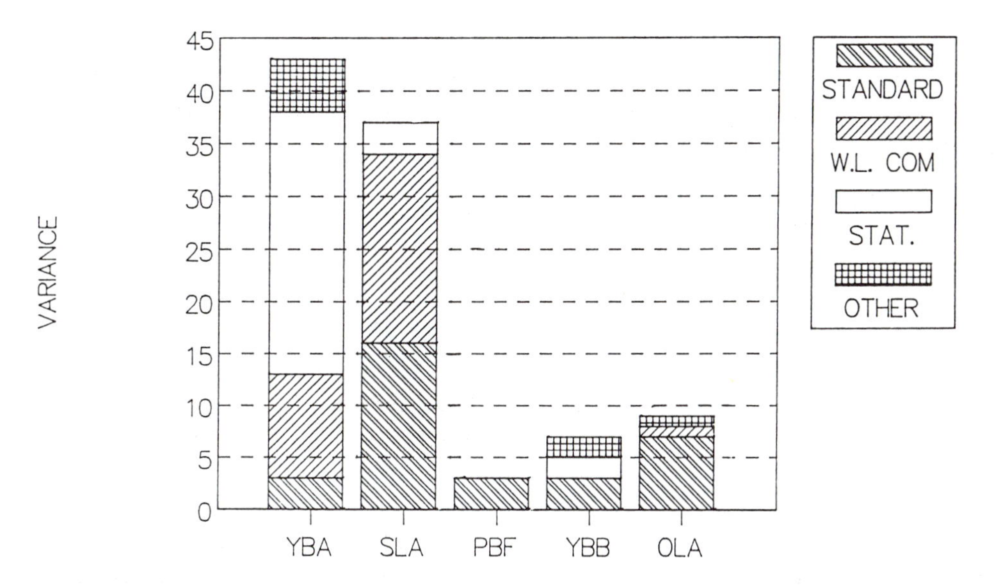

Figure 1. Comparison of variance = σ^2 (units of 10^{-20}) of the Rydberg constant

Legends:

YBA - Yale - N.B.S. Balmer α (reference 9)

SLA - Stanford Lyman α (two photon. See talk by T. Hänsch)

PBF - Paris 1S - 8D,10D (two photon. See talk by F. Biraben)

YBB - Yale - N.B.S. Balmer β (reference 9)

OLA - Oxford Lyman α (two photon - see ref. 3)

Sources of variance:

Standard - frequency or wavelength standard

W.L. COM - Comparison of measured wavelength with standard

Stat - Statistical errors

Other - Other sources of error, such as energy level shifts

3. Possible New Directions for Research

We consider, as an example, the determination of the Lamb shift in the ground state of hydrogen. The theoretical uncertainties are about a factor of ten smaller than the experimental error. [3] Thus an order of magnitude improvement in the experimental precision is necessary before laser spectroscopy can be said to be interesting. We consider a few possibilities:

1. Measurement of ratios. An example of this type of measurement was given by WIEMAN and HÄNSCH, [6] who successfully measured the ground state Lamb shift of H and D by comparing a laser at Balmer β at 4860 Å with another dye laser which was doubled to look at the two-photon transition 1S-2S, Lyman α. The measured

values of 8151(30) and 8188(30) MHz agree both with theory and the results of BOSHIER et al., which have a 40-fold higher precision. [3] This procedure has the advantage of eliminating the standard, but it involves the additional complexity of two simultaneous measurements.

2. Direct chain to the visible. This possibility involves improving the chain from the Cs frequency standard into the visible, preferably near an atomic hydrogen line. For example the 486 nm laser, useful for Balmer β or doubling for the 1S-2S line, is quite near the seventh harmonic of the He-Ne(CH_4) line at 3.39μ and is an attractive possibility for a direct link from Cs to a hydrogen line.

3. Use of improved standards in the infra-red and radio frequency regions. Currently, standards based on the osmium tetroxide and CH_4 absorptions at 10.8μ or 3.39μ, respectively, are at the level of a few parts in 10^{11} or better. [7] Measurement of a hydrogen line near one of these standards could improve the accuracy of the Rydberg constant by an order of magnitude. By measuring transitions between two high lying states in the radio frequency region, one has the potential accuracy of the Cs clock at a few parts in 10^{14}. [8]

In either case, by using the hydrogen lines in different wavelength regions, one would then have a new set of standards throughout the entire frequency spectrum. [9]

4. Exotic Atoms. Muonium

The purely leptonic hydrogen atom, muonium, consists of a positive muon and an electron. It is the ideal atom, free of the nuclear structure effects of H, D and T and also of the difficult, reduced mass corrections of positronium. An American-Japanese group has observed the 1S-2S transition in muonium to a precision somewhat better than a part in 10^7. [10] Because there were very few atoms available, the statistical errors precluded an accurate measurement. The "ultimate" value of this system is very great, being limited by the natural width of the 1S-2S line of 72 kHz, set by the 2.2 μsec lifetime of the muon.

5. Positronium

Positronium (e^+e^-) is a purely leptonic system, free of nuclear structure effects, but suffers from reduced corrections in the worst possible case of equal masses. This makes the system difficult to treat, since quantum electrodynamical calculations start from an infinite nuclear mass and treat reduced mass effects as a perturbation.

Many experiments have been performed on this system, since its historic discovery by DEUTSCH.[11] Table II shows a sample of these measurements, with the level of agreement between theory and experiment shown. It is a mixed bag, with several cases of highly significant difference between Q.E.D. theory and experiment.

Table II. Positronium. Comparison of experimental theory, and deviation between the two, expressed in experimental standard deviations

Quantity	Value	Dev(σ)
1S hfs	203.3991 Th	
(GHz)†	203.3849 Exp	12
	203.3870 Exp	
n=2, fs(MHz) S ↔ P†		
J=1 ↔ J=2	8625 Th	
	8620 Exp	1.9
J=1 ↔ J=1	13 011 Th	
	13 001 Exp	2.4
J=1 ↔ J=0	18 496 Th	
	18 504 Exp	0.8
Decay Rate	7.038 30 Th	
of O-Ps	7.051 6 Exp	10
(per μs)†		
1S - 2S	1233 607 202 Th	
Interval	1233 607 143 Exp	5.4
(MHz)†		

† Ref. 14

How should one view these disagreements? Certainly one should never be complacent. Nevertheless, it seems likely that in most cases, the problem lies in the difficulty of the Q. E. D. calculations, which have not been carried out to a high enough order. Perhaps, a totally new type of calculation is needed. [12] In the case of the 1S-2S interval, there is some doubt about the correctness of the experimental values and remeasurements are underway. [13,14]

6. The Electron-Positron system. A New Elementary Particle?

We have seen that the electron-positron system, positronium, shows irregularities in its ground and excited states. However, we have questioned whether or not the discrepancies between theory and experiment have any real physics. On the other hand, there are remarkable phenomena in this system at energies of a few hundred KeV, as shown by the heavy ion experiments at G. S. I. in Darmstadt, Germany (see Figure 2). [15] At the risk of being accused of introducing irrelevant material in a focussed symposium, I shall discuss these interesting events here. The reason is there has been little or no systematic investigation of the e^+e^- system at these energies and the results are provocative. It is possible that, unlike most other studies of the hydrogen atomic system with a well-understood Hamiltonian, these events could represent really new physics. [12] On the other hand, the author will explore the possibility that these bizarre events have a conventional explanation.

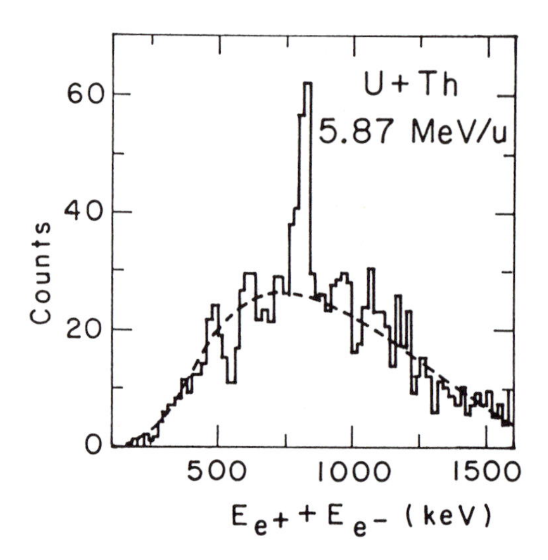

Figure 2. Distribution of the sum of coincident electron and positron kinetic energies in heavy ion collisions at G.S.I. (ref. 15)

The G. S. I. group measured the sum of the kinetic energies of electron-positron pairs produced in heavy ion collisions at GeV energies. The remarkable feature of these experiments was a sharp peak in the energy distribution, summed over the e^+e^- pair (Fig. 2). This lead to speculation as to the origin of this mysterious peak. A large number of papers have been written on the subject. [15]

The original discussion of this problem was based on the solution of the ground state solution of the Dirac equation for the hydrogenic atom consisting of a point nucleus of charge Z and a single electron:

$$E = [1 - (\alpha Z)^2]^{1/2}\, mc^2$$

At $Z > 1/\alpha = 137$, this solution "dives" into the electron sea of the positron continuum and no real solution exists. This has been interpreted as an unstable state which, if vacant, can be filled by "autoionization" from the positron continuum. [16]

In 1969, Gershtein and Zeldovich noted that a slow collision would cause the diving of the lowest bound level into the positron continuum. This caused great excitement and a rash of papers and experiments attempting to calculate the properties and observe these "escape" positrons. [16] The G. S. I. experiments grew out of this early stage of development.

A remarkable paper at this stage of the field was by RAFELSKI, MÜLLER and GREINER [17], who predicted a sharp line in the positron continuum, as a result of a reaction between the colliding nuclei. This would produce a time delay which would allow the spontaneous emission of the positron with a sharp line shape. Early observations of this positron line seemed to verify this prediction. The fact that there was only a single energy of the emitted positron, independent of the charge of the colliding nuclei, seemed to rule out the nuclear physics explanation, [15] and furthermore lead to a new possibility.

This was considered by COWAN et al. [18] who hypothesized "a common source of the monoenergetic positrons in the two-body decay of a previously undetected particle... A clear signal for a neutral particle could be provided by the detection of a monoenergetic electron in coincidence with the peak positrons." This particle, at first assumed to be the axion, a pseudoscalar of zero spin, has had various names, one of which is the "X_0". For the second time, a remarkable discovery, the observation of electron-positron pairs [15,19] (see Figure 2), confirmed a speculative hypothesis dramatically. The experimenters, although phrasing their interpretation cautiously, could hardly resist considering the possibility that they had indeed observed "a neutral particle which decays into an e^+e^- pair... The signature for such a process, when viewed from the particle rest frame, would be back-to-back colinear emission of monoenergetic electrons and positrons.... Features associated with the electron-positron decay of a slowly moving particle appear to be reflected in the observations involving electron-positron coincidences." [19] A new flurry of experiments, this time by particle physicists, started a frantic search for the X_0 but it never was found. Another problem with the new data was that the e^+e^- coincidence peak no longer occurred at a single energy. It has been seen at three different energies, and even more than one energy for a given collision pair. [15]

Many discussions have been in the literature about the nature of this new particle. Some have assumed that it was the axion; others have ruled out the standard axion, but have not ruled out non-standard axions. In fact, the particle physicists concluded that they had ruled out a new elementary particle. It seems rather improbable now that such a particle exists at all. [20]

The theorists, who never lack for ingenuity and inventiveness, no longer seem to be writing about the X_0 particle. However, there have been several discussions about a new phase of quantum electrodynamics, an even more radical interpretation of the G. S. I. results. [21]

Many other attempts to observe this new excited state of positronium have been made. Indications of correlated, equal energy, two-photon decay with a summed energy of 1062 keV have been found by DANZMANN et al. [22] in the same reaction as produced the electron-positron pairs. [15, 18, 19] The authors believe that this line, at a new, fourth energy, may belong to the same neutral system which produced the correlated electron-positron pairs. [22]

Many groups have searched directly for this entity by Bhabha electron-positron scattering, the inverse process to the decay of the postulated entity. Although this is the definitive experiment in principle, the technical problems are great. As of the present time, no group has yet produced convincing results. [23]

7. The Conventional Explanation of the G. S. I. Results

The author and A. Robatino have pointed out that the sharp positron spectrum resembles electron spectra found in atomic collisions by Niehaus and coworkers. [24,25] The quantum mechanics in both cases is analogous. In our point of view, the sharpness of the spectra arises from interferences arising at avoided crossings of potential curves of the molecules formed by the collision partners. In particular, such a model is consistent in a natural way with the multiple summed energies found by the G. S. I. experimenters. [15,16,19] The molecular model predicts very different angular distributions than those of the particle model. [26] The more recent discovery of electron positron pairs is equally consistent with the molecular model, as with more exotic explanations. [26]

A recent analysis of the G. S. I. experiments shows that the design of the experiments did not decide between the particle explanation [18,19] and the conventional, atomic physics model of the collisions. [27] Furthermore, the kinematic data of these experiments is incomplete. The sharpness in the summed energy of the electron positron pair is not unambiguous evidence of a back-to-back decay mechanism of a particle or other entity. Neither the equality of the e^+, e^- energies nor the equal but opposite momenta are directly observed. These experiments give too little information about the angular distributions or the relative energies of the electron-positron pair.

It seems to the author that a conventional explanation, which follows the principle of scientific parsimony, is preferable as a point of departure, before postulating new entities, involving totally new principles of physics. As experimenters still pursue with undiminished enthusiasm the possibility of a new physics, the investigation of the atomic alternative awaits the definitive test. Such a test would involve a study of the angular distribution of the electrons and/or positrons with a different geometry than that used by the G. S. I. EPOS group. [15]

Such an investigation would produce more than positive or negative results, proving or disproving the particle (or new Q. E. D. phase) explanations. If the

G. S. I. results in fact arose from molecular interactions, this would be the beginning of a new field of atomic physics, involving for the first time excitation from the negative energy, positron continuum into the positive energy, electron continuum. [27]

References

1. Facets of Physics, V.W. Hughes and D. A. Bromley, Eds., (Academic Press, N.Y., 1970) as told by Gregory Breit on the occasion of his retirement at Yale University. Breit goes on to say, "The meaning of Einstein's remark is also especially appropriate in the present period, which shows the effects of pushing, elbowing and self-advertising that physicists learned during the war years." Needless to say, Breit's statement has some relevance to the current situation in physics
2. T.W. Hänsch, I.S. Shahin, and A.L. Schawlow, Nature **235**, 63 (1972)
3. M.G. Boshier et al., Nature **330**, 463 (1987) and talk by D.N. Stacey
4. S. Triebwasser, E.S. Dayhoff and W.E. Lamb, Phys. Rev. **89**, 98 (1953)
5. See talks by F. Pipkin and Yu. L. Sokolov in this book. See also G. W. F. Drake, J. Patel and A. van Wijngaarden, Phys. Rev. Lett. **60**, 1002 (1988) for He^+
6. C. Wieman and T. Hänsch, Phys. Rev. **A22**, 192 (1980)
7. A. Clairon, A. Van Lerberghe, Ch. Breant, Ch. Salomon, G.Camy and Ch. J. Borde, J. de Phys. **42**, C8-127 (1981) [OsO_4]; S.N. Bagaev, V.P. Chebotaev, Sov. Phys.-Usp.(USA) **29**, 82 (1986) [Ch_4]
8. D. Kleppner, Bull. Amer. Phys. Soc. **20**, 1458 (1976); R. Hulet and D. Kleppner, Phys. Rev. Lett. **51**, 1430 (1983); J. Liang, M. Gross, P. Goy, and S. Haroche, Phys. Rev. **A33**, 4437 (1986). See the talk by M. Gross in this session
9. For detailed discussion of this proposal, see Laser Spectroscopy VIII (Eds. W. Persson and S. Svanberg, Springer-Verlag, 1987), article by P. Zhao, W. Lichten, J.C. Bergquist and H.P. Layer, p. 12; also P. Zhao et al., submitted to the Physical Review; also the talk by F. Biraben
10. S. Chu, A. P. Mills, Jr., A. G. Yodh, K. Nagamine, Y. Miyake, and T. Kuga, Phys. Rev. Lett. **60**, 101 (1988)
11. M. Deutsch, Phys. Rev. **82**, 455 (1951); Prog. Nucl. Phys. **3**, 131 (1953)
12. See the discussion in Physics Today, July 1982, p. 17. The author also is indebted to Peter Mohr and Steven Weinberg for explanations of the possible significance of the positronium results in this and the next section
13. S. Chu, private communication

14. For further discussion and references, see the speakers on the session on positronium and muonium in this book

15. For a recent review and guide to the literature, see T.E. Cowan and J.S. Greenberg in Physics of Strong Fields edited by Walter Greiner (Plenum, 1987, NY) p. 111

16. For a history of early history of this field, see S.J. Brodsky and P. J. Mohr in Structure and Collisions of Ions and Atoms, I Sellin, Ed., (Springer-Verlag, New York, 1978), chapter 2

17. J. Rafelski, B. Muller and W. Greiner, Z. Phys. **A285**, 49 (1978)

18. T. Cowan et al., Phys. Rev. Lett. **54**, 1761 (1985)

19. T. Cowan et al., Phys. Rev. Lett. **56**, 444 (1986)

20. For reviews, see M. Suzuki, Phys. Lett. **B175**, 364 (1986); L.M. Krauss and M. Zeller, Phys. Rev. **D34**, 3385 (1986); A.B. Balantekin, Nucl. Instr. Meth. Phys. Res. **B24/25**, 273 (1987)

21. B. Müller, J. Reinhardt, W. Greiner and A. Schäfer, J. Phys. **G12**, L109 (1986); C.Y. Wong and R.L. Becker, Phys. Lett. **B182**, 251 (1986) who examine new types of resonances in the positronium system; R.D. Peccei, T.T. Wu and T. Yanigada, Phys. Lett. **172B**, 435 (1986); D.Y. Kim and M.S. Zahir, Phys. Rev. **D35**, 886 (1987); D.G. Caldi and A. Chodos, Phys. Rev **D36**, 2876 (1987); W.J. Ng and Y. Kikuchi, Phys. Rev. **D36**, 2880 (1987); J.B. Kogut, E. Dagotto and A Kocic, Phys. Rev. Lett. **60**, 772 (1988)

22. K. Danzmann, W.E. Meyerhof et al., Phys. Rev. Lett. **59**, 1885 (1987)

23. For example, see K.A. Erb, I.Y. Lee and W.T. Milner, Phys. Lett. **B181**, 52 (1986); K. Maier et al., Z.Phys. A-Atomic Nuclei **326**, 527 (1988); A.P. Mills, Jr. and J. Levy, Phys. Rev. **D36**, 707 (1987); R. Peckhaus, Th.W. Elze, Th. Happ, and Th. Dresel, Phys. Rev. **C36**, 83 (1987); T.F. Wang, I. Ahmad, S.J. Freedman, R.V.F. Janssens, and J.P. Schiffer, Phys. Rev. **C36**, 2136 (1987); U. von Wimmersperg et al., Phys. Rev. Lett. **59**, 266 (1987); H. Tsertos et al., Phys. Lett. **B207**, 273 (1988)

24. W. Lichten and A. Robatino, Phys. Rev., Lett. **55**, 135 (1985); W. Lichten, A.I.P. Conference Proceedings **136**, 319 (1985); W. Lichten and A. Robatino, Phys. Rev. Lett. **55**, 135 (1985)

25. R. Morgenstern, A. Niehaus, and U. Thielmann, Phys. Rev. Lett. **37**, 199 (1976); G. Gerber and A. Niehaus, J. Phys. B: Atom. Molec. Phys. **9**, 123 (1970); A.C. Kessel, R. Morgenstern, B. Müller, and A. Niehaus, Phys. Rev. **A20**, 804 (1979)

26. A more detailed discussion is given in a paper by W. Lichten, submitted to Phys. Rev. D

27. W. Lichten, unpublished

Doppler-Free Two-Photon Spectroscopy of Hydrogen Rydberg States: Remeasurement of R_∞

M. Allegrini, F. Biraben, B. Cagnac, J.C. Garreau, and L. Julien*

Laboratoire de Spectroscopie Hertzienne de l'ENS**,
4, Place Jussieu, Tour 12 1er étage,
F-75252 Paris Cedex 05, France
* Permanent Address: Istituto di Fisica Atomica e Molecolare del CNR, Pisa, Italy
**Laboratoire associé au CNRS: UA18

1. Introduction

In the last few years, the application of very high resolution laser spectroscopy to atomic hydrogen has allowed the determination of the Rydberg constant with an increasing precision up to a few parts in 10^{10} /1-7/. Because R_∞ plays a key role in atomic physics, quantum electrodynamics and metrology, it is desirable to improve even further this precision ; experiments to this aim are currently being performed in various laboratories. In this work we report on latest measurements of R_∞ which are in fact limited by the precision of the wavelength standard in the optical domain (1.6×10^{-10}).

The method we use is Doppler free two-photon laser spectroscopy, applied to the atomic hydrogen transitions from the metastable 2S state to the Rydberg nD states (n = 8, 10, 12) /8/.

Compared to the method based on the study of the 2S-3P /4/ or 2S-4P /5/ one-photon transitions, our method takes advantage of the narrow linewidths of the Rydberg levels (~ 300kHz for the 10D level). From this point of view, the 1S-2S two-photon transition with a natural linewidth of 1.3 Hz offers in principle the best experimental resolution. However, this transition is affected by the uncertainty on the 1S Lamb shift, while the 2S Lamb shift has been measured with a very high precision and the nD Rydberg levels have negligible Lamb shifts. Thus the measurement of the 1S-2S frequency /6,7/ provides an experimental value of the 1S Lamb shift rather than an independent value of the Rydberg constant.

The Hydrogen Atom Editors: G.F. Bassani · M. Inguscio · T.W. Hänsch

2. Experimental method and apparatus

The apparatus consists of two major parts : (i) the vacuum chamber where the beam of metastable 2S atoms is produced, excited to the nD levels and detected (Fig.1) ;(ii) the optical system with the lasers used for the excitation and for the control of the transition frequency (Fig.2).

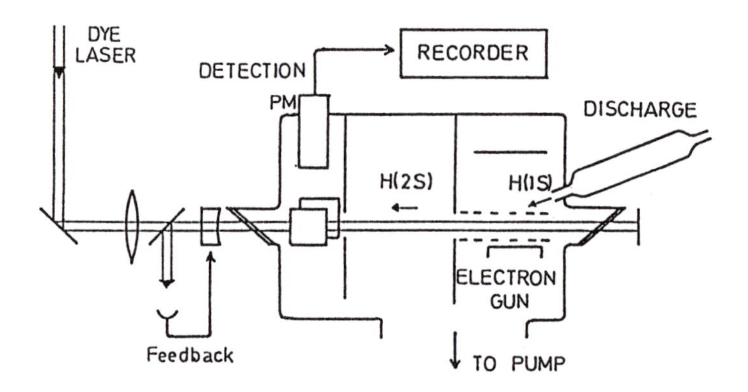

Figure 1 : Metastable beam apparatus

The metastable atomic beam is produced in two steps : molecular hydrogen flowing in a pyrex tube is dissociated by a radiofrequency discharge and the resulting ground state effusive beam enters in a first vacuum chamber ; then atoms are excited to the 2S state by electronic bombardment. Because of the inelastic collisions with electrons, the atomic beam is deviated by about 20°. The optical absorption takes place in a second vacuum chamber evacuated by two cryogenic pumps and where electric and magnetic fields are reduced as much as possible. In this chamber the metastable atomic beam is collinear with two counterpropagating laser beams, in order to reduce the line broadening due to the finite transit time of the atoms in the laser beams. Metastable atoms are detected in a third chamber where an electric field is applied to mix the 2S with the 2P levels and the resulting Lyman α radiation is measured.

The light source is a home made CW ring LD 700 dye laser, pumped by a Kr^+ laser. In the range 730-780 nm (wavelength of the two-photon 2S-nD transitions for $n \geqslant 8$) it provides a power of about 1W on single mode operation. The frequency stabilization is made by locking the laser to an external auxiliary Fabry-Perot cavity indicated FPA in Fig.2 ; the resulting linewidth is about 50 kHz.

A Fabry-Perot cavity (shown in Fig.1) having its optical axis coincident with the metastable atomic beam provides a standing wave that induces the two-photon transitions. This cavity is locked to the laser frequency in order to increase the intensity of the standing wave to ~ 50W in each propagation direction.

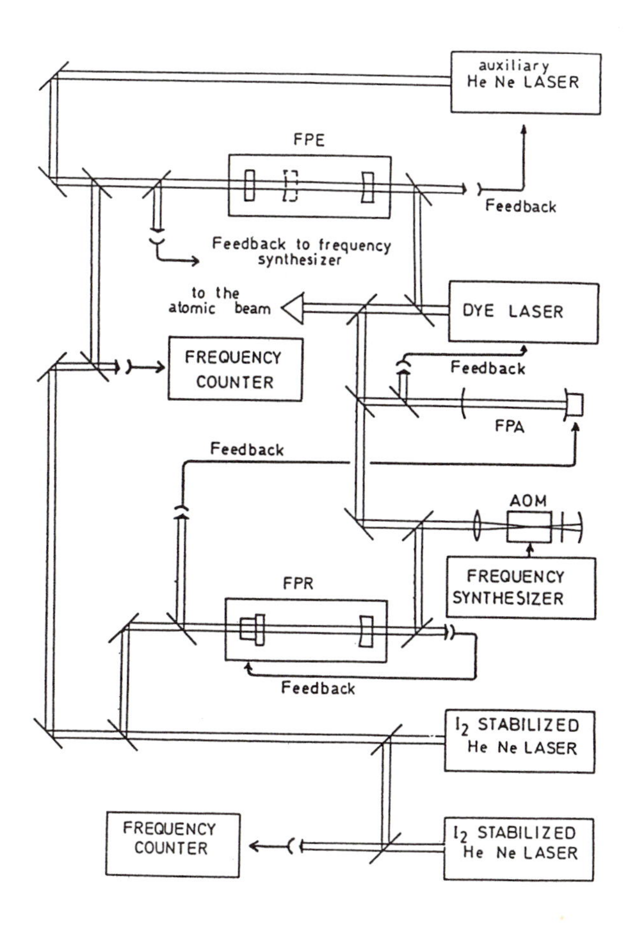

Figure 2 : Experimental set-up for the control and the measurement of the dye laser frequency

To sweep the dye laser its beam is split and the secondary beam is driven into an acousto-optic device. The frequency-shifted beam is reflected back into the acousto-optic crystal so that one of the emerging beams is shifted twice. This beam then enters a reference Fabry-Perot cavity (indicated as FPR in Fig. 2) of very high finesse, whose length is locked to an I_2 - stabilized

He-Ne laser. The frequency of the shifted infrared beam is locked to this reference Fabry-Perot cavity whose length is fixed. By changing the acousto-optic modulation frequency, which is provided by a computer-controlled frequency synthesizer, we can therefore precisely control the dye laser frequency over a range of 250 MHz centered at any desired frequency.

3. Study of the line profiles

After a two-photon excitation from the 2S metastable state, about 95% of atoms in the nD states undergo radiative cascade to the 1S ground state . The two-photon transition can then been detected by observing the corresponding decrease of the 2S beam intensity. Figure 3 shows a recording of the 2S intensity when the laser wavelength is swept through the $2S_{1/2}(F=1) - 10D_{5/2}$ transition of the hydrogen atom (the signal has been averaged over 10 scans). The signal amplitude corresponds to a 11% decrease of the metastable beam intensity. The experimental linewidth (in terms of total two-photon transition frequency) is 1.25 MHz, to be compared to the natural width of the transition which is 296 kHz.

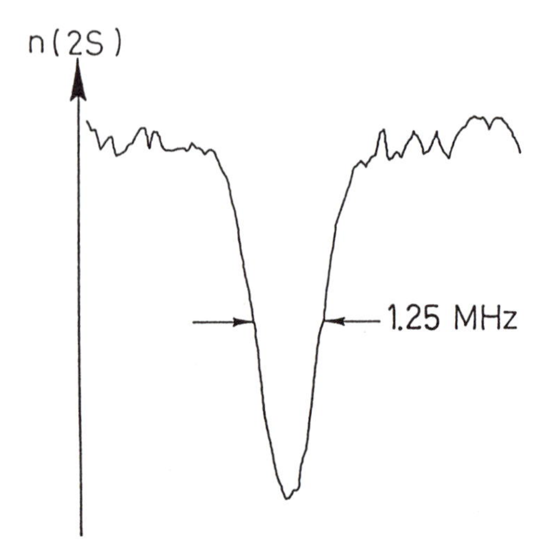

Figure 3 :
Recording of the $2S_{1/2}(F=1)-10D_{5/2}$ two-photon transition in hydrogen observed as a decrease of the metastable beam intensity n(2S).

There are several possible causes for the broadening and shift of the signal :

(i) The laser linewidth is responsible for a line broadening of 100 kHz.

(ii) Second order Doppler effect : for an atomic beam of 3.2 km/s mean velocity, the second-order Doppler effect decreases the line frequency by 44 kHz and broadens it by about 60 kHz.

(iii) Finite transit time /9/ : the geometry of an atomic beam gives a maximum line-broadening of 14 kHz corresponding to the largest possible angle of atomic trajectories with respect to the laser beams.

(iv) Light-shift and saturation : The laser beam waist inside the excitation chamber is $w_0 \simeq 600$ μm. For a 50W light power in each propagation direction, the light shift for an atom at rest in the center of the laser beam is about 560 kHz. Because the light power seen by each atom depends on its trajectory and varies along it, the light shift also contributes to the line broadening. For the same light power inside the excitation chamber 1m long, there is also a broadening effect due to the saturation of the two-photon transition (excitation rate 1.7×10^5 s^{-1}, transit time 2.8×10^{-4} s).

The line broadening due to light-shift and saturation is shown in Fig.4 where the experimental linewidth of the $2S_{1/2} - 10D_{5/2}$ transition in deuterium is reported versus the light intensity transmitted through the excitation cavity.

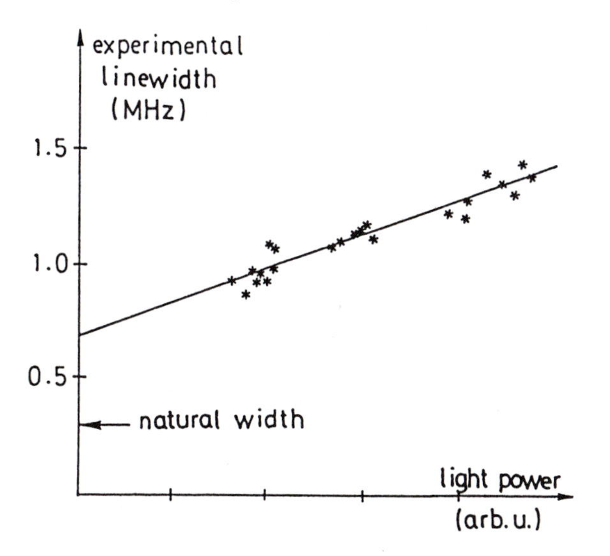

Figure 4 : Example of the variation of the experimental linewidth as a function of the light power inside the excitation cavity. Data are relative to the $2S_{1/2} - 10D_{5/2}$ transition in deuterium ; the arrow indicates the natural linewidth of the transition

A numerical calculation of the line profiles due to the combined effect of the natural lifetime, the light-shift and the saturation has been performed taking into account all possible trajectories of atoms inside the metastable beam. Actually, the study of experimental linewidths shows there are some other stray effects responsible for the broadening of the lines. We have considered their contribution by making a convolution of the line profile with a gaussian curve.

As an example, Fig.5a shows the fit of an experimental signal relative to the transition $2S_{1/2} - 10D_{5/2}$ in hydrogen with the theoretical line profile obtained after this convolution. The parameters of the fit are the line position, the light power and the gaussian broadening. The difference between the experimental and the fitted theoretical profiles is plotted in Fig. 5b which shows that there is no systematic error in the fit.

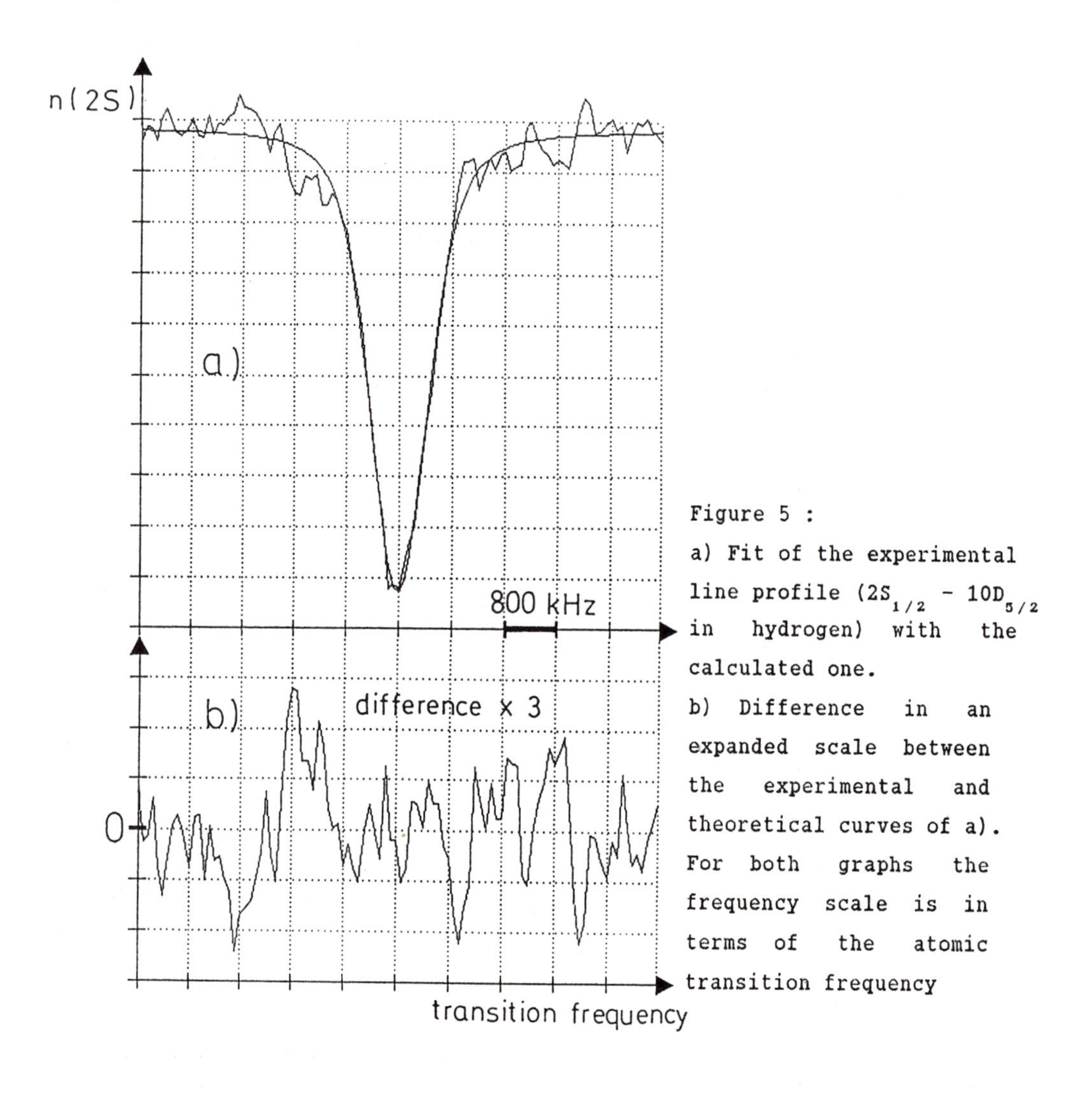

Figure 5 :
a) Fit of the experimental line profile ($2S_{1/2}$ - $10D_{5/2}$ in hydrogen) with the calculated one.
b) Difference in an expanded scale between the experimental and theoretical curves of a). For both graphs the frequency scale is in terms of the atomic transition frequency

The line position (relative to the frequency determined by the reference Fabry-Perot cavity) obtained from the fit is then investigated as a function of the light power (see Fig.6) ; extrapolation to zero light power gives the value corrected for light shifts.

Great care has been taken in the experimental set-up to avoid any stray field. In order to evaluate the effect of residual electric fields we have

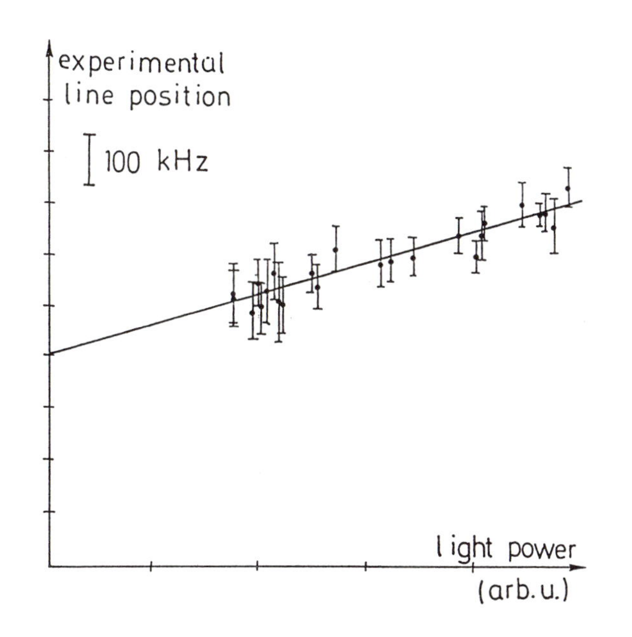

Figure 6 :
Example of an extrapolation of the two-photon line position versus the light power ($2S_{1/2}$-$10D_{5/2}$ transition in hydrogen)

excited the transitions to higher n because the Stark effect contribution to the line broadening varies as n^2. In Fig. 7 the signal for the 2S-20D transition is reported ; even on the assumption that the total broadening 950 kHz is due solely to the Stark effect and to the light shift, we obtain a stray field of 2mV/cm. As an example, the effect of such a low field on the $2S_{1/2}$- $10D_{5/2}$ transition is a broadening of $\sim$ 100kHz and a shift of $\sim$ 1kHz.

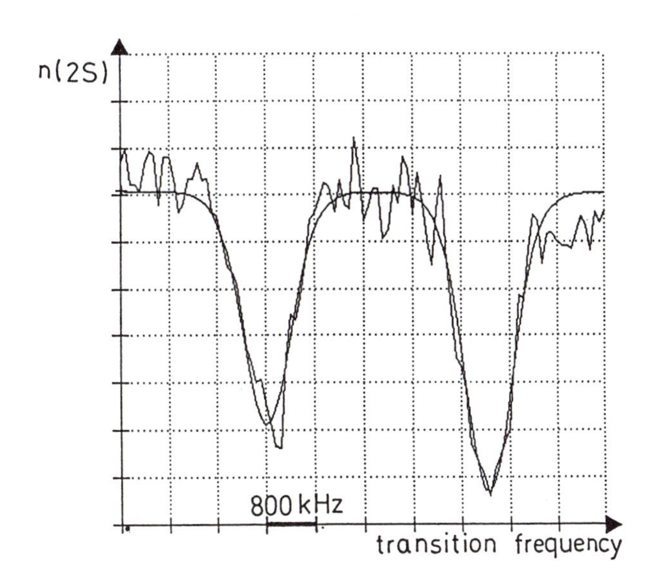

Figure 7 :
Same fit as that of Fig.5a) for the two fine structure components of the 2S-20D transition

4. Measurement of the transition wavelength

The absolute frequency position of the two-photon transition is measured by comparing the infrared dye laser wavelength with an I_2- stabilized He-Ne reference laser at 633 nm (see Fig.2). The key of the wavelength comparison is a nonconfocal etalon Fabry-Perot cavity (indicated as FPE in Fig.2) kept under a vacuum better than 10^{-6} mbar. This optical cavity is built with two silver-coated mirrors, one flat and the other spherical (R = 60 cm), in optical adhesion to a zerodur rod. Its finesse is ~ 60 at 633 nm and ~ 100 at 778 nm. An auxiliary He-Ne laser as well as the dye laser are mode-matched and locked to this Fabry-Perot cavity. Simultaneously the beat frequency between the auxiliary and etalon He-Ne lasers is measured by a frequency counter.

The frequencies at the red and infrared radiations inside the etalon Fabry-Perot cavity are determined by the resonant condition /10/

$$\nu = \frac{c}{2L} \left(N + \phi + \Phi \right)$$

where L is the cavity length, N is an integer number, ϕ is the reflective phase shift for light of frequency ν and Φ is the Fresnel phase shift

$$\Phi = \frac{1}{\pi} \cos^{-1} \left(1 - \frac{L}{R} \right)^{1/2}$$

The phase shift ϕ due to the mirror coatings is eliminated by the method of virtual mirrors /10/ by using two rods of different lengths (50cm and 10cm) for the Fabry-Perot. The Fresnel phase shift Φ is determined by measuring the frequency interval between the fundamental TEM_{00} mode and the first transverse mode TEM_{01} (or TEM_{10}). In principle TEM_{01} and TEM_{10} are degenerate by symmetry. However, due to imperfections in the mirror curvature, they are not exactly degenerate. Depending on the alignment of the laser beam with respect to the FPE optical axis, we may excite various superpositions of these two modes which give different results for the relative frequency of the first transverse mode and the TEM_{00} mode. We have made measurements for various alignments and the result is shown in Fig.8 which reports the dependence of this frequency difference upon the orientation of the transmitted light spots. The Fresnel phase-shift we need is then given by the mean value /11/ of the curve of Fig.8.

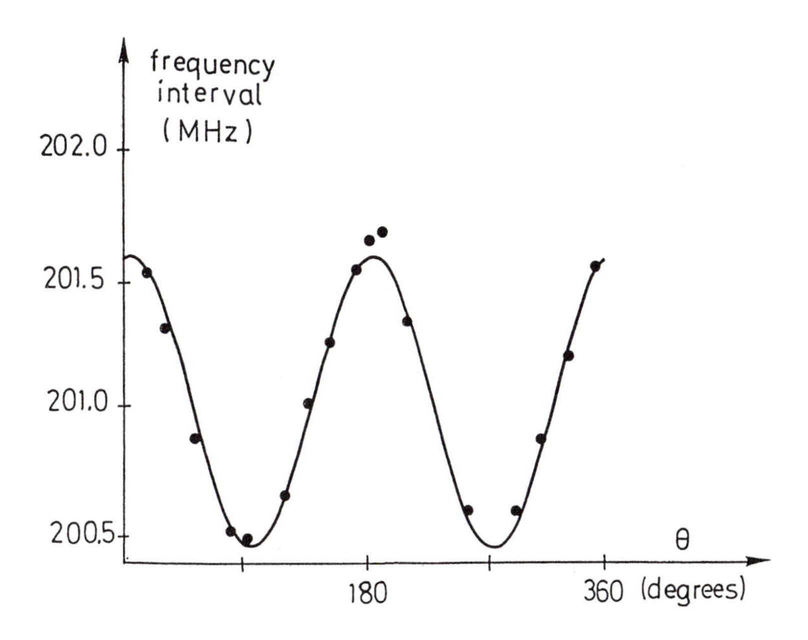

Figure 8 : Frequency interval between the fundamental mode and the first transverse mode of the Fabry-Perot etalon, measured as a function of the orientation of the plane of incidence of the auxiliary He-Ne laser. Similar behaviour has been observed for the dye laser radiation

Finally we have compared our reference I_2-stabilized He-Ne laser with that at the "Institut National de Métrologie" which had been previously compared with the standard He-Ne lasers of the "Bureau International des Poids et Mesures". As a result, the frequency of our reference laser relative to the lasers of the BIPM is known with a precision better than 10^{-11}.

5. Results

In both atomic hydrogen and deuterium we have studied the three transitions $2S_{1/2}-8D_{5/2}$, $2S_{1/2}-10D_{5/2}$ and $2S_{1/2}-12D_{5/2}$. The frequencies measured, corrected for hyperfine splittings and for the second-order Doppler effect, are reported in Table I.

Table I

Transition	Frequency measured (MHz)	Rydberg constant -109737 (cm^{-1})
$2S_{1/2}-8D_{5/2}$ in H	770649561.764 (38)	.3157130 (55)
$2S_{1/2}-8D_{5/2}$ in D	770859253.058 (38)	.3157158 (55)
$2S_{1/2}-10D_{5/2}$ in H	789144886.620 (39)	.3157136 (55)
$2S_{1/2}-10D_{5/2}$ in D	789359610.444 (39)	.3157124 (55)
$2S_{1/2}-12D_{5/2}$ in H	799191727.593 (40)	.3157120 (55)
$2S_{1/2}-12D_{5/2}$ in D	799409185.185 (40)	.3157149 (55)

Using the theoretical work of Erickson /12/ for the 2P-nD interval and either the experimental value of the 2S Lamb-shift in hydrogen /13/ or the theoretical value of the 2S Lamb-shift in deuterium ($\mathcal{S}$ = 1059.229 MHz) /14/, we can deduce six independent determinations of the Rydberg constant. These values are in excellent agreement each with other. Our final result is :

$$R_\infty = 109\ 737.\ 315\ 7136\ (186)\ cm^{-1}.$$

The main source of error is due to the determination of the frequency of the standard laser /15/. In fact our precision with respect to this standard is less than 5×10^{-11}.

The various experimental errors are evaluated in Table II.

Table II : Estimated errors (parts in 10^{11})

Electron-to-proton mass ratio	1.1
Lambshift and energy level calculations	1.5
Stark effect	0.5
Statistical	1.4
Uncertainty of the theoretical line shape	2.0
Second-order Doppler effect	2.0
Ageing of the mirrors	1.8
Fresnel phase shift measurement	1.2
Comparison between the He-Ne lasers	1.0
Root mean square error	4.4
Standard laser	16.0
Final error	16.6

In Fig.9 our result is compared to those recently obtained by other groups, either from the Balmer transitions or from the 1S-2S transition assuming the theoretical value of the 1S Lamb-shift.

6. Conclusion

We have applied the Doppler-free two-photon technique to atomic hydrogen Rydberg states, and have achieved a new determination of the Rydberg constant. A more refined analysis of our results is at present in progress. As it stands, the main limitation of our determination is due to the frequency of the I_2 - stabilized He-Ne laser. This result shows that a new frequency standard is needed in the optical range. Moreover, since the hydrogen atom has transitions both in the visible and microwave ranges, the Rydberg constant may play a key role in the direct comparison between microwave and optical frequencies, as it has been pointed out in various occasions.

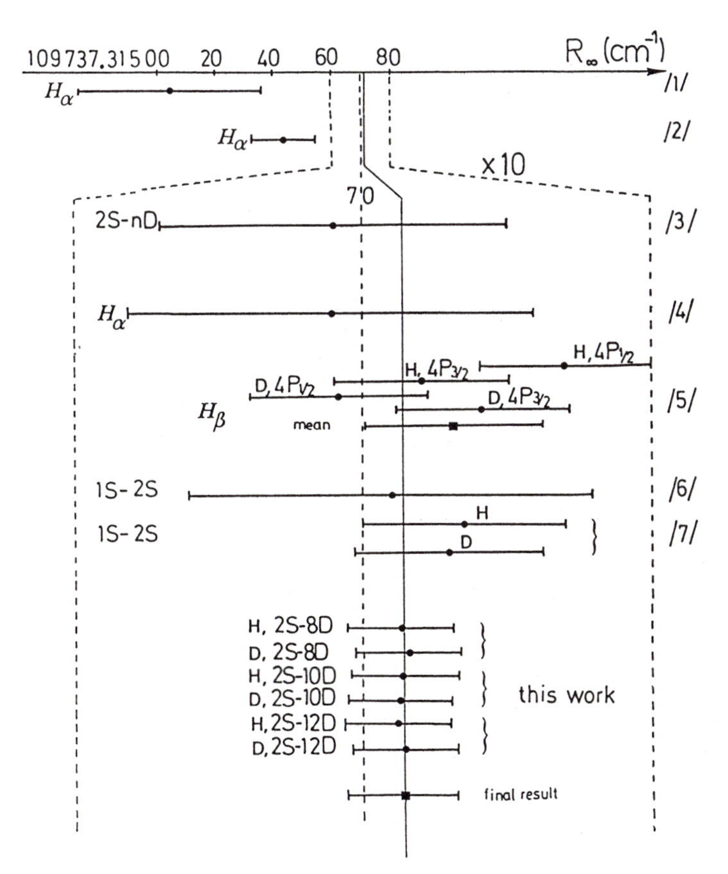

Figure 9 : Comparison of our result with those recently obtained by other groups

The authors are grateful to Prof. G.W. Series for critical reading of the manuscript. This research is supported in part by the "Bureau National de Métrologie".

REFERENCES

1. J.E.M. Goldsmith, E.W. Weber and T.W. Hänsch, Phys. Rev. Lett. 41, 1525 (1978)
2. S.R. Amin, C.D. Caldwell and W. Lichten, Phys. Rev. Lett. 47, 1234 (1981)
3. F. Biraben, J.C. Garreau and L. Julien, Europhys. Lett. 2, 925 (1986)
4. P. Zhao, W. Lichten, H.P. Layer and J.C. Bergquist, Phys. Rev. A34, 5138 (1986)
5 P. Zhao, W. Lichten, H.P. Layer and J.C. Bergquist, Phys. Rev. Lett. 58, 1293 (1987); and Proceedings of the VIIIth International Conference on Laser Spectroscopy (Are, June 1987): Laser Spectroscopy VIII edited by W. Persson and S. Svanberg (Springer Verlag, 1987) p.12
6. R.G. Beausoleil, D.H. McIntyre, C.J. Foot, E.A. Hildum, B. Couillaud and T.W. Hänsch, Phys. Rev. A35, 4878 (1987)
7. M.G. Boshier, P.E.G. Baird, C.J. Foot, E.A. Hinds, M.D. Plimmer, D.N. Stacey, J.B. Swan, D.A. Tate, D.M. Warrington and G.K. Woodgate, Nature 330, 463 (1987)
8. F. Biraben and L. Julien, Opt. Comm. 53, 319 (1985); and L. Julien, F. Biraben and B. Cagnac in Proceedings of Journée Thématique du B.N.M. (Paris, Sept. 1985): Bulletin du B.N.M. 66, 31 (1986)
9. F. Biraben, M. Bassini and B. Cagnac, J. Physique 40, 445 (1979)
10. H.P. Layer, R.D. Deslattes and W.G. Schweitzer, Appl. Opt. 15, 734 (1976)
11. C. Fabre, R.G. Devoe and R.G. Brewer, Opt. Lett. 11, 365 (1986)
12. G.W. Erickson, J. Phys. Chem. Ref. Data 6,, 831 (1977)
13. V.G. Pal'chickov, Yu.L. Sokolov and V.P. Yakovlev, Pis'ma Zh. Eksp. Teor. Fiz. 38, 347 (1983) (JETP Lett. 38, 418 (1983))
14. G.W. Erickson and H. Grotch, Phys. Rev. Lett. 60, 2611 (1988)
15. D.A. Jennings, C.R. Pollock, F.R. Petersen, R.E. Drullinger, K.M. Evenson, J.S. Wells, J.L. Hall and H.P. Layer,Optics Lett. 8, 136 (1983)

Two-Photon Transitions Between Discrete States

A. Quattropani[1] *and N. Binggeli*[2]

[1]Institut de Physique Théorique, EPFL,
Ecublens, CH-1050 Lausanne, Switzerland

[2]Institut de Physique Appliquée, EPFL,
Ecublens, CH-1050 Lausanne, Switzerland

In this note, we discuss different approximation schemes for the evaluation of the two-photon transition rate between discrete states. Non relativistic atomic hydrogen is used as a test of the reliability of the methods. We consider a one particle system described by a Hamiltonian H_0, whose eigenstates and eigenvalues are denoted by $|n>$ and E_n, respectively. In the gauge with div$\mathbf{A}$ = 0, the interaction of the particle with the electromagnetic field has the usual form

$$h_J = \frac{-e}{mc}\mathbf{A}\cdot\mathbf{p} + \frac{1}{2m}\left(\frac{e}{c}\right)^2\mathbf{A}^2 \tag{1}$$

Since we are interested in the two-photon transition rate, we will consider a vector potential

$$\mathbf{A}_J = \hat{\mathbf{e}}_1\, A_{01} \exp[-i\,\omega_1 t] + \hat{\mathbf{e}}_2\, A_{02} \exp[-i\,\omega_2 t] + \text{c.c.} \tag{2}$$

The dipole approximation in (2) is justified in all the cases that we will consider [1]. As a consequence of this approximation the term proportional to $\mathbf{A}^2$ in the interaction does not contribute to the two-photon transition rate, which takes in the velocity gauge the well known form

$$W_J^{(2)}(m,n) = \frac{2\pi e^4\,|A_{01}A_{02}|^2}{\hbar^4 c^4}\,|\Omega_J(m,n)|^2\,\delta(\Delta\omega)\ , \tag{3}$$

where

$$\Omega_J(m,n) = (1+P_{12})\sum_{\mu} \frac{<m|\hat{\mathbf{e}}_1\cdot\mathbf{p}/m|\mu><\mu|\hat{\mathbf{e}}_2\cdot\mathbf{p}/m|n>}{\omega(\mu)-\omega(n)-\omega_2}\ . \tag{4}$$

In (3) A_{01}, A_{02} are the amplitudes of the vector potentials with frequencies ω_1, ω_2 and polarization unit vectors $\hat{\mathbf{e}}_1$, $\hat{\mathbf{e}}_2$, the operator P_{12} interchanges $(\hat{\mathbf{e}}_1, \omega_1)$ with $(\hat{\mathbf{e}}_2, \omega_2)$, $\omega(\mu) = E_\mu/\hbar$, $\Delta\omega = \omega(m) - \omega(n) - \omega_1 - \omega_2$, and $\{|\mu>\}$ represents a complete set of

The Hydrogen Atom Editors: G.F. Bassani · M. Inguscio · T.W. Hänsch

eigenstates of H_0 with eigenvalues E_μ. Since rot$\mathbf{A} = 0$ in the dipole approximation, the vector potential can be eliminated by a gauge transformation. In the so called length gauge J_o with $A_{J_o} = 0$, the interaction Hamiltonian takes the form

$$h_{J_o} = \frac{e}{c} \mathbf{x} \cdot \frac{\partial}{\partial t} \mathbf{A}_J \tag{5}$$

The transition rate in the length gauge can be obtained from (3,4) by the substitution $\mathbf{p} \cdot \hat{e}_i / m \rightarrow \omega_i \mathbf{x} \cdot \hat{e}_i$. It is important to notice that the transition rates between eigenstates of the same free Hamiltonian H_0, evaluated in different gauges are gauge independent, i.e.

$$W_J^{(2)} = W_{J_o}^{(2)} . \tag{6}$$

For the special case of non relativistic Hydrogen, the multiphoton transition rate can be obtained exactly using methods based on Green function techniques, which avoid summations over intermediate states. This approach was introduced in order to treat time independent problems, and later extended to time dependent ones [2]. In the Green function method, the evaluation of the infinite sums over intermediate states is reduced to the solution of a linear differential equation. For systems other than Hydrogen, this method can also be used, but the associated differential equation has to be integrated numerically. The two-photon transition rate can also be evaluated exactly by performing explicitly the summation over the intermediate states.

We present exact calculations of the two-photon transition rates between discrete states of the non relativistic Hydrogen atom and we compare these results with those obtained with various approximation schemes. We plot in Fig. 1 the 1s-3s resonant transition amplitude D_1^3 as a function of one of the photon frequencies [3], where

$$D_1^3 [J_0] = \frac{3}{2} \frac{2\pi R}{a_0^2 \omega_1 \omega_2} \Omega_{J0} (1s, 3s) \tag{7}$$

and a_0 and R are the Bohr radius and the Rydberg frequency, respectively.

The transition amplitude exhibits dramatic structure over the entire frequency range. Besides the resonance enhancement for $\omega_1 = 2\pi R [1 - 1/\mu^2]$ with $\mu = 2, 3$, the plot clearly shows the two-photon transparency. We refer to [3, 4] for the behavior of the transition rate $W^{(2)}$ to higher excited s and d states and for the dependence of $W^{(2)}$ on the polarization of the incoming radiation. A detailed discussion of the transparencies for different transitions is also given in a recent paper by Florescu et al [4]. In Table 1, we reproduce some of their results, showing in particular that the position of the transparencies are almost independent of the final state.

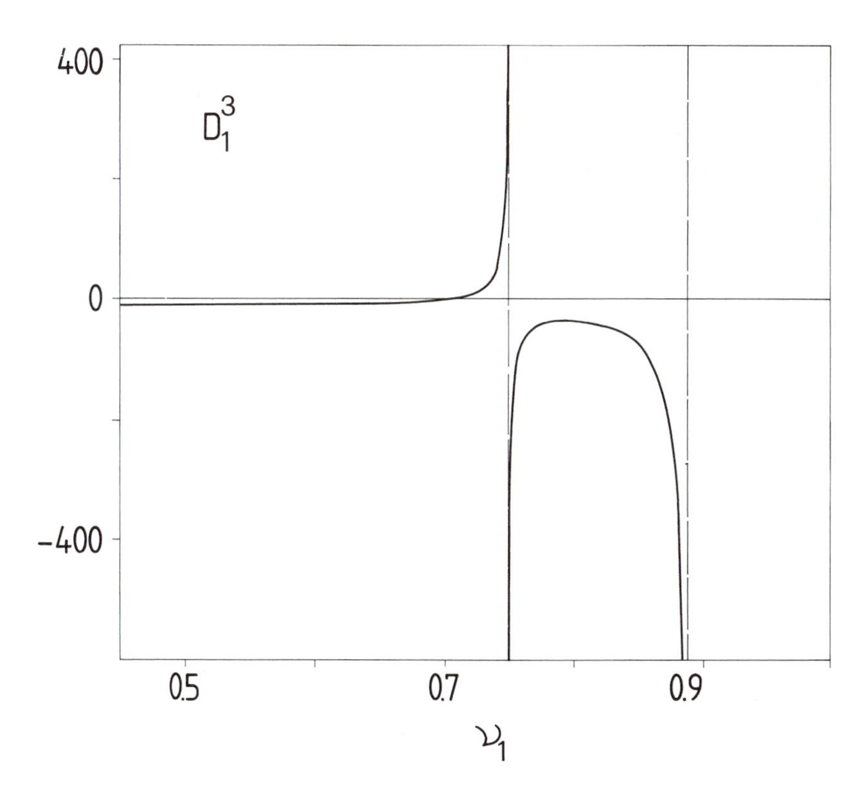

Fig. 1. Two-photon resonant transition amplitude D_1^3 for 1s-3s transitions in Hydrogen has a function of ν_1, for $\frac{4}{9} \leq \nu_1 < \frac{8}{9}$ [Rydberg units]

Frequency interval	0.3750 0.7500	0.7500 0.8899	0.8899 0.9375	0.9375 0.9600	0.9600 0.9720	0.9720 0.9796
1s-3s	0.6936					
1s-4s	0.6912	0.8714				
1s-4d		0.8381				
1s-5s	0.6905	0.8707	0.9299			
1s-5d		0.8307	0.9215			
1s-6s	0.6901	0.8705	0.9295	0.9560		
1s-6d		0.8274	0.9192	0.9529		
1s-10s	0.6897	0.8704	0.9294	0.9557	0.9697	0.9780
1s-10d		0.8232	0.9163	0.9510	0.9676	0.9767
1s-20s	0.6895	0.8703	0.9293	0.9557	0.9697	0.9780
1s-20d		0.8216	0.9160	0.9505	0.9672	0.9766
1s-∞s	0.6894	0.8703	0.9293	0.9557	0.9697	0.9780
1s-∞d		0.8211	0.9157	0.9503	0.9671	0.9765

Table 1. Two-photon transparency frequencies for several 1s-ns and 1s-nd transitions (frequencies in Rydberg units), from Ref. [4]

We turn now to the evaluation of $W^{(2)}$ using various approximation schemes. Problems that one has to consider in an approximate calculation are the rate of convergence of the sum over intermediate states, the importance of the contribution from the continuum and the gauge independence of the final result. For frequencies out of the condition of resonance enhancement, a heuristic approximation generally adopted, is to consider only a limited number of intermediate states, choosing those close to initial and final states because of their smaller energy denominators. As an example, we present in Table 2 the resonant 1s-3s two-photon transition amplitude in atomic Hydrogen, for various values of the incident frequencies, in the length and velocity gauges. We notice that both gauges give the same final result, but the respective contribution from the intermediate states are dramatically different.

OMEGA(1) = 0.44444, OMEGA(2) = 0.44444

N	D[J0,N]	D[J,N]
2	3.96259	2.08965
3	-10.83427	2.08965
4	-7.45560	1.31014
5	-6.54089	0.99402
6	-6.14107	0.83003
7	-5.92520	0.73293
8	-5.79366	0.67034
9	-5.70690	0.62749
10	-5.64637	0.59682
11	-5.60233	0.57408
12	-5.56921	0.55674
13	-5.54364	0.54320
14	-5.52347	0.53243
15	-5.50726	0.52372
DD	-5.39815	0.46363
DC	2.18723	-3.67456
D	-3.21093	-3.21093

OMEGA(1) = 0.21389, OMEGA(2) = 0.67500

N	D[J0,N]	D[J,N]
2	9.20118	6.63866
3	-11.04357	6.63866
4	-6.71940	5.27370
5	-5.57604	4.73307
6	-5.08192	4.45579
7	-4.81685	4.29266
8	-4.65598	4.18792
9	-4.55015	4.11642
10	-4.47647	4.06533
11	-4.42292	4.02751
12	-4.38270	3.99869
13	-4.35168	3.97622
14	-4.32721	3.95835
15	-4.30756	3.94389
DD	-4.17555	3.84442
DC	2.50625	-5.51372
D	-1.66930	-1.66930

OMEGA(1) = 0.12389, OMEGA(2) = 0.76500

N	D[J0,N]	D[J,N]
2	-39.39279	-43.29640
3	-70.23253	-43.29640
4	-64.38017	-46.11053
5	-62.88897	-47.18464
6	-62.25544	-47.72621
7	-61.91877	-48.04184
8	-61.71562	-48.24332
9	-61.58251	-48.38032
10	-61.49007	-48.47796
11	-61.42303	-48.55011
12	-61.37274	-48.60499
13	-61.33400	-48.64774
14	-61.30347	-48.68171
15	-61.27898	-48.70916
DD	-61.11479	-48.89761
DC	2.91474	-9.30244
D	-58.20004	-58.20004

Table 2. Two-photon transition amplitude D_1^3 for 1s-3s transitions in Hydrogen for three different values of the incident photon frequency in the length and velocity gauges J_0 and J respectively. $D[J_0, N]$ (D[J; N]) denotes the contribution from the discrete spectrum in the length (velocity) gauge up to the N-th intermediate state. The total contribution from the discrete and continuum spectrum are denoted by DD and DC, respectively

Robinson [5] has shown how to obtain gauge-independent results for the two-photon transition rates in cases when only a limited number of intermediate states are used. Following Robinson, we first transform the transition amplitudes in a more compact form

$$D[J_0] = \frac{3}{2\, v_1\, v_2} \sum_{\mu} F_\mu\, C_\mu \tag{8a}$$

and

$$D[J] = -\frac{3}{2\,\nu_1\,\nu_2} \sum_{\mu} F'_{\mu}\, C_{\mu} \tag{8b}$$

where

$$C_{\mu} = (1 + P_{12}) \langle m \,|\, \hat{e}_1 \,.\, \mathbf{x}\,/a_0 \,|\, \mu \rangle \; \langle \mu \,|\, \hat{e}_2 \,.\, \mathbf{x}\,/a_0 \,|\, n \rangle \, [\nu(\mu) - \nu(n) \; - \nu_1] \tag{9}$$

$$F_{\mu} = \frac{\nu_1\,\nu_2}{[\nu(\mu) - \nu(n) - \nu_2] \;\; [\nu(\mu) - \nu(n) - \nu_1]} \tag{10a}$$

$$F'_{\mu} = \frac{[\nu(\mu) - \nu(m)] \;\; [\nu(\mu) - \nu(n)]}{[\nu(\mu) - \nu(n) - \nu_2] \;\; [\nu(\mu) - \nu(n) - \nu_1]}\,. \tag{10b}$$

Using energy conservation, i.e. $\nu(m) - \nu(n) = \nu_1 + \nu_2$, one has

$$F'_{\mu} = 1 - F_{\mu}. \tag{11}$$

From the equality of the transition amplitudes in different gauges one obtains the sum rule

$$\sum_{\mu} C_{\mu} = 0\ . \tag{12}$$

The sum rule (12) can be used in order to have a gauge independent approximation scheme. The transition amplitude $D[J_0]$ is split into two parts :

$$D[J_0] = \sum_{\mu}^{(c)} F_{\mu} C_{\mu} + \sum_{\mu}^{(u)} F_{\mu} C_{\mu}\ , \tag{13}$$

in the first term, the summation extends over states which are exactly known or for which a good approximation is available (characterized states), while the states in the second summation are supposed to be unknown (uncharacterized states). In the approximation scheme proposed by Robinson, the coefficients F_{μ} under the sum over the uncharacterized states in (13) are replaced by a constant coefficient F_K, which is obtained from F_{μ} replacing $\nu(\mu)$ with an averaged frequency $\nu(K)$. Using then the sum rule (12), we obtain an approximate expression for the transition amplitude :

$$D[R; J_0] = \sum_{\mu}^{(c)} (F_{\mu} - F_K)\; C_{\mu}\,. \tag{14}$$

In the velocity gauge an analogous expression holds for the transition amplitude, and it can be shown [5] that the approximate results obtained in the length and velocity gauges are equal.

The approximation can be tested for the 1s-2s transition in Hydrogen. All p-states with $\mu > 2$ are taken as uncharacterized, the average frequency $\nu(K)$ is determined by fitting (14) to the exact transition amplitude $D_1^2\ [J_0]$ from [2] at $\nu_1 = 0.375$, leading to $\nu(K) = 0.0171$. In Table 3 we compare at various frequencies the exact amplitude D_1^2 with the amplitudes $D_1^2\ [R, J_0\ 2]$ and $D_1^2\ [J_0, 2]$, where only the 2p intermediate state is included. We notice that the error in $D_1^2\ [R, J_0\ 2]$ is less than 2%, and much smaller than the error in $D_1^2\ [J_0, 2]$. The gauge invariant approximation partially overcomes the major difficulty encountered in any approximate calculation; this scheme, however, relies on free parameters which must be determined independently and hence its application is not straightforward.

ν_1	$D_1^2\ [J_o]$	$D_1^2\ [R; J_o\ 2]$	% error	$D_1^2\ [J_o\ 2]$	% error
0.3750	- 11.7805			17.8785	(61.8)
0.5250	- 14.7319	- 14.8339	(0.69)	- 21.2839	(44.5)
0.6750	- 41.1484	- 41.8616	(1.73)	- 49.6624	(20.7)
0.6875	- 49.6878	- 50.5207	(1.68)	- 58.5113	(17.8)
0.7000	- 62.6595	- 63.6348	(1.56)	- 71.8331	(14.6)
0.7125	- 84.5252	- 85.6713	(1.36)	- 94.0971	(11.3)
0.7250	- 128.683	- 130.037	(1.05)	- 138.712	(7.79)
0.7375	- 262.165	- 263.772	(0.61)	- 272.722	(4.03)
0.7475	- 1334.33	- 1336.18	(0.14)	- 1345.37	(0.83)

Table 3. Comparison of theoretical transition amplitudes for various ν_1 using only the 2p-state as intermediate state. $D_1^2\ [J_o]$ is the exact theoretical value, $D_1^2\ [R; J_o\ 2]$ is from the modified Eq. (14) and $D_1^2\ [J_o\ 2]$ is from Eq. (8a), where in both cases only the 2p intermediate state has been included. The errors relative to the exact results are given in parenthesis. The average frequency $\nu(K) = 0.0171$ is determined by setting $D_1^2\ [R; J_o\ 2] = D_1^2\ [J_o\ 2] = - 11.7805$ for $\nu_1 = 0.375$. (From ref. [1])

Finally, we propose an alternative approach based on the representation of the Hamiltonian over a finite basis set. In general, the lowest variational states are close to the true states, but this is not the case for the high energy eigenstates, in particular for those with positive energy representing the continuous spectrum. In our approach, we still use the finite set of eigenfunctions to evaluate the sum over the intermediate states, including the continuum. Since the exact result can in principle be obtained with an infinite basis set, the accuracy of the procedure can be checked through the convergence of the transition rate versus basis size. For Hydrogen, we find fast convergence using real exponentional or Gaussian radial basis functions. As an example, we show in Table 4 the 1s-3s two-photon transition rate, at selected frequencies. The calculations were performed for both beams polarized in the z-direction, in the length and velocity gauges. Convergence tests showed that the basis size (24 Gaussian functions) is sufficient to provide results with 1% accuracy.

ν_1	LENGTH GAUGE D(1s-3s)	VELOCITY GAUGE D(1s-3s)	EXACT D(1s-3s)
0.4444	-3.2108	-3.2104	-3.2109
0.6750	-1.6699	-1.6693	-1.6693
0.7000	0.9832	0.9838	0.9847
0.7250	11.211	11.212	11.216
0.7475	226.71	226.71	226.81
0.7650	-58.184	-58.183	-58.200
0.8000	-38.301	-38.300	-38.310
0.8250	-46.571	-46.569	-46.580
0.8500	-74.412	-74.409	-74.420
0.8750	-219.98	-219.96	-219.98
0.8860	-1116.0	-1116.0	-1117.2

Table 4. 1s-3s Two-photon transition amplitude calculated with variational states. The radial basis set for s an p states are of the form $\varphi_s = c_\alpha \exp - \alpha r^2$ and $\varphi_p = c'_\alpha\, r \exp - \alpha r^2$ respectively. Twenty four different exponents α, with $10^{-3} \leq \alpha \leq 10^{-5}$ (a.u.) in geometrical sequency have been used.

Variational energies : (Rydberg units)

s-states - 1.0000 / - 0.1111

p-states - 0.2500 / - 0.1111 / - 0.0625/- 0.0397 / - 0.0214 / + 0.0821 / ...

1. See e.g. A. Quattropani and R. Girlanda, Rivista Nuovo Cimento 6, 1 (1983).
2. A. Dalgarno and J. T. Lewis, Proc. Roy. Soc., London, Ser. A, 233, 70 (1956); J. Gontier, N.K. Rahman and M. Trahn, Nuovo Cimento D, 4, 1 (1984); S. Baroni and A. Quattropani, Nuovo Cimento D, 5, 89 (1985), and references quoted therein.
3. F. Bassani, J. J. Forney and A. Quattropani, Phys. Rev. Letters 39, 1070 (1977); A. Quattropani, F. Bassani and Sandra Carillo, Phys. Rev. A25, 3079 (1982).
4. J. H. Tung, A. Z. Tang, G. J. Salamo and F. T. Chan, J. Opt. Soc. Am B3, 837 (1986); Viorica Florescu, Suzana Patrascu and O Stoican, Phys. Rev. A36, 2155 (1987).
5. C. Wilse Robinson, Phys. Rev. A26, 1482 (1982).

Two-Photon Spectroscopy of Hydrogen 1S–2S

D.N. Stacey

Clarendon Laboratory, Parks Road, Oxford, OX1 3PU, U.K.

1. Introduction and Overview

1.1 The interest in 1S - 2S; early experiments

In 1976, reviewing the field of two-photon spectroscopy, BLOEMBERGEN and LEVENSON wrote: " The 1S - 2S transition in atomic hydrogen is among the most important in physics, and it can be resolved by the Doppler-free two-photon technique. The experiment is a tremendous challenge....." [1]. More than a decade later, this remains a fair summary. The special nature of the 1S - 2S transition was pointed out independently by several authors, notably BAKLANOV and CHEBOTAYEV [2,3], who carried out a detailed study, and CAGNAC et al [4]. By a remarkable combination of properties it brings the ground state of hydrogen into the class of atomic levels which can be studied by the methods of high resolution laser spectroscopy. In recent years, experiments on the 1S - 2S transition in hydrogen and its isotopes have been used to test quantum electrodynamics, to determine the Rydberg constant and to give a value of the electron-proton mass ratio. It is true that so far all these objectives have been accomplished at least as well by other means; 1S - 2S experiments bristle with technical difficulties, and the first continuous wave (cw) measurements were reported only recently [5,6]. Now that this milestone has been passed, however, the potential for improvement is enormous.

To see the experiments in context, one must appreciate that high resolution spectroscopy of atomic hydrogen has largely been confined to transitions between excited states because of the experimental problems associated with studying the ground level. There is a 10 eV energy gap between the ground level and the nearest excited levels (in the n = 2 manifold); this corresponds to a wavelength of 121.5 nm, in the vacuum ultraviolet. To carry out a Doppler-free 1S - 2P experiment one would have to generate tunable narrow-band radiation at this wavelength; even then, the radiative lifetime of the upper level would give rise to a homogeneous linewidth of 100 MHz and one would still face the problem of finding a suitable frequency standard at 121.5 nm. The way in which measurements on the 1S - 2S transition can overcome these difficulties was shown in a remarkable series of experiments by HÄNSCH and his

The Hydrogen Atom Editors: G.F. Bassani · M. Inguscio · T.W. Hänsch

coworkers [7,8,9] to measure the ground state Lamb shift. These were carried out during the period 1974–80 at Stanford. The use of two-photon spectroscopy takes the wavelength of the radiation required to 243 nm, into the range in which it is possible to generate tunable radiation from laser light by standard non-linear methods. As regards the line-width, the two-photon technique allows one to eliminate first-order Doppler broadening; but in addition, because the 2S level is metastable, the transition has a natural width of only 1.3 Hz, so that the resolution limit is of order 1 part in 10^{15}! Of course, this limit has not yet been remotely approached; in these early Stanford experiments, the use of pulsed excitation led to linewidths (at 243 nm) of order 200 MHz. Finally, the problem of calibration was solved by using another transition in hydrogen, $H\beta$ (n = 2 – n = 4), as a reference. The 243 nm radiation required for the 1S – 2S excitation was generated by frequency doubling, so that the fundamental wavelength was at 486 nm, which coincides with $H\beta$. This transition was excited (using the fundamental laser radiation) in a separate hydrogen discharge, and the small frequency differences between one quarter of the 1S – 2S transition and components of $H\beta$ were measured. If Bohr theory were exact, these differences would be zero, so the uninteresting gross structure of hydrogen was eliminated from the measurements. This ingenious approach allowed the Lamb shift to be determined without the need for an accurate value of the Rydberg constant. Experiments were carried out in hydrogen and deuterium, eventually giving results with quoted uncertainties of about 30 MHz [9].

1.2 The drive towards higher precision

The precision achieved in the early experiments was limited primarily by the use of pulsed techniques. The fact that two non-linear processes – generation of 243 nm radiation and two-photon excitation – are involved makes it much more difficult to carry out a cw experiment; nevertheless, one must do so if one is to begin to realise the potential of the 1S – 2S transition. An uncertainty of tens of megahertz is much too large if the ground state Lamb shift is to provide a significant test of QED. The radio-frequency measurements of the $2^2S_{\frac{1}{2}}$ – $2^2P_{\frac{1}{2}}$ interval of about 1 GHz as first performed by LAMB and RETHERFORD [10] have now reached a quoted accuracy of the order of a few kilohertz [11,12,13]. The ground state Lamb shift is about 8 GHz, so one is aiming for considerably better than 100 kHz on a ground state result to provide a comparable test. It would be useful to carry out such measurements since there is little prospect of dramatic improvement in the 2S – 2P measurements; the uncertainty is already less than 10^{-4} of the 2P line-width.

However, there are other reasons for improving the precision of 1S – 2S measurements. For example, a measurement of the absolute frequency (in contrast to one in terms of $H\beta$) is sensitive to the value of the Rydberg constant. More generally, one can expect to get the most out of measurements of transition frequencies in hydrogen-like systems by combining them both with each other and with data from other types of experiment. It is therefore likely that the study of the 1S - 2S transition will play an increasingly important role in understanding fundamental systems and the determination of fundamental constants.

For several years, considerable effort was put into attempts to develop a successful cw experiment [14,15]. Interest also developed in the coherent multiple pulse approach suggested by BAKLANOV and CHEBOTAYEV [16]. In the meantime, the objective of making absolute measurements of the transition frequency was also pursued, and in 1985 BARR et al [17] reported measurements of the frequencies of transitions in molecular tellurium in the vicinity of 486 nm which could act as references for the two-photon work. This was shortly followed by two papers from the Stanford and Southampton groups [18,19] in which the new tellurium standards were used in pulsed experiments to measure the absolute 1S - 2S frequency for the first time. However, even before these appeared, the first observation of a cw 1S - 2S signal was made at Stanford by FOOT et al [20]. This became the basis of a measurement in terms of the tellurium standard [5], and the cw experiment at Oxford gave its first results shortly afterwards [6]. The groups adopted different methods to generate the 243 nm radiation; Stanford chose sum frequency mixing, while at Oxford frequency doubling was used, a newly developed crystal (β-barium borate, or BBO) playing a crucial role. The advantage of the frequency doubling approach is that it opens the way to a very precise cw measurement of the 1S Lamb shift via a comparison with Hβ as in the earlier pulsed experiments. Although this comparison has always been the main objective of the Oxford group, the work so far reported has been carried out using tellurium calibration. This limited the precision because the tellurium standards are simply not good enough to match the accuracy of a cw experiment. Nevertheless, the work at Oxford has given the most precise values of the Lamb shifts of the ground levels of hydrogen and deuterium, with uncertainties of less than 1 MHz. Alternatively, if one accepts that the calculation of the Lamb shift is reliable at this level of accuracy, the experiment can be interpreted as a measurement of the Rydberg constant, as accurate as the best of recent determinations based on other transitions [21].

1.3 The present position

No further measurements have been reported, though the research is being actively pursued at Munich, Novosibirsk, Oxford, Southampton and Yale. The frequency doubling scheme is basically simpler and more straightforward than other ways of producing radiation at 243 nm, and has the advantage that the fundamental is close to Hβ; it will probably become the standard method of exciting the transition. The results obtained so far are soon likely to be superseded as techniques are refined and frequency standards improve, so that this chapter is being written during an exciting period of rapid development. What the outcome will be in terms of physics it is too early to tell. As a test of QED, hydrogen has the disadvantages of uncertainty in the nuclear size correction (some 50 kHz in the 1S Lamb shift) and a low value of the expansion parameter $Z\alpha$. This is underlined by recent work in He^+ [22] and highly ionized atoms [23]. Of course, high resolution spectroscopists dream of eventually realising experimentally the 1.3 Hz linewidth of the transition; however, major investments of time and effort would be needed to achieve this, and we are concerned here with the first steps along the way, and with more immediate objectives. In section 2 the experiments carried out to date are discussed, particularly that carried out at Oxford since it is the most recent and most accurate (details are from BOSHIER [24] and BOSHIER et al [25]). Section 3 contains concluding remarks.

2. Review of 1S - 2S Experiments

2.1 Basic Structure

A 1S - 2S experiment can be roughly divided up as follows:

- Production of ground state hydrogen atoms, and their transport to an interaction region.
- Generation of 243 nm radiation, beam handling, enhancement/amplification.
- Excitation and detection system.
- Calibration.

The way in which 243 nm radiation was produced - specifically, the distinction between pulsed and cw experiments - represents the most obvious subdivision of work reported so far, and we shall discuss these categories separately. However, future cw experiments are likely to be primarily distinguished by the methods adopted to detect and calibrate the two-photon signals. We consider calibration separately at the outset.

2.2 Calibration

As has already been pointed out, two different methods of calibration have been used in the work so far reported. One can measure in terms of $H\beta$, and in this way determine the ground state Lamb shift. If instead one uses a frequency standard, one can determine the 1S Lamb shift by allowing for the other contributions to the measured 1S - 2S interval; this requires a value of the Rydberg constant from another source. Alternatively, if one assumes the 1S Lamb shift known from theory at the level of accuracy of the experiment, one can regard the measurement as a determination of the Rydberg constant. All the most recent work has been in terms of the external standards provided by the tellurium transitions, but these introduce an uncertainty which in the case of the cw experiments dominates the final error. This situation must be avoided in future work, probably by a change in calibration technique. A frequency doubled cw experiment offers the possibility of returning to a comparison with Hβ or making measurements in terms of an improved external standard at 486 nm, and work towards both these objectives is in progress.

2.3 Pulsed experiments

These cannot compete with cw experiments in terms of the accuracy of measurement, so we shall discuss them very briefly, and primarily insofar as they have a bearing on cw work. With pulsed excitation, there is no difficulty in generating enough 243 nm radiation to observe strong signals with easily realizable number densities of atomic hydrogen. The problem is the uncertainty introduced by the distorting effects of pulsed dye amplifiers which affect the shape and position of the observed 1S – 2S resonance. These effects have been responsible for the main contribution to the error in all the pulsed experiments except the first (the dominant uncertainty in this case was due to the Doppler broadening of Hβ; in subsequent work, the Doppler-free techniques of saturation or polarization spectroscopy were used).

The detection scheme for the earlier pulsed experiments in which the atoms were excited in a cell [7,8,9] made use of collisional transfer of excitation between the 2S and 2P levels. The intensity of Lyman-α radiation (at 121.5 nm) resulting from the subsequent decay to the ground level was measured as a function of laser frequency. This is a convenient method, but not suitable for the highest precision; by its nature it prevents the experiment from being carried out in a collision-free environment. Also, its efficiency is hard to estimate; signal levels in these experiments were significantly lower than expected, and it is possible that radiative trapping of Lyman-α allowing other relaxation mechanisms to depopulate the excited levels may have been responsible. The simplicity of the method has led to its being used also for both cw experiments carried out so far, but there are obvious advantages in using excitation from a ground state beam with subsequent detection of metastables. In one of the two most recent pulsed experiments [18] the excitation occurred in a beam, and in both [18,19] a new detection system was adopted. The Doppler-free two-photon excitation to the 2S level was followed by ionization by a further 243 nm photon, and the ions could be detected with high efficiency.

2.4 CW Experiments

2.4.1 Feasibility

In contrast to the pulsed experiments, the main problem for many years in cw experiments was to observe any signal at all. However, this is not apparent from simple feasibility estimates, as we now show. We assume for simplicity that one excites the atoms using two counter-propagating plane-polarized gaussian beams each with the same power P, and take the bandwidth of the light to be small compared with the width Γ of the resonance. Let the waist size of the beam be w_0, and suppose that we detect atoms excited over a length L of the beam much less than the Rayleigh length. Then if the atomic number density is N, the peak excitation rate in the detection volume is

$$R \approx \frac{1}{\pi\epsilon_0{}^2c^2\hbar^4}\ \frac{L}{w_0{}^2}\ \frac{4}{\Gamma}\ N\ M^2\ P^2$$

where

$$M^2 = \left|\ \sum_j \frac{\langle i|ez|j\rangle\ \langle j|ez|f\rangle}{\omega_{ji} - \omega}\ \right|^2 ,$$

$|i\rangle$ and $|f\rangle$ respectively representing the 1S and 2S states. Evaluating this expression, one finds

$$R \approx 2.7 \times 10^{-7}\ (L/w_0{}^2\Gamma)\ N\ P^2\ W^{-2}\ m^4\ s^{-2}.$$

Practical considerations limit the choice of w_0, L and N. We take the values appropriate to the Oxford experiment [6], $w_0 = 100\mu m$, L = 10 mm, with a typical pressure in the interaction region of around 100 mtorr. Then with $\Gamma/2\pi \sim 10$ MHz we have a signal count rate of

$$S = \eta R \sim 1.3 \times 10^{10}\ P^2\ W^{-2}\ s^{-1}.$$

for fully dissociated hydrogen, assuming that every excitation is followed by a detectable Lyman-α photon. The efficiency η of the detection system (solid angle, filter) has been taken to be 10^{-3}. From the evidence of the pulsed work, we can expect our estimate of S to be too large (and with hindsight, from the results of both cw experiments, it appears that the factor under some conditions can be as much as 1000). Even so, we can hope to generate 1 mW or so of 243 nm radiation and use an enhancement cavity to give a value of $P \sim 10$mW and a signal $S \sim 1000$ counts s^{-1}.

The real difficulty lay elsewhere. Until quite recently, the problem with the frequency doubling approach was the lack of suitable crystals. Lithium formate, the material used in the early pulsed work, is inefficient and subject to uv damage. Urea appeared on the scene around 1979, and was reported to have much superior properties [26]; however, it proved enormously difficult to obtain a crystal of good enough optical quality. The material is fragile, hard to grow and to polish. It also suffers surface damage when generating uv. Nevertheless, the first observation of a cw two-photon signal by frequency doubling was made with urea, following significant advances in polishing techniques at the Clarendon. However, urea was soon abandoned in favour of the newly-developed material BBO [27,28], which has about the same efficiency as urea and is now available in crystals of excellent quality with no degrading of performance in use. We now describe the Oxford experiments, which were carried out with a single crystal of BBO, since they are the most accurate yet reported.

2.4.2 The Oxford Experiment

Flowing molecular hydrogen or deuterium was dissociated in a radio-frequency discharge operating at about 30 MHz. For some of the experiments, a 5% H, 5% D, 90% He mixture was used to reduce pressure shifts as in [5]; measurements were made in the range 20 to 300 mtorr and extrapolated to zero pressure. Efficient transport of dissociated hydrogen atoms from a discharge to an interaction region is important in a cw experiment, but despite much experimentation it remains a somewhat empirical science and recombination may be responsible for some of the apparent loss in signal discussed earlier. A short length of Teflon tubing was used, looped to prevent discharge light from reaching the interaction region.

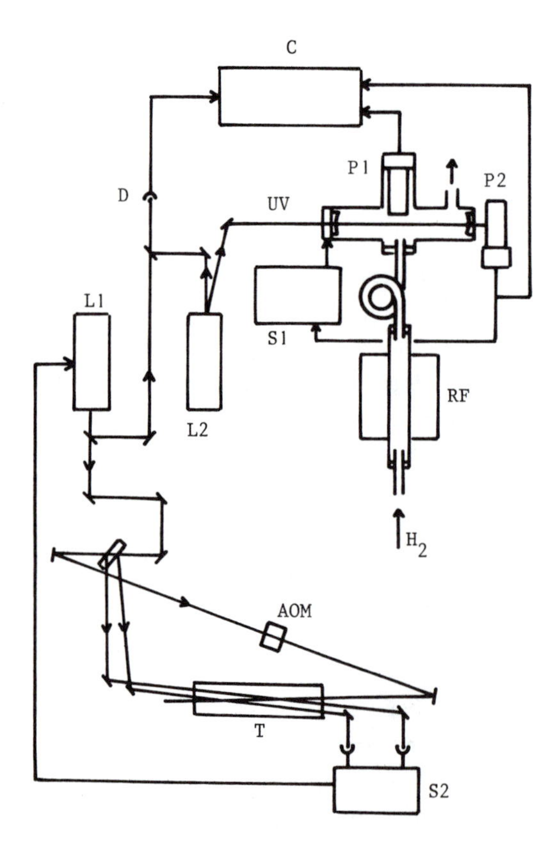

Figure 1 Apparatus of Oxford experiment [6]. L1, L2: tunable dye lasers. UV: ultra violet radiation (243 nm). RF: radio-frequency dissociation of flowing molecular hydrogen. P1: signal photomultiplier (Lyman-α detector). P2: photomultiplier for cavity locking and signal normalisation. S1: cavity length servo-control. C: computer. AOM: acousto-optic modulator. T: heated quartz cell containing tellurium. S2: laser frequency servo-control. D: fast photodiode

The BBO crystal was mounted intra-cavity in a krypton-pumped Coumarin 102 ring dye laser, and produced typically 2 mW at 243 nm. The crystal was phase-matched by angle tuning. The large walk-off angle of about 5° gave a beam with a very elongated intensity profile (aspect ratio 19:1); however, an optical system was devised which transferred this non-gaussian beam into the TEM_{00} mode of an enhancement cavity in the hydrogen cell well enough to give intracavity powers P greater than 20 mW. Signals were up to 25,000 counts s^{-1}, depending on conditions, with background (largely scattered 243 nm light) typically 1000 s^{-1}. Signal/noise was thus never a problem in the measurements. Profiles were recorded of the stronger hyperfine component of the transition for both hydrogen and deuterium; a typical recording is shown in figure 2.

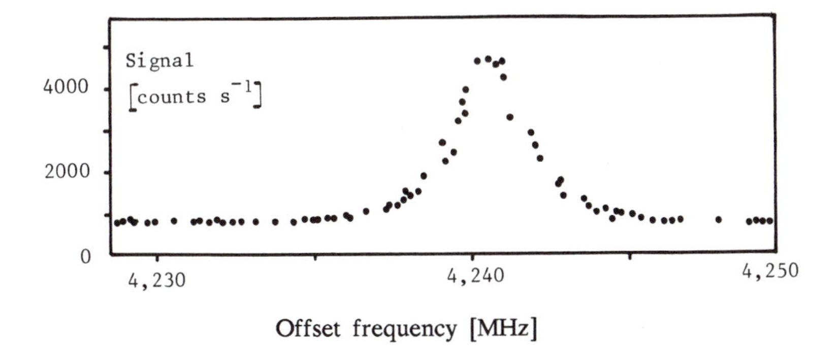

Figure 2 The stronger component of the 1S–2S two photon transition in deuterium. The signal is the normalised Lyman-α fluorescence observed as a function of the frequency difference between lasers L1 and L2 (fig. 1) when L1 is locked to the appropriate transition (b_2) in $^{130}Te_2$. The measured offset frequency is 20 MHz greater than the true value because of the shift introduced by the acousto-optic modulator. The pressure in the deuterium cell was 270 mtorr

It was found that the normal method of scanning the frequency doubling laser by rotating the tipping Brewster plate in the reference cavity was insufficiently smooth over the small ranges required; a satisfactory alternative was found to be to shift electronically the reference point used for locking the reference cavity. Further refinements to the lasers were unnecessary, because their performance did not limit the accuracy of the measurements. The calibration procedure was responsible for most of the uncertainty, and was the least satisfactory aspect of the experiment as we now discuss.

After initial experiments with a tellurium cell which later turned out to be contaminated, we borrowed the actual cell which had been calibrated and reproduced the conditions under which it had been operated at the NPL. Even with this system, the three main contributions to the uncertainty in measuring the 1S – 2S frequencies in hydrogen and deuterium were attributable to the method of calibration, as follows (all in kHz):

	H	D
Statistical uncertainty	152	140
Uncertainty in locking point on Te line	320	160
Uncertainty in Te frequencies	660	660

Other effects are negligible compared with these, but are discussed in [24] and [25]. The statistical uncertainty is mainly due to small temperature fluctuations of the tellurium cell; these had periods of order minutes, and gave corresponding shifts of the calibration frequencies. The tellurium lines are relatively wide (~ 20 MHz), but even so the uncertainty in the locking point could be reduced considerably by the use of a more sophisticated locking system. However, the main problem is the uncertainty in the frequencies of the transitions themselves. The final results for the 1S - 2S frequencies are:

$$f(1S - 2S) = 2,466,061,414.12(75) \text{ MHz (H)};$$

$$2,466,732,408.45(69) \text{ MHz (D)}.$$

In deriving values of the Lamb shifts of the 1S levels from these results, one needs to subtract the Dirac energy. To calculate this requires in turn a value of the Rydberg constant. We use that of ZHAO et al [21], who measured on Hβ; the reference frequency in their work was a He-Ne laser which has been directly compared with that employed in the NPL calibration of the tellurium lines. Some of the errors in the two measurements are therefore systematic; effectively, the 1S - 2S frequency is being compared with Hβ as in the early pulsed experiments, albeit via three steps. We obtain the following values for the 1S Lamb shifts, which we take to be the sum of all QED contributions plus the correction for finite nuclear size [29]:

$$\text{H:} \quad \Delta\nu = 8,172.94(85) \text{ MHz (theory } 8,173.05(10) \text{ MHz)},$$

$$\text{D:} \quad \Delta\nu = 8,184.10(80) \text{ MHz (theory } 8,184.09(12) \text{ MHz)}.$$

The theoretical results quoted here and below make use of the most recent calculations of the reduced mass and recoil corrections [30,31,32] and values of the fundamental constants (see [24,25]). We stress once more that systematic and random uncertainties due to the calibration procedure completely dominate the quoted experimental errors; detailed examination of the data suggests that the total contribution from other sources is less than 50 kHz, despite the use of cell excitation and the need to extrapolate to zero pressure. Even though the results so far do not provide a critical test of QED, therefore, there is the prospect of considerable improvement. However, if we accept the evidence provided by the n = 2 Lamb shift experiments that the theory is very unlikely to be incorrect at the level of our experimental error, we can interpret the 1S - 2S frequency measurements as determinations of the Rydberg constant; we obtain

$$R_\infty = 109{,}737.31573(3)\ \mathrm{cm}^{-1}.$$

This is as precise as the best previous measurement [21]. Finally, we can derive the isotope shift in the 1S - 2S transition; this is

$$\delta\nu\ (\mathrm{D - H}) = 670{,}994.33(64)\ \mathrm{MHz}\ (\text{theory } 670{,}994.39(12)\ \mathrm{MHz}).$$

Work is in progress at Oxford to improve this result by wide-band heterodyning techniques which avoid the need for frequency standards.

2.4.3 Comparison with Sum-frequency Mixing

In the Stanford experiment [5,20], 243 nm radiation was generated by mixing 351 nm light from an argon ion laser with a dye laser output at 790 nm in a crystal of KDP. This gave a little more uv output power than the Oxford frequency doubling system, but the method has the disadvantage that calibration of the two-photon signal is less straightforward. In the Oxford experiment, it was possible to measure the frequency difference between the fundamental from the frequency doubling laser and the tellurium calibration lines simply by optical heterodyning at 486 nm. The tellurium cell was placed in a saturation spectrometer, a second dye laser being locked to the appropriate line for hydrogen or deuterium. The method lends itself readily to a comparison with Hβ. At Stanford, calibration was carried out at 243 nm, so the light from the reference laser had to be frequency doubled. The difficulties raised by this procedure prevented any results being obtained for deuterium. However, the experiment did give the first cw measurement of the hydrogen ground state Lamb shift; the result was 8173.3(1.7) MHz.

Comparisons of the results of the work at Oxford, Stanford and elsewhere are shown in figs. 3 and 4.

3. Conclusion

The first point to make is that the results quoted above show that the present priority is to improve the comparison of the 1S - 2S frequency with a reference. This alone could bring the ground state Lamb shift into serious contention as a critical test of QED. Three ways suggest themselves, all of which are being pursued:

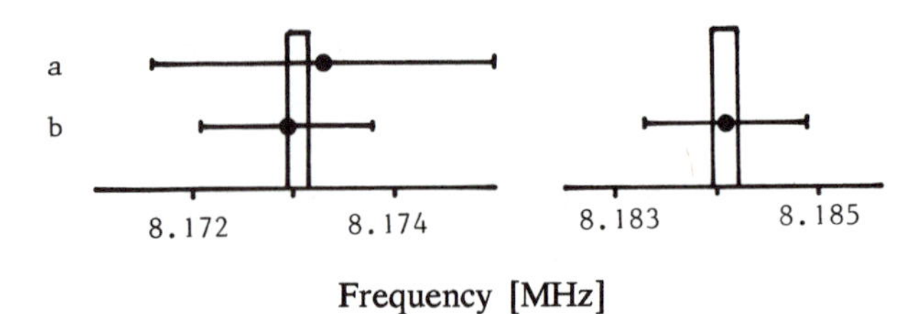

Figure 3 Comparison of measured 1S Lamb shifts with theory [24,25], the limits of which are shown as vertical lines. (a) Reference [5], (b) present work [6]. No earlier work of comparable precision exists for deuterium

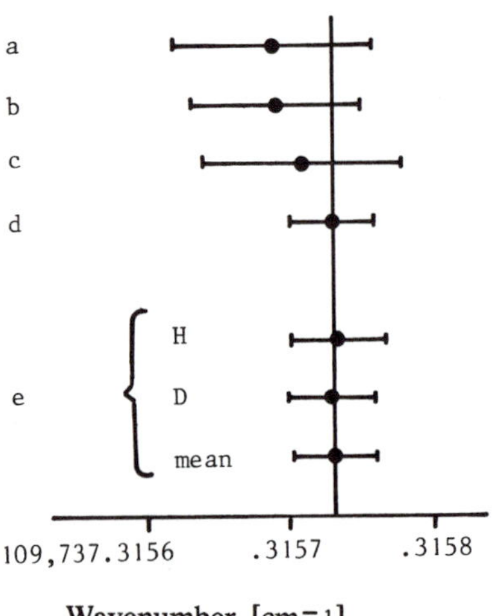

Figure 4 The present results for the Rydberg constant compared with other recent values. (a) Reference [36], (b) reference [35], (c) reference [5], (d) reference [21], (e) present work [6]. The vertical line is drawn through the mean of the new results

1. More refined work on tellurium lines. This is not a route towards much higher precision, but a more thorough study of lines currently limiting the accuracy of 1S - 2S experiments is clearly desirable.
2. A frequency chain providing an accurate standard in the vicinity of 486 nm without the need for interferometry.
3. Direct comparison of 1S - 2S with other hydrogen transitions, particularly $H\beta$ as in the earlier pulsed work. This is the immediate goal of the Oxford group.

There are other obvious refinements. The Oxford experiments were carried out using commercial lasers, with frequency jitter around the 1 MHz level; there are well established techniques which can reduce this substantially. Cell excitation must ultimately be abandoned; although it is not at present a limiting factor, it becomes increasingly unsatisfactory as techniques develop and precision improves. For a comparison with Hβ, it would be desirable to work with cold beams for both parts of the experiment. In Oxford, we are building a system in which the metastables created in the 1S - 2S excitation are used in a measurement of the sharp 2S - 4S transition. Such an approach should permit a very direct comparison of frequencies.

The main significance of the work so far carried out at Oxford is that it demonstrates the feasibility of making measurements of the 1S - 2S transition with a frequency doubled cw system. This opens the way to considerable improvements in precision, but the interpretation of the measurements will depend on what other information is available - the transition frequency is sensitive to QED effects, the proton size, and the values of fundamental constants. Cynics might point out that after well over a decade of work on the 1S - 2S transition, the results have produced no real surprises. A more optimistic view is that it is only over the last year or two that it has become possible to study the transition with the standard techniques of high resolution cw laser spectroscopy, and the next decade will bring dramatic developments. We shall see.

The Oxford group is grateful to the Science and Engineering Research Council for supporting this work, including the award of a Senior SERC Fellowship to the present author.

1. N. Bloembergen and M.D. Levenson: in High-Resolution Laser Spectroscopy, ed. by K. Shimoda, Springer Ser. Topics in Applied Phys., Vol. 13 (Springer-Verlag, Berlin, Heidelberg, New York 1976)
2. E.V. Baklanov and V.P. Chebotayev: Opt. Comm. 12, 312 (1974)
3. E.V. Baklanov and V.P. Chebotayev: Opt. Spectrosc. 38, 215 (1975)
4. B. Cagnac, G. Grynberg, F. Biraben: Jour. de Physique 34, 84 (1973)
5. R.G. Beausoleil, D.H. McIntyre, C.J. Foot, B. Couillaud, E.A. Hildum, T.W. Hänsch: Phys. Rev. A35, 4878 (1987)
6. M.G. Boshier, P.E.G. Baird, C.J. Foot, E.A Hinds, M.D. Plimmer, D.N. Stacey, J.B. Swan, D.A. Tate, D.M. Warrington, G.K. Woodgate: Nature 330, 463 (1987) [work first reported at the Eighth International Conference on Laser Spectroscopy, Are, Sweden, June 1987]
7. T.W. Hänsch, S.A. Lee, R. Wallenstein, C. Wieman: Phys. Rev. Lett. 34, 307 (1975)
8. S.A. Lee, R. Wallenstein, T.W. Hänsch: Phys. Rev. Lett. 35, 1262 (1975)
9. C. Wieman and T. W. Hänsch: Phys. Rev. A22, 192 (1980)
10. W.E. Lamb and R.C. Retherford: Phys. Rev. 72, 241 (1947)
11. D.A. Andrews and G. Newton: Phys. Rev. Lett. 37, 1254 (1976)
12. S.R. Lundeen and F.M. Pipkin: Phys. Rev. Lett. 46, 232 (1981)

13. V.A. Pal'chikov, Yu.L. Sokolov and V.P. Yakovlev, Metrologia 21, 99 (1985)
14. B. Couillaud, T.W. Hänsch, S.G. MacLean: Opt. Comm. 50, 127 (1984)
15. C.J. Foot, P.E.G. Baird, M.G. Boshier, D.N. Stacey, G.K. Woodgate: Opt. Comm. 50, 199 (1984)
16. E.V. Baklanov and V.P. Chebotayev: Sov. J. Quantum Electron. 7, 1253 (1977)
17. J.R.M. Barr, J.M. Girkin, A.I. Ferguson, G.P. Barwood, P. Gill, W.R.C. Rowley, R.C. Thompson: Opt. Comm. 54, 217 (1985)
18. E.A. Hildum, U. Boesl, D.H. McIntyre, R.G. Beausoleil, T.W. Hänsch: Phys. Rev. Lett. 56, 576 (1986)
19. J.R.M. Barr, J.M. Girkin, J.M. Tolchard, A.I. Ferguson: Phys. Rev. Lett. 56, 580 (1986)
20. C.J. Foot, B. Couillaud, R.G. Beausoleil, T.W. Hänsch: Phys. Rev. Lett. 54, 1913 (1985)
21. P. Zhao, W. Lichten, H.P. Layer, J.C. Bergquist: Phys. Rev. Lett. 58, 1293 (1987)
22. G. W. F. Drake, J. Patel, A. van Wijngaarden: Phys. Rev. Lett. 60, 1002 (1988)
23. C. T. Munger and H. Gould: Phys. Rev. Lett. 57, 2927 (1986)
24. M.G. Boshier, D. Phil. Thesis, Oxford University (unpublished, 1988)
25. M.G. Boshier, P.E.G. Baird, C.J. Foot, E.A. Hinds, M.D. Plimmer, D.N. Stacey, J.B. Swan, D.A. Tate, D.M. Warrington, G.K. Woodgate: Phys. Rev. A (to be published)
26. J.-M. Halbout, S. Blit, W. Donaldson, C.L. Tang: IEEE J. Quant. Electr. QE-15, 1176 (1979)
27. Chen Chuangtian, Wu Bochang, Jiang Aidong, You Guiming: Scientia Sinica B28, 235 (1985)
28. K. Kato: IEEE J. Quant. Electr. QE-22, 1013 (1986)
29. W.R. Johnson and G. Soff: At. Data Nucl. Data Tables 33, 405 (1985)
30. G. Bhatt and H. Grotsch: Phys. Rev. A31, 2794 (1985)
31. G. Bhatt and H. Grotsch: Phys. Rev. Lett. 58, 471 (1987)
32. G. Bhatt and H. Grotsch: Ann. Phys. (N.Y.) 178, 1 (1987)
33. J.E.M. Goldsmith, E.W. Weber, T.W. Hänsch: Phys. Rev. Lett. 42, 1525 (1978)
34. B.W. Petley, K. Morris, R.E. Shawyer: J. Phys. B13, 3099 (1980)
35. F. Biraben, J.C. Garreau, L. Julien: Europhys. Lett. 2, 925 (1986)
36. P. Zhao, W. W Lichten, H.P. Layer, J.C. Bergquist: Phys. Rev. A34, 5138 (1986)

1S–2S Transition-Frequency Calibration

A.I. Ferguson, J.M. Tolchard, and M.A. Persaud
Department of Physics, University of Southampton,
Southampton, SO9 5NH, U.K.

1. INTRODUCTION

Recent experiments on atomic hydrogen have shown that it is now possible to make measurements with a reproducibility far exceeding the accepted absolute precision of the available visible frequency standards. For example, the recent work of ALLEGRINI et al. [1] has shown that the overall systematic error of their measurements of the Rydberg states in hydrogen is about 4 parts in 10^{11}. Their measurements have had to be referred to the visible iodine stabilised HeNe laser which has an internationally accepted absolute accuracy of only 1.6 parts in 10^{10}. If any progress is to be made in making precision measurements of fundamental atomic constants and in tests of QED, new visible frequency standards and new ways of comparing optical frequencies will have to be developed.

In this contribution we speculate on some new techniques of optical frequency comparison by use of modulated lasers, and on a possible new optical frequency standard based on the methane stabilised HeNe laser. We start by reviewing our own work in providing a convenient secondary frequency standard based on $^{130}Te_2$ transitions in the vicinity of Balmer β in hydrogen and deuterium. We also report on a new and improved measurement of the hydrogen 1S to 2S transition frequency using a pulse amplified laser.

2. TELLURIUM CALIBRATION AT 486 nm

The $^{130}Te_2$ spectrum is extremely rich in the blue region of the spectrum. In collaboration with the NPL we have calibrated two reference lines in $^{130}Te_2$ using Doppler-free saturation spectroscopy [2]. These transitions have been referred to the iodine stabilised HeNe laser using the NPL 1m plane-plane interferometer. A typical $^{130}Te_2$ spectrum in the region of Balmer β in hydrogen is shown in figure 1. The reference line is indicated in this figure. This line lies some 1.4 GHz below one quarter the centroid of the 1S to 2S transition frequency. The lines are approximately 20MHz wide. Some of the problems in using this standard and in

The Hydrogen Atom Editors: G.F. Bassani · M. Inguscio · T.W. Hänsch

using interferometry include phase shift corrections on coatings, illumination effects, prismatic imbalances and diffraction effects. Furthermore the cell has to be used at a carefully controlled temperature in order to control the pressure shift. Care has to be taken in using this reference since, for example, we have found that one cell (belonging to Oxford) from the same source has a blue shift of some 1 MHz. This is thought to be due to a leak caused by heating the cell too rapidly. The $^{130}Te_2$ calibration is probably usable to an accuracy of about 4 parts in 10^{10}. We understand that the NPL are going to undertake a more careful study of the $^{130}Te_2$ spectrum which may improve the accuracy somewhat [3].

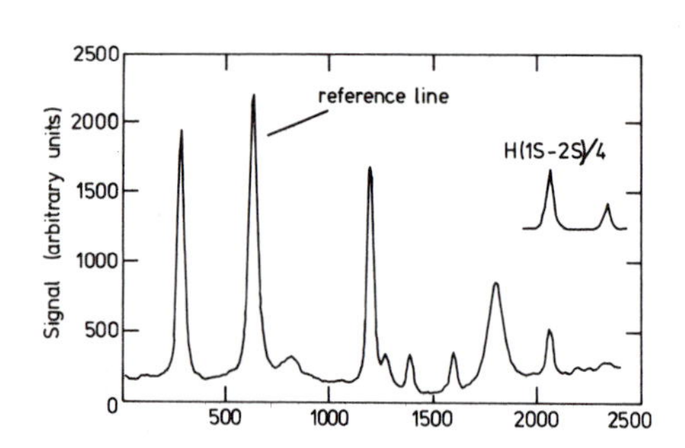

Fig 1.

Doppler-free saturation spectrum of $^{130}Te_2$ in the region of Balmer β in hydrogen shown as a scan over 2.5 GHz. Also shown is an inserted spectrum of the hydrogen 1S to 2S transition at one quarter of the transition frequency.

3. PULSED SPECTROSCOPY OF 1S-2S

The earliest approaches to the observation of the 1S to 2S transition in atomic hydrogen relied on the use of pulsed light sources. This was first done by Hänsch and his co-workers who used pulsed oscillators and amplifiers [4]. This was later improved by amplifying a continuous-wave dye laser [5]. One of the problems of the use of pulsed amplifiers has been the frequency chirp associated with the transient nature of the gain. This effect has been reduced by spectrally narrowing the amplified radiation in a confocal filter at both Stanford [6] and Southampton [7]. In these experiments a continuous-wave C102 dye laser was amplified in a three or four stage amplifier pumped by either an excimer laser [6] or a frequency tripled Nd:YAG laser [7]. The resulting linewidth of a few hundred MHz was spectrally filtered to about 30 - 40 MHz using a confocal filter. The hydrogen 1S to 2S transition frequency was compared with the $^{130}Te_2$ transitions which were calibrated by Southampton/NPL.

The results of these experiments were in reasonable agreement although the Stanford group reported rather smaller errors. Subsequent, continuous-wave experiments at both Stanford and Oxford show poor agreement with the Stanford pulsed result. This has led to speculation that the frequency chirp in pulse amplified experiments is so difficult to characterise that pulse amplifiers cannot be used for precision measurements.

We have undertaken an experiment to try to improve the performance of pulse amplifier experiments. The system is shown schematically in figure 2. It consisted of a continuous-wave C102 dye laser amplified in three stages by a frequency tripled Q-switched Nd:YAG laser. The output energy was approximately 2.0 mJ in a 150 MHz linewidth and was up-shifted from the continuous-wave laser by 60 MHz caused by the frequency chirp. This light was then spectrally filtered in a confocal interferometer with a finesse of 40 and a free spectral range of 300 MHz. The linewidth of the filtered radiation was approximately 16 MHz.

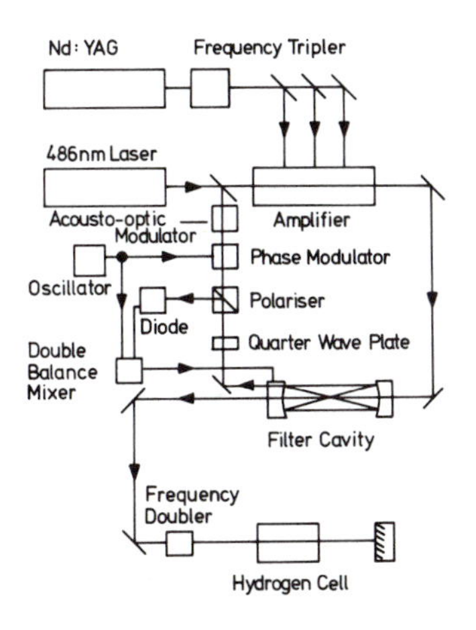

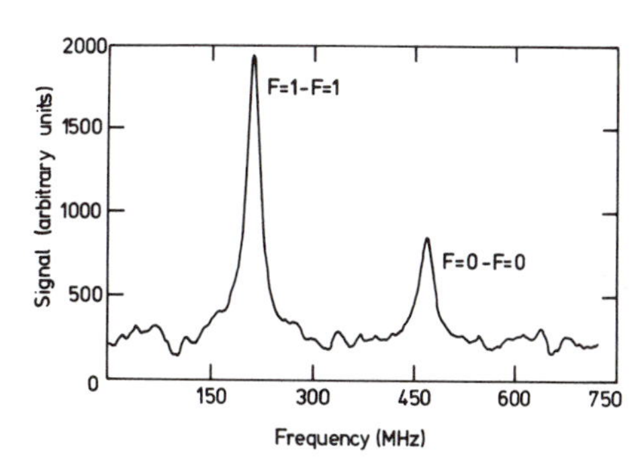

Fig 3 Spectrum of the hydrogen 1S to 2S transition taken using pulse amplified radiation.

Fig 2 Schematic diagram of the apparatus used to observe the 1S to 2S transition in hydrogen using a pulse amplified laser.

The novel aspect of this experiment was that the confocal cavity was locked to continuous-wave radiation which was frequency shifted by an acousto-optic modulator such as to centre the filtering cavity onto the chirped amplified radiation. This reduced the residual amplifier shift to -2(1) MHz. The dominant contribution to this shift resulted from the cw light being injected off-axis into the cavity. Because the filter cavity had a high finesse we used a phase modulation scheme for locking. Indeed, we normally locked the dye laser to the filtering cavity and scanned the spectrum by scanning the filter cavity.

The filtered radiation was then frequency doubled in urea and the 1S to 2S signal was detected by two-photon resonant, three-photon ionisation. A typical spectrum is shown in figure 3 and shows a linewidth of 18 MHz at 486 nm. This spectrum has been compared with the $^{130}Te_2$ reference lines and we have measured a 1S to 2S transition frequency of 2466061407(9) MHz. This compares with 2466061414.13(79) MHz reported for the best continuous-wave measurement [8]. A full report on this measurement is in preparation.

It is clear that the best measurements on the 1S to 2S transition in hydrogen will be accomplished using continuous-wave lasers. However, there is a range of exotic atoms such as muonium, positronium and anti-hydrogen where high energy, high resolution pulsed laser sources will be needed. We believe that with care, precise measurements can be made with pulse amplified radiation. We hope to apply some of these techniques to a study of muonium at the Rutherford Appleton Laboratory (RAL) in a collaboration led by Zu Putlitz and Jungmann at Heidelberg, Hughes at Yale, Baird at Oxford and Barr and myself at Southampton. The RAL muon source is some two orders of magnitude more intense than that used by Chu and collaborators at KEK in Japan [9].

4. COHERENT MULTIPLE PULSE SPECTROSCOPY

The idea of using a train of coherent pulses for the observation of the 1S to 2S transition in hydrogen was first suggested by BAKLANOV et al. in 1976 [10]. The observation of Doppler-free spectra using a coherent pulse train from a synchronously pumped dye laser was demonstrated by ECKSTEIN et.al [11].

In this scheme a mode-locked synchronously pumped dye laser is used to provide a train of tunable light pulses at a precise repetition rate in the region of 100 MHz. The output from the dye laser is passed through the sample and then retro-reflected from a mirror. The distance between the sample and the mirror is adjusted such that a pulse passing through the sample in one direction collides with its neighbour in the train going in the opposite direction. This produces a standing-wave field where a Doppler-free excitation can take place.

The Doppler-free excitation can best be understood in the frequency domain where the train of pulses appears as a comb of modes, all equally spaced in frequency and covering a spectral region which is roughly equal to the inverse of the pulse duration of a single pulse in the train. For a typical dye laser this might be 500 GHz. If one of the modes is in two-photon resonance with the sample then the adjacent modes higher in frequency and lower in frequency will also be in resonance, as will the modes next higher and lower in frequency, and so on. Thus when one mode is in resonance many of the modes will contribute to the signal. It is easy to show that the signal strength that is expected is the same as that obtained from a single frequency laser of the same average power. A resonance condition also exists when the sum frequency of two adjacent modes equals the atomic resonance frequency.

If we have a two-level sample we expect to see a series of resonances separated by half the inverse of the repetition rate of the laser as the carrier frequency is scanned. If a second transition is within the bandwidth of the laser then this too will give rise to a series of resonances. The resulting spectrum is rather like that obtained from a Fabry-Perot interferometer with overlapping orders. However, in the mode-locked case the modes are precisely equally spaced in frequency.

The mode-locked laser can be used for comparing frequency differences. For example the hydrogen-deuterium isotope shift can be accurately measured using this technique. In this scheme the 671GHz shift can be spanned by a mode-locked dye laser. By scanning a frequency interval of only one mode spacing all the information on the isotope shift can be obtained. An alternative method of using this technique is to lock the mode-locked laser carrier frequency to the 1S to 2S transition frequency. At 486 nm the dye laser will appear as a comb of modes. These modes can be heterodyned against a single frequency laser locked to Balmer β or to a $^{130}Te_2$ reference.

The main features of the coherent multiple pulse technique are: that it is self-calibrating in frequency; that the signal strength is the same as that obtained with a single mode laser of the same average power and that frequency doubling efficiency can be good.

These features are exploited in an experiment which is taking place in our laboratory. A schematic diagram of this experiment is shown in figure 4. The system consists of an all-lines violet mode-locked Kr^+ ion laser operating at a repetition rate of about 250 MHz which synchronously pumps a C102 dye laser. The dye laser typically produces about 300 mW of average power and pulse durations of about 3 psec. This is frequency doubled to 243 nm in a crystal of β-barium borate to produce in excess of 2 mW average power. The output from the second harmonic crystal is then mode-matched into an ultra-violet enhancement cavity. The free spectral range of this cavity is adjusted to match the repetition rate of the pulse train in such a way that the pulses collide in the centre of the cavity. We have obtained enhancements in excess of x20.

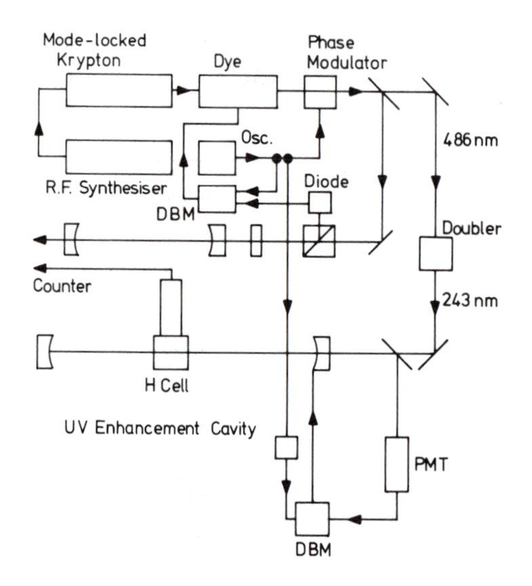

Fig 4 Diagram of the apparatus being used to observe the 1S to 2S transition using a train of ultrashort pulses from a mode-locked dye laser.

Frequency stabilisation and scanning is accomplished by use of a confocal cavity of free spectral range matched to the dye laser repetition rate. Phase modulated sidebands are put on to the mode spectrum of the mode-locked pulse train and used to lock the laser to the reference cavity. The frequency modulation technique is also used to lock the ultra-violet enhancement cavity to the mode-locked pulse train.

Two-photon absorption is detected by observing collisionally induced Lyman α fluorescence in a cell containing atomic hydrogen. This experiment has been set up and we shall report on the results soon.

The mode-locked pulse train is one of a range of ways of comparing optical frequencies. A second technique which we have been investigating is the use of a frequency modulated (FM) dye laser. This has similarities to the mode-locked laser in that we are using the precise nature of the mode spacing when intracavity modulation is applied. In the case of the FM laser phase modulation is applied and in the case of the mode-locked laser amplitude modulation is applied.

5. FM LASER SPECTROSCOPY

An ideal FM laser is a laser which produces an output of constant amplitude but whose instantaneous frequency is sinusoidally modulated about a central carrier frequency. Thus the electric field can be described as

$$E(t) = (E_0/2)\exp(i\omega_0 t + i\Gamma \sin\Omega t) + c.c. \quad (1)$$

In this case ω_0 is the carrier frequency, Ω is the modulation frequency and Γ is the modulation index which describes the region over which the modulation takes place.

This ideal FM spectrum can be Fourier transformed into the frequency domain to give a spectrum of equally spaced modes with a Bessel function amplitude distribution. These equally spaced modes can be used for comparing optical frequencies by heterodyning a reference laser, unknown laser and FM laser on a nonlinear detector. Three beats can be observed ie the beats between the reference laser and one of the modes of the FM laser, the beats between the unknown laser and one of the modes of the FM laser and the mode spacing of the FM laser. The separation between the reference and unknown laser can hence be deduced.

We have developed FM lasers based on a commercial ring laser (Coherent 699-21). In this case all the intracavity etalons are removed and replaced by a lithium niobate phase modulator. This modulator can be resonantly driven at a frequency close to the cavity mode spacing. A simple theory of FM operation of a laser suggests that the modulation index is given by [12]

$$\Gamma = \left(\frac{\delta}{\pi} \right) \left(\frac{\Omega}{|\omega_m - \Omega|} \right), \quad (2)$$

where δ is the single pass phase retardation of the phase modulator, ω_m is the cavity mode spacing.

In figure 5 we plot the measured FM bandwidth of a dye laser with a three plate birefringent tuning element as a function of the modulation frequency in the region of matching modulation frequency to cavity mode spacing. There is reasonable agreement between the results and the predictions of equation (2). Bandwidths in excess of 500 GHz have been achieved with this system. For smaller detuning the FM spectrum becomes distorted and FM mode-locking is observed. Wider bandwidth operation has been observed by using a two plate birefringent filter as a tuning element. Bandwidths in excess of 2 THz have been obtained in this case.

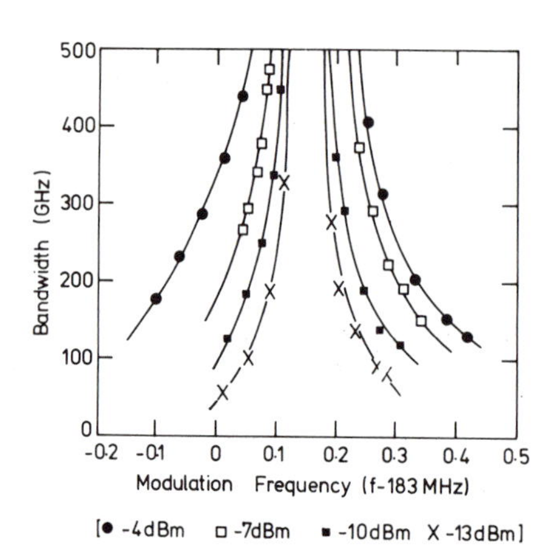

Fig 5 Bandwidth of an FM dye laser tuned by a three plate birefringent filter as a function of the detuning between the passive mode spacing and the driving frequency applied to the phase modulator.

It is important to be able to test the FM laser spectrum to see how closely it approaches that of an ideal FM oscillator. Mode analysis of the FM spectrum is too tedious. We have developed methods of testing the quality of the FM spectrum which involve nonlinear mixing of the FM beams. If we phase shift a part of the FM beam by θ we obtain a modified field,

$$E'(t) = (E_0/2)\exp(i\omega_0 t + i\Gamma \sin(\Omega t + \theta)) + c.c.$$

If we have a nonlinear process which mixes the beam represented by $E(t)$ and $E'(t)$ we get

$$E(t)E'(t) \propto E_0^2 \exp(i2\omega_0 t + i2\Gamma \cos(\theta/2) \sin(\Omega t + \theta/2)) + c.c. \qquad (3)$$

This represents another FM oscillation centred at a carrier frequency $2\omega_0$ with modulation index $2\Gamma\cos(\theta/2)$. When θ is chosen to equal π we obtain a single frequency at twice the carrier frequency.

We have demonstrated this effect in two experiments [13,14]. One is in single frequency ultraviolet generation by sum frequency mixing and the other is in Doppler-free two-photon spectroscopy, with the FM laser.

Schematic diagrams of the two experiments are shown in figure 6. In the case of the sum frequency experiment a delay is introduced by splitting part of a beam, going through a delay line and recombining the beams with a lens in a nonlinear crystal. The sum frequency is generated at the bisector of the two beams and is mode analysed by a UV interferometer. In figure 7 we plot some UV spectra taken for an FM bandwidth of about 10 GHz for different values of the phase change $\Delta\theta = \theta - \pi$. When $\Delta\theta = 0$ a single frequency UV spectrum is obtained indicating a clean FM spectrum from the dye laser. As $\Delta\theta$ is increased sidebands appear at the modulation frequency. We have assigned an effective modulation index Γ' which agrees very well with the expression given in equation (3).

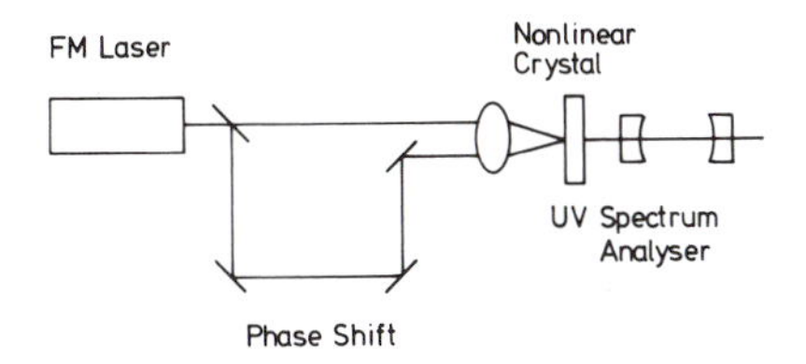

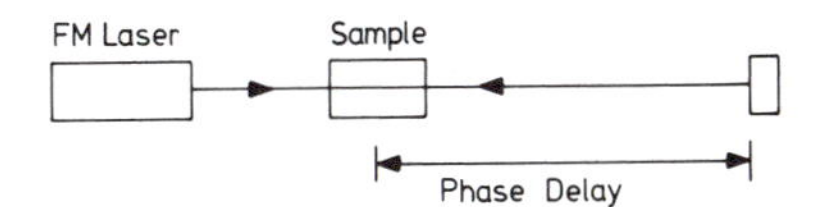

Fig 6 Two experimental arrangements for demonstrating the quality of an FM spectrum a) sum frequency generation in a nonlinear crystal followed by mode analysis of the generated ultraviolet b) Doppler-free two-photon spectroscopy.

The experimental arrangement for the demonstration of Doppler-free two-photon spectroscopy is shown in figure 6 (b). In this case the phase delay is accomplished by adjusting the distance between the sample and the retro-reflecting mirror. In figure 8 (a) we show a Doppler-free two-photon spectrum of the 3S to 4D transition in sodium taken with a single frequency dye laser. Figure 8 (b) shows the identical spectrum obtained with an FM laser. It can be seen that the two sets of spectra are almost identical except that the Doppler background is smaller in the case of the FM laser. This is explained by the fact that the FM bandwidth exceeds the Doppler width and hence only a small part of the FM laser can contribute to the Doppler background.

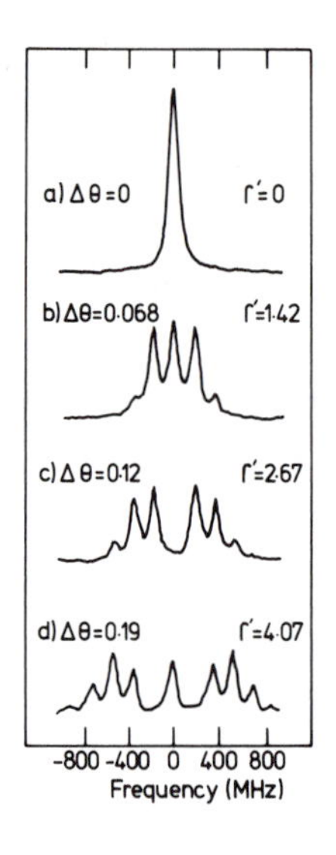

Fig 7 The spectrum of the sum frequency of the FM dye laser for different values of the phase delay.

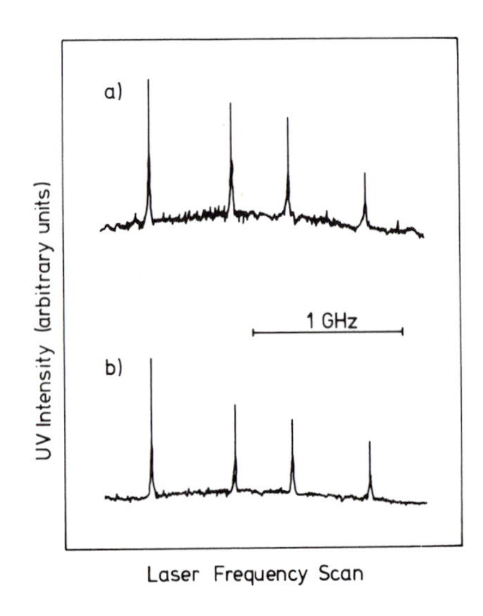

Fig 8 a) Doppler-free two-photon spectrum of the sodium 3S to 4D transition taken with a single frequency dye laser.
b) The same transition observed using an FM dye laser.

6. ABSOLUTE FREQUENCY CALIBRATION

We have developed techniques for the comparison of frequency intervals in excess of 2 THz using modulated dye lasers: either mode-locked or FM. This has still not addressed the problem of obtaining an absolute frequency standard in the visible which can be referred back to the Cs standard of time.

An exciting proposal for generating a more stable visible reference has been to use the harmonics of the methane stabilised HeNe laser. This laser has been developed to a stage where it is stable to better than a few parts in 10^{12} and can be easily referred to the Cs standard[15]. The harmonics of the methane stabilised laser lie in the visible region. For example the seventh harmonic lies at 485 nm. There are a number of schemes for generating the seventh harmonic of the methane stabilised laser. If this could be done it would be only 2.1 THz away from one quarter the 1S to 2S transition frequency in hydrogen. This opens up the possibility of using an FM dye laser to compare a laser locked to the seventh harmonic of the methane stabilised HeNe and one locked to the 1S to 2S transition.

Indeed lasers locked to the harmonics of the methane stabilised HeNe laser may prove to be of importance in the next generation of fundamental measurements in the visible region. It is possible to envisage schemes of harmonic generation and mixing covering the region from 848 nm down to 212 nm with the reference frequencies separated by 88 THz. This would mean that there would be nowhere in the visible or ultraviolet region of the spectrum more than 44 THz away from a reference laser.

7. CONCLUSION

The rapid progress in recent years in the spectroscopy of the hydrogen atom has renewed pressure for a much better optical frequency standard. This in itself would not be enough to solve the measurement problem. New techniques of comparing optical frequencies are needed. We have developed methods of modulating lasers which can be used for frequency differences in excess of 2THz.

New optical frequency standards based on harmonics of methane stabilised lasers will mean that we will never be more than 44 THz away from a reference frequency. New techniques of making frequency interval measurements of this magnitude will then be needed.

8. ACKNOWLEDGEMENTS

This work has been supported by the Science and Engineering Research Council.

9. REFERENCES

1. M. Allegrini, F.Biraben, B.Gagnac, J.C. Garreau and L. Julien: (Proceedings of this conference)

2. J.R.M. Barr, J.M. Girkin, A.I. Ferguson, G.P. Barwood, P. Gill, W.R.C. Rowley, R.C. Thompson: Opt. Commun. 54, 217 (1985)

3. P. Gill and B.W. Petley (Private Communication)

4. T.W. Hänsch, S.A. Lee, R. Wallenstein, C. Wieman: Phys. Rev. Lett. 34, 307 (1975)

5. C.E. Wieman, T.W. Hänsch: Phys. Rev. Lett. 36, 1170 (1976)
   C.E. Wieman, T.W. Hänsch: Phys. Rev. A, 22, 192 (1980)

6. E.A. Hildum, U. Boesl, D.H. McIntyre, R.G. Beausoliel, T.W. Hänsch: Phys. Rev. Lett. 56, 576 (1986)

7. J.R.M. Barr, J.M. Girkin, J.M. Tolchard, A.I. Ferguson: Phys. Rev. Lett. 56 580 (1986)

8. M.G. Boshier, P.E.G. Baird, C.J. Foot, E.A. Hinds, M.D. Plimmer, D.N. Stacey, J.B. Swan, D.A. Tate, D.M. Warrington and G.K. Woodgate: Nature 330, 463 (1987)

9. S. Chu, A.P. Mills, A.G. Yodh, K. Nagamine, Y. Miyake, T. Kuga: Phys. Rev. Lett. 60, 101 (1988)

10. Y. Baklanov, V.P. Chebotayev: Appl. Phys. 12, 97 (1977)

11. J.N. Eckstein, A.I. Ferguson, T.W. Hänsch: Phys. Rev. Lett. 40, 847 (1978)

12. S.E. Harris, O. McDuff: IEEE J. Quantum Electron. QE-1, 245 (1965)

13. S.R. Bramwell, A.I. Ferguson, D.M. Kane: Opt. Commun. 61, 87 (1987)

14. S.R. Bramwell, A.I. Ferguson, D.M. Kane: Opt. Lett. 12, 666 (1987)

15. N.G. Basov, M.A. Gubin, V.V. Nikitin, A.V. Nikul'chin, E.D. Protsenko, D.A. Tyurikov, A.S. Shelkovnikov: Sov. J. Quantum Electron. 17, 545 (1987)

High Resolution Spectroscopy of Hydrogen

T.W. Hänsch

Max-Planck-Institute for Quantum Optics,
D-8046 Garching, Fed. Rep. of Germany and
Sektion Physik, University of Munich, D-8000 Munich, Fed. Rep. of Germany

1. Introduction

The hydrogen atom continues to hold fascinating challenges for spectroscopists. As the simplest of the stable atoms it has long permitted unique confrontations between basic theory and experiment. Very large further improvements in spectral resolution and accuracy appear feasible with the help of emerging new methods for stabilizing and measuring optical frequencies, together with advanced laser techniques for preparing, slowing, and manipulating atoms. Such prospects make more accurate quantum electrodynamic computations of hydrogen levels highly desirable. If theory is correct, measurements and comparisons of transition frequencies in hydrogen will yield precise new values of fundamental constants, in particular the Rydberg constant, the electron mass, and the charge radius of the proton. It would be much more exciting, of course, if we found out that theory fails at some level of scrutiny.

G. SERIES [1] has reminded us how spectroscopy of hydrogen has played a central role in the history of atomic physics and quantum mechanics. Seemingly minute discrepancies between theory and observation have more than once stimulated major conceptual breakthroughs. A good example is the spectral profile of the red Balmer-α line which presented one of the most tantalizing problems of atomic physics for more than a decade, in the 1930s and 40s. Unfortunately, the intricate fine structure always appeared blurred and smeared out by the Doppler-effect due to the random thermal motion of the light hydrogen atoms. Suspected deviations from the relativistic Dirac theory were only confirmed after the war with the observation of the 2S Lamb shift by radiofrequency spectroscopy of an atomic beam, a discovery that led to the development of quantum electrodynamics by Feynman, Schwinger, and Tomonaga.

2. Doppler-Free Spectroscopy and the Rydberg Constant

Major advances in optical high resolution spectroscopy became possible only in the early 1970s with the advent of monochromatic broadly tunable dye lasers [2] and powerful techniques of Doppler-free laser spectroscopy [3]. In the earliest successful method, saturation spectroscopy, the output of a tunable laser is split into a saturating beam and a probe beam which are sent in opposite directions through a cell with the absorbing gas. The saturating beam is strong enough to excite a noticeable fraction of the absorbing atoms so that it bleaches a path through the absorber. The probe beam can detect this bleaching if the laser is tuned to the center of a Doppler-broadened line so that both beams can interact with the same atoms, those that are standing still or are at most moving sideways.

Fig. 1 shows an early saturation spectrum of the hydrogen Balmer-α line, recorded in this way at Stanford [4], together with the seven theoretically predicted fine structure components and the Doppler-

The Hydrogen Atom Editors: G.F. Bassani · M. Inguscio · T.W. Hänsch

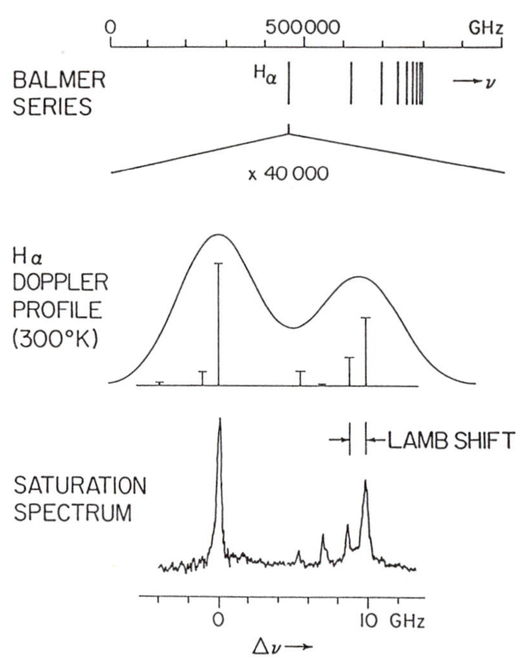

Fig. 1. Doppler-free saturation spectrum of the hydrogen Balmer-α line compared to theoretical fine structure and Doppler profile at room temperature [4].

broadened line profile at room temperature. We were thrilled to resolve the 2S Lamb shift for the first time directly in the Balmer-α spectrum. By measuring the absolute wavelength of the strongest $2P_{3/2}$-$3D_{5/2}$ component, we determined a new tenfold improved value of the Rydberg constant in 1974 [5]. This constant not only determines the binding energy between electron and nucleus, but plays an important role in its relationship to other fundamental physical constants [6].

This first laser measurement of the Rydberg constant has meanwhile been surpassed by numerous other experiments, studying several optical transitions in hydrogen and deuterium by different techniques of Doppler-free laser spectroscopy. Fig. 2 gives a summary of six recent measurements carried out at Yale [7,8], Paris [9,10], Stanford [11] and Oxford [12]. The accuracy of all these experiments is now limited by wavelength metrology. The most recent Rydberg measurement (D) has reached an accuracy of better than 2 parts in 10^{10}, determined only by the precision of the iodine-stabilized helium neon laser which is still the most accurate optical wavelength standard. This problem is the impetus behind recent efforts to measure the Rydberg constant by observing microwave transitions between circular Rydberg states [13]. Nonetheless, with its newly reached precision, the Rydberg constant is already the most accurately known of the fundamental constants except for the g-factor of the electron.

Other contributions in this volume give detailed accounts of the work at Yale, Paris, and Oxford. This article will therefore focus on the past experiments of our laboratory at Stanford and on our current and future program at the Max-Planck-Institute for Quantum Optics in Garching.

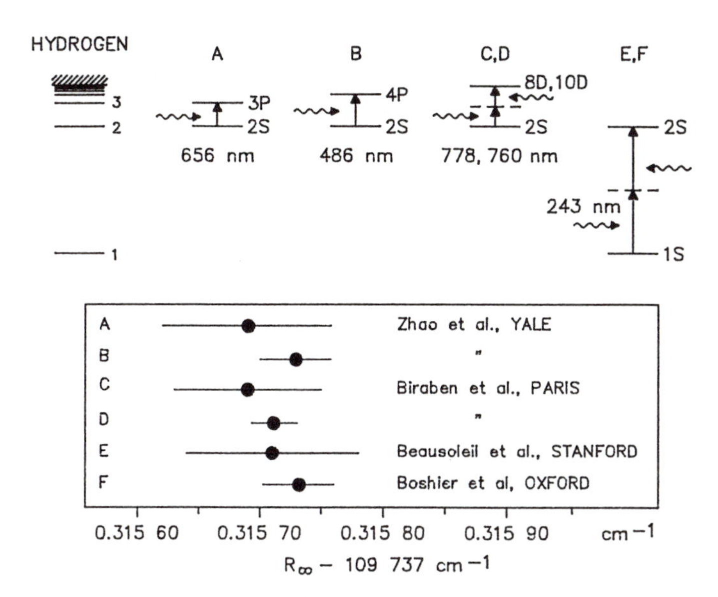

Fig. 2. Recent measurements of the Rydberg constant by laser spectroscopy of hydrogen and deuterium. Data points A and B are derived from the wavelength of the Balmer-α [7] or Balmer-ß line [8], observed by quenching of a beam of metastable 2S atoms with crossed dye laser beams. Data C and D are obtained by Doppler-free two-photon spectroscopy of 2S-8..12S transitions, recorded by quenching of a metastable atomic beam [9,10]. The Rydberg values E and F have been measured by Doppler-free two-photon spectroscopy of the 1S-2S transition in a gas cell [11,12].

3. Two-Photon Spectroscopy of the 1S-2S Transition

Even the earliest saturation spectra of the Balmer-α line approached the resolution limit imposed by the short lifetime of the excited levels. We have therefore long been concentrating our experimental efforts on the 1S-2S transition with its much narrower width [11,14-17]. The 1/7 second lifetime of the 2S state implies a natural line width of only about 1 Hz, or an ultimate resolution better than 10^{-15}.

Although one-photon transitions from 1S to 2S are forbidden, the 2S level can be excited by simultaneous absorption of two photons of 243 nm wavelength. If these photons come from opposite directions, first-order Doppler-broadening cancels without any need to select slow atoms, as first recognized by V.P. Chebotaev and coworkers [18]. The main difficulty in two-photon spectroscopy of hydrogen 1S-2S has long been the lack of a good tunable laser near 243 nm. Until recently, such experiments had to rely on frequency doubled pulsed dye laser systems with rather large instrumental linewidth and cumbersome frequency chirping problems [14-17].

In all but one [17] of the 1S-2S experiments to date, the atoms have been observed in a gas cell. Molecular hydrogen is dissociated in a gas discharge, and the atoms travel by gas flow and diffusion into the observation cell. The signal is observed by monitoring collision induced vacuum ultraviolet Lyman-α photons.

In 1S-2S spectra of hydrogen and deuterium, recorded by C. WIEMAN [16] in this way, the observed linewidth remained as large as 100 MHz. Nonetheless, the isotope shift could be measured well enough to yield first experimental evidence for a relativistic correction to the nuclear recoil effect. This correction was known theoretically, but was considered too small to be observable.

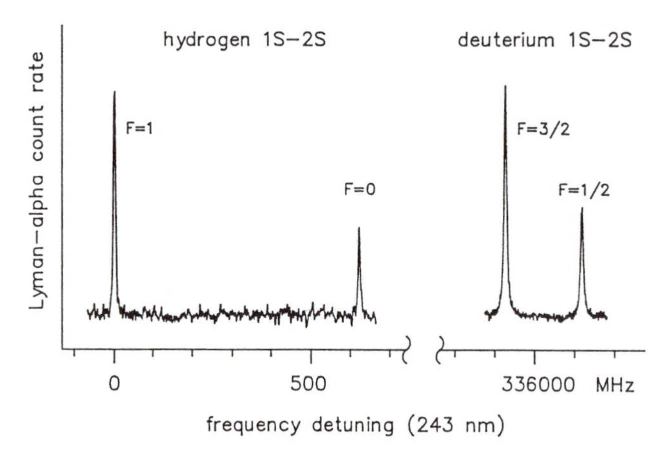

Fig. 3. Continuous wave Doppler-free two-photon spectra of hydrogen and deuterium 1S-2S [11].

Despite such successes, it was obvious that only a continuous wave laser source can do justice to the extremely sharp 1S-2S transition. Production of intense cw radiation near 243 nm remained long an elusive goal. Satisfactory power levels of several mW were first achieved by B. COUILLAUD et al. [19] by summing the frequency of a 351 nm argon laser and a 790 nm dye laser in a crystal of KDP. In the first cw experiment with this source [11], the power in the observation cell was further enhanced with a standing wave build-up cavity. Fig. 3 shows two-photon spectra recorded in this way. Although the resolution is much superior to the earlier pulsed spectra, it remains limited to a few MHz by laser frequency jitter, collision effects, and transit-time broadening. A further at least millionfold improvement in resolution should ultimately be achievable.

In 1986, R. BEAUSOLEIL and D. McINTYRE completed their thesis research at Stanford with an absolute frequency measurement of the F = 1 component of hydrogen 1S-2S [11,20]. As frequency reference they employed a 486 nm cw dye laser, locked to a narrow absorption line of $^{130}Te_2$ vapor. This line was chosen near a reference line, calibrated to within 4 parts in 10^{10} by A. FERGUSON et al. [21]. Its second harmonic coincides very nearly with the resonance frequency of the hydrogen two-photon transition, so that the frequencies can be precisely compared by observing a radio frequency beat signal.

With a careful study of pressure shifts in pure hydrogen and in a dilute mixture of hydrogen in helium, the absolute wavelength was determined to within 7 parts in 10^{10}. The corresponding value of the Rydberg constant (data point E in Fig. 2) is in good agreement with the other recent Rydberg measurements.

4. The Ground State Lamb Shift

If the Rydberg constant is considered known, a measurement of the 1S-2S frequency permits a different interpretation: it can yield an experimental value for the Lamb shift of the 1S ground state, or, more precisely, for the difference of the Lamb shifts of the 1S and 2S levels. Measurements of the 2S Lamb shift have been carried to very high precision [22] in order to test quantum electrodynamic theory. The Lamb shift of the 1S ground state is eight times larger, but cannot be measured by radio-frequency techniques, since there is no nearby P state.

Wieman was able to determine the 1S Lamb shift to within a few parts per thousand by comparing the 1S-2S frequency with the n = 2-4 Balmer-ß transition, recorded simultaneously with the blue fundamental dye laser radiation [16]. By comparing 1S-2S, in effect, with the Balmer-ß measurement at Yale, the new Stanford experiment [11] gave a 1S Lamb shift of 8173.3(1.7) MHz, in good agreement with the theoretical value of 8172.94(9) MHz. With an uncertainty of 2 parts in 10^4, it is about an order of magnitude away from challenging the best current calculations. Unlike measurements of the 2S Lamb shift, comparisons of optical transition frequencies in hydrogen need not be limited by poor resolution due to a short-living P state, and they may soon provide the most stringent tests of quantum electrodynamics for a bound system.

5. Towards Higher Resolution

At Garching, R. Kallenbach and C. Zimmermann have begun work on a new experiment, aimed at an initial thousandfold improvement of the resolution of the 1S-2S transition. Together with J. Sandberg, they have developed a new highly monochromatic ultraviolet light source, based on second harmonic generation in a crystal of barium beta borate [23]. Light from a 486 nm ring dye laser is injected into a doubly resonant build-up cavity surrounding the crystal. This cavity enhances the intensities of fundamental and second harmonic radiation, and it provides a clean TEM_{00} output mode despite a considerable walk-off in the angle tuned crystal. Several mW have in this way been generated with an input power of 300 mW.

The rms linewidth of the dye laser has so far been reduced to 300 Hz relative to a reference cavity with the help of an intracavity ADP phase modulator and a fast servo system which compensates for small rapid optical path fluctuations in the liquid dye jet [24]. A perhaps even more elegant alternative is the external laser frequency stabilizer [25] which compensates for phase and frequency noise after the light has left the laser cavity. J. Hall and coworkers [26] have recently reduced the linewidth of a commercial ring dye laser to sub-Hz levels with such a device.

Even with such a highly monochromatic laser, the natural line width of the 1S-2S transition can only be approached if the atom can interact with the light field for a sufficiently long time and without being perturbed by collisions. As a first step towards this end we have constructed a hydrogen atomic beam apparatus. The atoms are cooled to a temperature of about 6 K with the help of an escape nozzle mounted on a liquid helium cryostat [27], as illustrated in Fig. 4. A coaxial ultraviolet standing wave field is maintained inside a build-up cavity surrounding the beam source. Excitation over an interaction length of several cm reduces transit broadening to a few kHz. Atom cooling also reduces relativistic second order Doppler-shifts from 70 kHz at room temperature to about 1 kHz, and it increases the excitation probability per atom in proportion to the square of the mean interaction time.

We are exploring other means of slowing the atoms for future experiments with even higher resolution. One promising approach may be laser radiation pressure cooling [28]. Unfortunately, such cooling requires vacuum ultraviolet Lyman-α radiation near 121.5 nm. On the other hand, a hydrogen atom can be slowed from 6 K to a few mK by resonant scattering of only about 100 photons. A train of suitable pulsed Lyman-α radiation could, for instance, be produced by injecting 364.5 nm light from

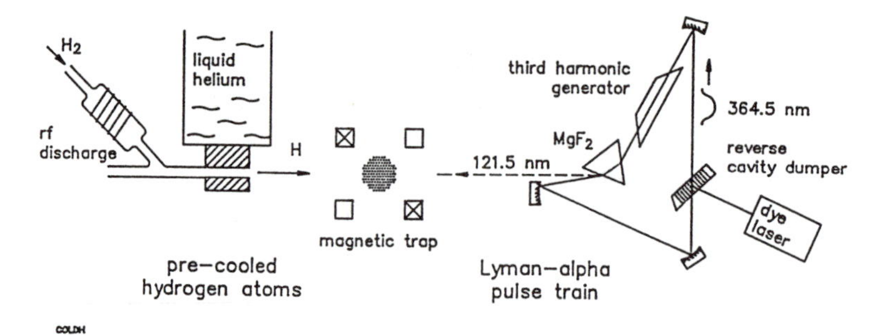

Fig. 4. Hydrogen atomic beam apparatus and scheme for laser cooling and magnetic trapping of hydrogen atoms.

a dye laser into an optical ring resonator, so that it passes repeatedly through a gas cell which produces some third harmonic radiation, as illustrated in Fig. 4.

A magnetic trap can be employed to suspend the cold atoms so that they do not escape or fall down due to gravity. Hydrogen atoms cooled with a dilution refrigerator have recently been trapped magnetically by N. MASUHARA et al. [29] and by R. VAN ROIJEN et al. [30], as also reported elsewhere in this volume. Even though trapped hydrogen atoms are perturbed by the magnetic field, high resolution two-photon spectroscopy of 1S-2S should still be possible, since upper and lower levels have very nearly equal magnetic moments so that first- and second-order Zeeman shifts remain small.

For ultimate precision one could let the cold atoms escape from the trap so that they can be observed in free fall due to gravity. Two-photon optical Ramsey spectroscopy of an "atomic fountain" has been analyzed in a semiclassical model [31]. On their parabolic trajectories the atoms are passing twice through the same standing wave laser field, so that the total excitation probability shows interference effects between the two excitation amplitudes. In contrast to ordinary Ramsey spectroscopy, the fastest atoms in a fountain spend the longest time between the two interactions and therefore produce the narrowest interference fringes. This difference has an interesting consequence when averaging over a broad atomic velocity distribution: The central interference signal appears very nearly as a simple Lorentzian resonance with just the natural linewidth. Such an experiment can hence reach the same resolution as if the atoms were standing still in the laser field. A linewidth of 1 Hz requires a fountain of just a few cm height.

6. Optical Frequency Metrology

Long before such ultimate resolution is reached, new approaches for measuring and comparing optical frequencies are urgently needed.

One interesting reference standard may be the methane stabilized helium neon laser at 3.39 μm. Its infrared frequency can be compared directly with the microwave cesium frequency standard with the help of a relatively short frequency chain [32]. An accuracy of 1 part in 10^{12} or better appears feasible for a transportable secondary standard.

At Garching, we are exploring several different approaches for synthesizing the seventh harmonic of a 3.39 μm helium neon laser, which coincides within about 2 THz with the 0.486 μm fundamental wavelength for spectroscopy of hydrogen 1S-2S. The most straightforward scheme is a laser frequency chain employing tunable solid state lasers and nonlinear optical crystals for frequency mixing.

An alternative, proposed by R. Kallenbach, employs two tunable dye lasers, operating at the fourth and sixth harmonic of the 3.39 μm helium neon laser frequency f. The frequency ratio of 3:2 is assured by comparing the second harmonic of the laser at 6f with the third harmonic of the laser at 4f. The frequency difference is simultaneously maintained at 2f by monitoring a beat signal between the sum frequency 4f+f and the difference frequency 6f-f. Summing the frequencies 6f and f finally produces 0.485 μm radiation at 7f.

A conceptually even simpler approach uses just one optical parametric oscillator, pumped by a dye laser or diode laser at 4f and oscillating at the two frequencies f and 3f. The signal frequency f is enforced by injection locking with light from the 3.39 μm reference laser. The pump frequency is adjusted so that the idler frequency agrees with the third harmonic of the reference laser. The seventh harmonic is then generated by simply summing idler and pump frequency.

A much more general and appealing scheme would be a true optical frequency divider, as could be realized by synchronizing the cyclotron motion of an electron with a laser field [33,34].

Fig. 5 illustrates a less elegant but perhaps more feasible scheme for an optical difference frequency divider and frequency synthesizer [35]. Each stage can be compared to a differential gear box: It produces a new frequency f_3 exactly halfway in between two input frequencies f_1 and f_2. To this end, it contains a tunable laser, whose second harmonic is phase locked to the sum frequency f_1+f_2. By connecting several such stages in cascade, the original frequency difference is successively halved until it becomes accessible to a microwave frequency counter. If a laser frequency f and its second harmonic are used as the starting frequencies, a beat signal at $f/2^n$ can be observed after n stages. Only 16 stages would be sufficient to link 486 nm radiation to the 9 GHz cesium frequency standard.

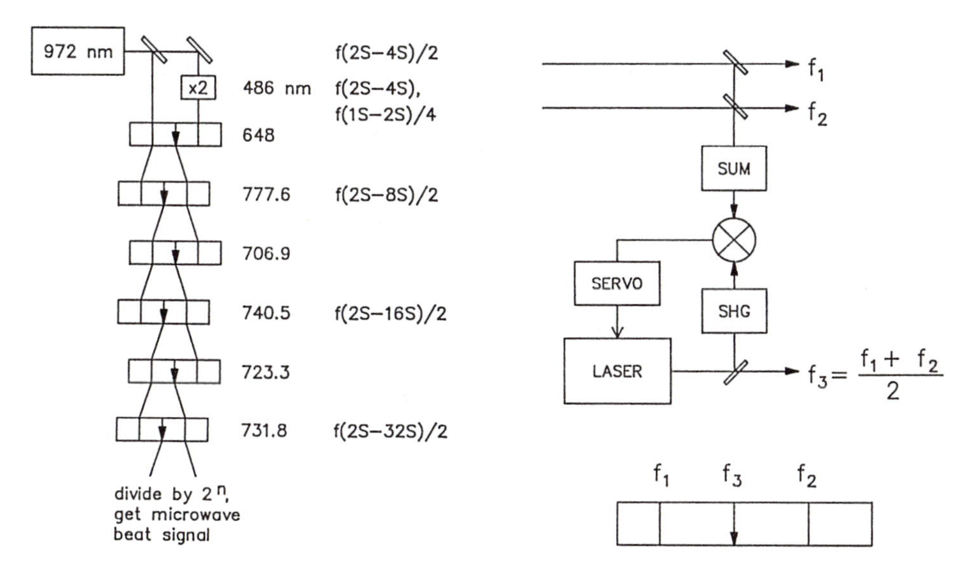

Fig. 5. Optical difference frequency divider and synthesizer of hydrogen transition frequencies [35].

The exact path can be chosen for convenience or so as to synthesize a desired optical frequency. Very few stages are sufficient to synthesize a number of hydrogen transition frequencies as illustrated in Fig. 5. Such a scheme represents then an "artificial" hydrogen atom which can be used to compare different hydrogen transition frequencies with extreme precision, and without the need to make absolute frequency measurements.

7. Comparing Experiment and Theory

Such comparisons promise interesting tests of QED. Unfortunately, however, the theory of hydrogen is no longer simple, once we try to predict its energy levels with adequate precision [36]. The quantum electrodynamic corrections to the Dirac energy of the 1S state, for instance, have an uncertainty of about 35 kHz, caused by numerical approximations in the calculation of the one-photon self-energy of a bound electron, and 50 kHz due to uncalculated higher order QED corrections.

The accuracy of computed energy levels is further limited by uncertainties of fundamental constants and by nuclear size and structure effects. The dominant contributor to the uncertainty of the 1S-2S frequency ($\simeq$500 kHz) is still the Rydberg constant. However, the electron/proton mass ratio, known within $2 \cdot 10^{-8}$ [37], introduces an additional uncertainty of about 30 kHz. The rms charge radius of the proton contributes at least another 50 kHz [38], assuming that it has been measured to within 2.5%; estimates of this accuracy still differ widely.

A measurement of the 1S-2S frequency alone can therefore give only a modest further improvement of the Rydberg constant before its interpretation is limited by other uncertainties. At one time it has been suggested to determine the proton-electron mass ratio from the 1S-2S isotope shift. However, after the most recent measurements by VAN DYCK et al. [37], the uncertainty of the isotope shift is now dominated by nuclear size effects.

Fortunately, some of these troublesome uncertainties scale in a known way with the principle quantum number n, and they can be eliminated by combining measurements of different transitions. To give an example, we could measure the isotope shifts Δf of two separate transitions such as 1S-2S and 2S-nS. By forming the linear combination $7\Delta f(2S\text{-}nS)-(1-8/n^3)\cdot\Delta f(1S\text{-}2S)$, we obtain a new composite frequency which no longer contains terms proportional to $1/n^3$. It is thus independent of nuclear size and can be calculated much more precisely than its constituents. A comparison of experiment and theory can then yield an improved electron-proton mass ratio.

Following the same strategy, we can find a composite optical frequency, $7f(2S\text{-}nS)-(1-8/n^3)\cdot f(1S\text{-}2S)$, which does not depend on the nuclear charge radius or on higher order QED corrections scaling with $1/n^3$. With improvements in the computation of the self energy and with an improved measurement of the electron mass, this frequency can provide a very precise Rydberg value. If Coulomb's law is valid within atomic dimensions, exactly the same Rydberg value must be obtained from any 2S-nS transition as well as from microwave spectroscopy of high Rydberg states, provided perturbing external fields can be sufficiently controlled.

A dimensionless frequency ratio, such as f(1S-2S)/f(2S-nS), on the other hand, is independent of the Rydberg constant. Its measurement can serve as a sensitive test of quantum electrodynamic level shifts and as a means to determine the size of the proton or deuteron, provided QED is correct.

*8. **Positronium, Muonium, and Anti-Hydrogen***

Additional tests of fundamental theory become possible if transitions in hydrogen are compared with those in other hydrogen-like atoms. The exotic positronium (e^+e^-) and muonium (μ^+e^-) are particularly interesting since they consist of leptons so that no cumbersome hadron structure corrections are necessary. Unfortunately, both atoms are unstable so that they cannot be studied with the same high resolution as hydrogen.

Positronium still presents rather formidable theoretical challenges [39], since it is a relativistic two-body system which cannot be approximated by the motion of a particle of reduced mass in a fixed Coulomb potential. In addition, QED calculations must include annihilation terms. First laser measurements of the 1S-2S two-photon transition frequency in positronium [40] give a result about five standard deviation lower than the theoretical predictions, after taking a recalibration of the tellurium reference line into account [41]. On the other hand, as yet uncalculated higher order terms could well account for this discrepancy.

Recently, first experimental results have also been reported for the 1S-2S two-photon transition of muonium [42]. With the relatively heavy muon mass this atom is much more similar to hydrogen so that its energy levels can be computed more readily, and the 2.2 μsec lifetime of the muon should permit a resolution of better than 100 kHz.

Another most interesting atom that should permit spectroscopy with extreme resolution is anti-hydrogen (p^-e^+). After the recent successful trapping of low energy anti-protons by G. Gabrielse and coworkers [43] it appears conceivable to produce and magnetically trap slow anti-hydrogen atoms. High resolution spectroscopy of anti-hydogen 1S-2S two-photon transitions should require just a few atoms inside the trap. By observing simultaneously hydrogen and anti-hydrogen, one can investigate whether there are any spectroscopic differences between matter and anti-matter. At very low temperature and with a vertical trap of sufficient length it might even be possible to determine whether anti-hydrogen senses a stronger gravitational force than hydrogen, as recently suspected [44].

* * *

"Ultra-spectroscopy" of atomic hydrogen as envisioned here can provide a strong motivation to develop and refine our spectroscopic instruments and techniques far beyond the present state of the art. If past history is any guide such efforts are likely to be rewarded by new applications and unexpected discoveries.

References

1) G.W. Series, ed., "The Spectrum of Atomic Hydrogen: Advances", World Scientific Publishing Co., Singapore, 1988
2) T.W. Hänsch, Appl. Opt. 11, 895 (1972)
3) T.W. Hänsch, I.S. Shahin, and A.L. Schawlow, Phys. Rev. Lett. 27, 707 (1971)
4) T.W. Hänsch, I.S. Shahin, and A.L. Schawlow, Nature 235, 63 (1972)
5) T.W. Hänsch, M.H. Nayfeh, S.A. Lee, S.M. Curry, and I.S. Shahin, Phys. Rev. Lett. 32, 1396 (1974)
6) D.H. McIntyre and T.W. Hänsch, Metrologia 25, 61 (1988)
7) P. Zhao, W. Lichten, H.P. Layer, and J.C. Bergquist, Phys. Rev. A34, 5138 (1986)
8) P. Zhao, W. Lichten, H.P. Layer, and J.C. Bergquist, Phys. Rev. Lett. 58, 1293 (1987)
9) M. Allegrini, F. Biraben, B. Cagnac, J.C. Garreau, and L. Julien, to be published
10) F. Biraben, J.C. Garreau, and L. Julien, Europhysics Lett. 2, 925 (1986)

11) R.G. Beausoleil, D.H. McIntyre, C.J. Foot, E.A. Hildum, B. Couillaud, and T.W. Hänsch, Phys. Rev. A35, 4878 (1987)
12) M.G. Boshier, P.E.G. Baird, C.J. Foot, , E.A. Hinds, M.D. Plimmer, D.N. Stacey, J.B. Swan, D.A. Tate, D.M. Warrington, G.K. Woodgate, Nature 330, 463 (1987)
13) J. Liang, M. Gross, P. Goy, S. Haroche, Phys. Rev. A33, 4437 (1986)
14) S.A. Lee, R. Wallenstein, and T.W. Hänsch, Phys. Rev. Lett. 34, 307 (1975)
15) S.A. Lee, R. Wallenstein, and T.W. Hänsch, Phys. Rev. Lett. 35, 1262 (1975)
16) C. Wieman and T.W. Hänsch, Phys. Rev. A22, 192 (1980)
17) E.A. Hildum, U. Boesl, D.H. McIntyre, R.G. Beausoleil, and T.W. Hänsch, Phys. Rev. Lett. 56, 576 (1986)
18) L.S. Vasilenko, V.P. Chebotaev, and A.V. Shishaev, JETP Lett. 12, 113 (1970)
19) B. Couillaud, L.A. Bloomfield, and T.W. Hänsch, Optics Comm. 35, 127 (1984)
20) R.G. Beausoleil, D.H. McIntyre, C.J. Foot, E.A. Hildum, B. Couillaud, and T.W. Hänsch, Phys. Rev. A, submitted for publication
21) J.R.M. Barr, J.M. Girkin, A.I. Ferguson, G.P. Barwood, P. Gill, W.R.C. Rowley, and R.C. Thompson, Opt. Commun. 54, 217 (1985)
22) S.R. Lundeen and F.M. Pipkin, Phys. Rev. Lett. 46, 232 (1981)
23) C. Zimmermann, R. Kallenbach, J. Sandberg, and T.W. Hänsch, to be published
24) R. Kallenbach, C. Zimmermann, D.H. McIntyre, and T.W. Hänsch, Opt. Commun., submitted for publication
25) J.L. Hall and T.W. Hänsch, Opt. Lett. 9, 502 (1984)
26) J.L. Hall, JILA, private communication
27) A. Hershcovitch, A. Kponou, T.O. Niinikoski, Rev. Sci. Instrum. 58, 547 (1987)
28) T.W. Hänsch and A.L. Schawlow, Opt. Commun. 13, 68 (1975)
29) N. Masuhara, J.M. Doyle, J.C. Sandberg, D. Kleppner, T.J. Greytak, H.F. Hess, and G.P. Kochanski, Phys. Rev. Lett. 61, 935 (1988)
30) R. Van Roijen, J.J. Berkhout, S. Jaakkola, and J.T.M. Walraven, Phys. Rev. Lett. 61, 931 (1988)
31) R.G. Beausoleil and T.W. Hänsch, Phys. Rev. A33, 1661 (1986)
32) G. Kramer, PTB, private communication
33) D.J. Wineland, J. Appl Phys. 50, 2528 (1979)
34) A.E. Kaplan, Opt. Lett. 12, 489 (1987)
35) D.H. McIntyre and T.W. Hänsch, to be published
36) D.H. McIntyre, E.A. Hildum, R.G. Beausoleil, B. Couillaud, C.J. Foot, and T.W. Hänsch, Phys. Rev. A, to be published
37) R.S. Van Dyck, Jr., F.L. Moore, D.L. Farnham, and P.B. Schwinberg, Bull. Am. Phys. Soc. 31, 224 (1986)
38) L.N. Hand, D.J. Miller, and R. Wilson, Rev. Mod. Phys. 35, 335 (1963)
39) T. Fulton, Phys. Rev. A26, 1794 (1982)
40) S. Chu, A.P. Mills, Jr., and J.L. Hall, Phys. Rev. Lett. 52, 1689 (1984)
41) D.H. McIntyre and T.W. Hänsch, Phys. Rev. A34, 4504 (1986)
42) S. Chu, A.P. Mills, Jr., A.G. Yodh, K. Nagamine, Y. Miyake, and T. Kuga, Phys. Rev. Lett. 60, 101 (1988)
43) G. Gabrielse, X. Fei, K. Helmerson, S.L. Rolston, R. Tjoelker, T.A. Trainor, H. Kalinowsky, J. Haas, and W. Kells, Phys. Rev. Lett. 57, 2504 (1986)
44) T. Goldman, R.J. Hughes, and M.M. Nieto, Scientific American 258, 48 (1988)

Trapped Atomic Hydrogen

D. Kleppner
Research Laboratory of Electronics, MIT, Cambridge, MA 02139, USA

1. INTRODUCTION

The rapid development of techniques for cooling and trapping atoms using laser light has created a new subfield of atomic physics. Research opportunities include the study of matter at ultra low temperature, ultra precise atomic spectroscopy and the study of light-matter interaction in a new quantum regime.

Unfortunately, the difficulty of generating UV light has so far excluded hydrogen from this research, though at least two groups are now undertaking to laser cool hydrogen. Nevertheless, interest in ultra cold hydrogen is considerable because of hydrogen's continuing appeal as a testing ground for theory. For example, the hydrogen constitutes an ideal candidate for studying low energy scattering phenomena and the weakly interacting Bose gas. As a candidate for ultra precise spectroscopy hydrogen remains pre-eminent for testing theory and evaluating the fundamental constants.

At MIT we have recently succeeded in trapping hydrogen and cooling it to the millikelvin regime using an approach far removed from laser cooling. Our method, which was developed in the course of research on spin-polarized hydrogen, is based on special cryogenic techniques, essentially low temperature "tricks" that one can play with hydrogen. These developments open the way to several new lines of research including very high resolution spectroscopy. Some consideration governing spectroscopy of the $1S \rightarrow 2S$ two-photon transition are discussed below, following a brief summary of the hydrogen trap.

The Hydrogen Atom Editors: G.F. Bassani · M. Inguscio · T.W. Hänsch

2. BACKGROUND: SPIN-POLARIZED HYDROGEN

Pursuit of the goal of observing Bose-Einstein condensation in atomic hydrogen has resulted in the creation of techniques for stabilizing hydrogen and storing it at temperatures below 1 K. Electronically spin-polarized hydrogen interacts by the predominately repulsive triplet potential and is consequently immune to the normal molecular recombination processes. The electron spins can be "pinned" by working in the temperature-field regime $\mu_o B > k_B T$, where μ_o is the Bohr magneton and k_B is the Boltzmann constant. At a field of 10 T, $\mu_o B/k_B = 6.6\,K$. At a temperature of 0.3 K, the system is essentially 100% electron polarized. Spin-polarized hydrogen is formed by allowing cold atoms to flow into a high magnetic field. The "high-field seeking" states ("$H\downarrow$") are attracted, the "low-field seeking" states ("$H\uparrow$") are repelled.

Helium-covered surfaces are crucial for cryogenically cooling and trapping hydrogen. The binding energy of H on He^4 is only 1.0 K. This is so low that hydrogen can be cooled to 0.1K or less without undue adsorption. As the atoms flow into the magnetic field they are thermalized by surface collisions and cannot escape.

Unfortunately, the density of $H\downarrow$ is limited by a three body recombination process of the form $H\downarrow + H\downarrow \rightarrow H_2 + H\uparrow$. The highest density achieved under controlled conditions is 4.5×10^{18}atoms·cm^{-3}, at a temperature of 0.4 K. This is a factor of twenty too low for Bose-Einstein condensation.

A large body of experimental and theoretical literature exists on the properties of $H\downarrow$ in the temperature regime surrounding 0.3K. Because this information is well documented [1-3], we shall not review it here but turn instead to a new direction of the research.

The temperature for Bose-Einstein condensation varies with density as $n^{2/3}$. Because density is limited by three-body recombination, the search for the transition leads naturally to lower temperatures. Unfortunately, at temperatures below 0.1 K, adsorption rapidly becomes prohibitive. To avoid this problem, Hess [4] suggested confining the atoms in a magnetic trap without any surfaces. The states confined are the "low-field seeking" states, ($H\uparrow$, electron spin "up"). These are the hyperfine states $(F=1, m=1)$ and $(F=1, m=0)$.

3. THE HYDROGEN TRAP

The trap consists of a configuration of coils designed to produce a magnetic field minimum at its center, but without having a zero in the field. The geometry, which is essentially cylindrical, employs a configuration proposed by Pritchard [5]. Transverse trapping is provided by a quadrupole magnetic field. Longitudinal trapping is provided by a coaxial coil ("pinch solenoid") at each end which provides a barrier against leakage and a bias field at the center. The trap is loaded from a pulsed hydrogen discharge located in a high field. The gas "falls" into the potential well of the trap by inelastic collisions which isolate a population of atoms at a temperature significantly below the ambient temperature. In the first experiment [6], for instance, the gas was found to be at 40 mK when the ambient temperature was 70 mK. The temperature was inferred from the rate of escape as one of the pinch coils was lowered. The escaping atoms were monitored by a low-field hyperfine resonance detector. Further details are given in reference [6]. The geometry is shown in Fig. 1, and a pictorial view is provided in Fig. 2.

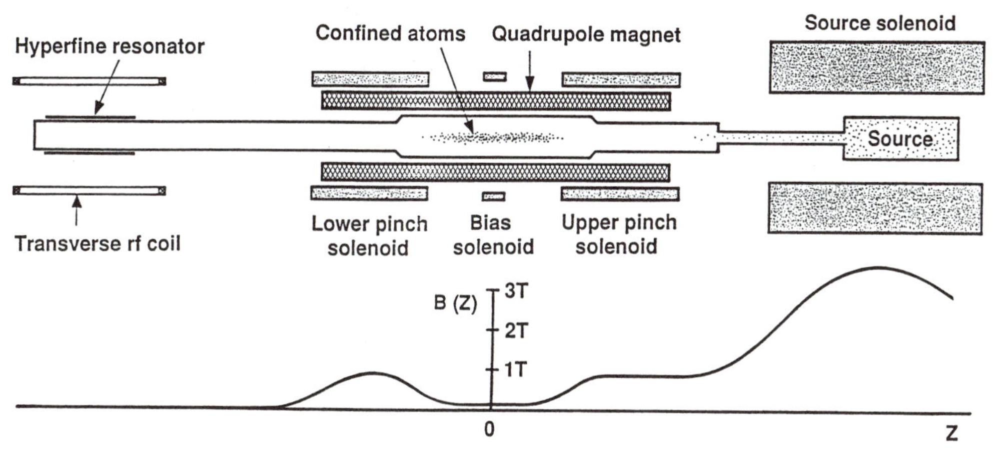

Figure 1. Schematic diagram of the MIT hydrogen trap, with magnetic field profile. (From ref. 7.)

The trapped gas can be further cooled by controlled evaporation as proposed by Hess [4]. The process is relatively efficient because the evaporating atoms have to surmount a high potential barrier and carry off energy substantially greater than the average thermal energy. Furthermore, the density in the trap stays approximately constant during the evaporation process because the loss of atoms is compensated by a decrease in temperature and the consequent decrease of the effective volume.

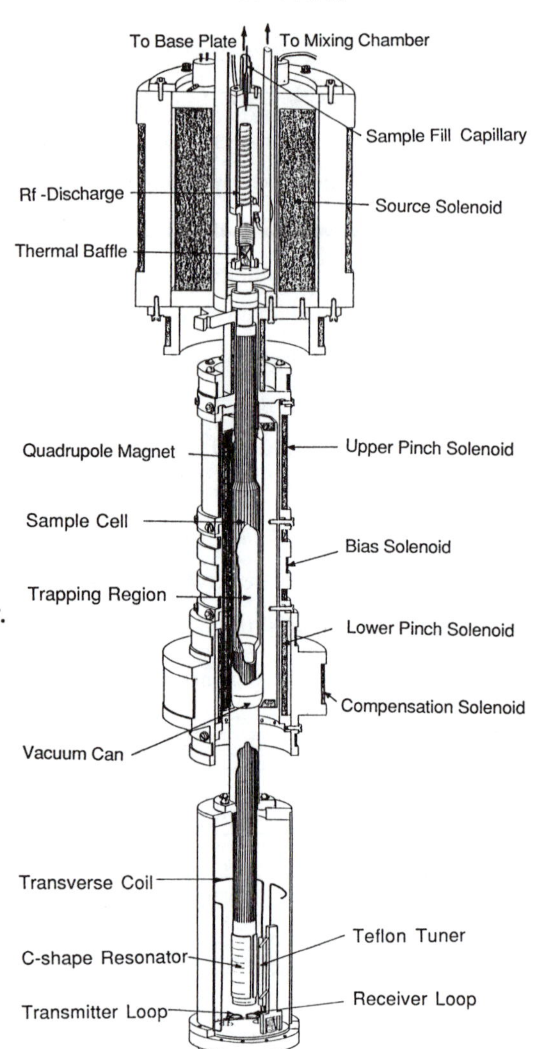

Figure 2. Pictorial diagram of the hydrogen trap. The magnets are immersed in liquid helium: the inner region of the trap is connected to a dilution refrigerator.

The trap is loaded in a time of a few seconds, and after a preselected delay its contents are dumped into the resonator region by lowering the field of one of the pinch solenoids. The signal from the hyperfine resonance detector provides a measure of the total number of atoms N in the trap. The stored atoms decay by dipole relaxation (described below) with a rate that is proportional to the density n. From values of N and n one can find the effective volume of the trap. The effective volume depends on the field geometry and the temperature. This roundabout route is used to find the temperature. In a recent experiment [7] the gas was cooled to a temperature of 3 mK at a density of $7.6\times10^{12}\,\mathrm{cm}^{-3}$. Further cooling to 1 mK was inferred from the data. It is evident, however, that a more direct method for studying the atoms is essential. A laser-based approach is attractive.

An important feature of laser cooling and/or trapping is the built-in diagnostic process method provides: the fluorescence constitutes a sensitive probe of the trapped gas. A number of schemes have been proposed for laser cooling hydrogen using the Lyman-α transition. However, the conventional Doppler cooling limit for this transition, $T = h\Gamma/2\,k_B$, is relatively high, approximately 15 mK (though it is now recognized that this does not constitute a fundamental limit). The recoil energy limit, $(h\gamma)^2/(2\,M_pc^2)$, is also high due to the high energy of the photon and the low mass of the atom: it corresponds to a temperature of 0.5 mK. Thus, laser cooling into the μK regime is a formidable task though, as Walraven has pointed out [8], there may be advantage to combining laser cooling with evaporation.

We have chosen optical excitation of the $1S \rightarrow 2S$ two-photon transition to study the trapped gas. Because the natural linewidth for this transition is only a few cycles, it is capable of yielding enormous spectral resolution. Consequently, two-photon excitation is capable of providing excellent momentum resolution for the gas and has the added attraction of offering a new possibility for carrying out ultra high resolution spectroscopy on hydrogen. It is the latter application which we discuss here. We note in passing that the two-photon transition is not well suited to laser cooling schemes: the 246 nm radiation is so effective at photoionizing the 2S state that the gas would be ionized before it is significantly cooled.

4. ULTRA HIGH RESOLUTION SPECTROSCOPY OF TRAPPED HYDROGEN

We summarize here some considerations affecting the potential accuracy of a spectroscopic measurement of the $1S \rightarrow 2S$ transition. We shall not dwell on the challenging problems of laser stabilization and optical frequency metrology, but only on the atomic considerations. In particular, we shall consider the major sources of line broadening and possible systematic shifts. We discuss below some of the factors which govern the accuracy of $1S \rightarrow 2S$ spectroscopy in the hydrogen trap.

NATURAL LINE WIDTH

The lifetime of the metastable state is $\tau_o = 1/7\,s$. The "Q" of the transition is 2.2×10^{15}. In principle, the resonance can be located to a small fraction of the linewidth.

LIFETIME IN THE TRAP

The trapped hydrogen is in the hyperfine state F=1, m=1. The gas can decay by dipole relaxation to a lower-lying hyperfine state and escape from the trap. The decay rate is given by $\Gamma_d = nG$, where n is the total density and the factor G is approximately temperature independent. Lagendijk, Silvera and Verhaar [9] have calculated $G = 1 \times 10^{-15}\,cm^3 \cdot s^{-1}$, which is in agreement with the observed value [7] $G = (1.2 \pm 0.5) \times 10^{-15}\,cm^3 \cdot s^{-1}$. With a density of $10^{12}\,cm^{-3}$, the lifetime is $10^3\,s^{-1}$. Dipole decay imposes a requirement on the supply rate of hydrogen to maintain the density, but does not limit the spectral resolution.

DOPPLER SHIFTS

First order Doppler broadening can be eliminated by using a standing wave geometry (i.e. oppositely running waves) to excite the two-photon transition. The fractional second order Doppler shift, $v^2/2c^2$, is less than 2×10^{-16} at a temperature of 1 mK.

ZEEMAN SHIFT

The $1S \rightarrow 2S$ transition obeys the selection rule $\Delta F = 0$, $\Delta m = 0$ and is almost field-independent. However, the g-factor for the bound electron is slightly less than for free space due to relativistic effects, and this gives the transition a small first-order field dependence. In the 1S state the g-factor is $g(1S) = g_e\,(1 - \alpha^2/3)$ [10]. The relativistic term is proportional to the binding energy so that $g(2S) = g_e\,(1 - \alpha^2/12)$. Thus, the field-dependence of the transition $1S \rightarrow 2S$, $(F=1, m=1; \Delta F=0, \Delta m=0)$ leads to a frequency shift

$$\nu_z = \frac{g_e}{2}\,\frac{\alpha^2\,\mu_o\,B}{4h} = 1.8 \times 10^5\ \text{Hz} \cdot \text{B(tesla)}\ .$$

If the mean field can be determined to 10^{-5}T (0.1 gauss), ν_z is known to 1.8 Hz and the frequency uncertainty due to ν_z can be expected to be small. However, the trapping process demands that the atom sample a spatially varying field, and that can result in line broadening.

The characteristic spread in magnetic field is given by $\Delta B = k_B T/\mu_o$. For a temperature of 1 mK, $\Delta B = 1.6\times10^{-3}$ T, and the spread in ν_z is $\Delta\nu_z = 300$ Hz. For a temperature of 30 μK the linewidth is 10 Hz. Thus, Zeeman-broadening in the trap is potentially significant. However, as described below, it is reduced by motional averaging effects.

The 2S state has a large electron polarizability due to the proximity of the 2P state. Motion through the magnetic field produces an electric field $E_m \cong (v/c) B$. This field can lead to the quenching of the 2S state, and a Stark shift. The quenching rate is $\Gamma_s \cong |V|^2/\Gamma_{2p}$, where $B^2 = |<2S|ezE|2P>|^2$, and Γ_{2P} is the radiative decay rate of the 2P state. At a temperature of 1 mK, $\Gamma_s \cong 10^{-3}\,s^{-1}$. The Stark shift is smaller by a factor of approximately ten, and is also negligible.

MOTIONAL AVERAGING

Zeeman broadening can be reduced by motional averaging providing that the rate at which the atom transverse the trap is large compared to the Zeeman frequency spread. The average linewidth is given by

$$\Delta \nu_z' \cong 2\,\pi\,(\Delta \nu_z)^2 \tau_c$$

where τ_c is the correlation time in the field. We can regard $\eta = (2\pi\Delta\nu_z\tau_c)^{-1}$ as a narrowing factor. For a harmonic (but anisotropic) trap, $\tau_c \cong \omega_t^{-1}$, where ω_t is the natural frequency in the trap. Taking $B(r) = B_o\, r^2/a^2$, where B_o is the magnetic field at some characteristic radius r = a, measured with respect to the trap minimum, then

$$\tau = (Ma^2/2\,\mu_o B_o)^{1/2},$$
$$\eta^{-1} = (\pi\alpha^2/2\sqrt{2})\,(a/h)\,(M\mu_o B_o)^{1/2}\,.$$

If we choose a to be the effective radius of the trapped sample, then $\mu_o = k_B T$. and we have $\eta^{-1} = (\pi\alpha^2 a/2\sqrt{2}h)(Mk_B T)^{1/2} = 4.3\,a\sqrt{T}$, where a is the radius in cm, and T is the temperature in mK.

Motional averaging puts a high premium on a small trapping volume - i.e. a "stiff" trap. For a = 0.1 cm, the narrowing factor is 2.5 at 1 mK, and 15 at 30 μK. In the latter case, the contribution to the linewidth due to the Zeeman effect is 0.7 Hz, which is unimportant.

OTHER CONSIDERATIONS

The ideal conditions for studying an atom is to have it at rest in free space, or in free fall as in a "fountain" experiment. Any process which confines an atom perturbs it. However, as has been shown, at ultra low temperatures the perturbations of hydrogen due to a magnetic trap are small. Furthermore, the trap provides an enormous advantage in density compared to atomic beams or fountains: density of $10^{11} \sim 10^{12} cm^{-3}$ is readily available. Thus, the trap is particularly attractive from the point of view of signal to noise ratio.

A number of schemes are available for observing optical resonance in the trap. A straightforward technique is to quench the 2S state with a small electric field (or possibly by collisional quenching) and observe the Lyman-α emission. Most of the quenched atoms will escape from the trap, but the loss rate is so small that its effect on the density would be negligible.

5. SUMMARY

The recently developed hydrogen trap provides an opportunity to extend the precision of hydrogen spectroscopy into new regimes of accuracy. Carrying forward such a program, however, requires developing methods for introducing laser light into the trap— not a simple matter in the ultra low temperature regime— and for observing the fluorescence. It also puts great demands on optical metrology, but since advancing the art of metrology is a central goal for this enterprise, and so such a struggle is not only inevitable, it is healthy.

ACKNOWLEDGEMENTS

The experimental work described here was carried out by N. Masuhara, J.M. Doyle, J.C. Sandberg, H.F. Hess, G.P. Kochanski, and T.J. Greytak. We thank T.W. Hänsch for advice and guidance on the generation of UV light. This research is sponsored by the National Science Foundation, Grant DMR-85-13769.

REFERENCES

1. T.J. Greytak and D. Kleppner, in *New Trends in Atomic Physics,* Proceedings of the Les Houches Summer School Session XXXVIII, edited by G. Greenberg and R. Stora (North Holland, Amsterdam, 1984).

2. I.F. Silvera and J.T. M. Walraven, in *Progress in Low Temperature Physics,* **10**, edited by D. Brewer (North Holland, Amsterdam, 1986), Chap. D.

3. D.A. Bell, H. F. Hess, G.P. Kochanski, S. Buchman, L. Pollock, Y.M. Xiao, D. Kleppner, and T. J. Greytak, Phys. Rev. B. **34**, 7670 (1986).

4. H.F. Hess, Phys. Rev. B. **34**, 3476 (1986).

5. D.E. Pritchard, Phys. Rev. Lett. **51**, 1366 (1983).

6. H.F. Hess, G.P. Kochanski, J.M. Doyle, N. Masuhara, D. Kleppner, and T.J. Greytak, Phys. Rev. Lett. **59,** 672 (1987).

7. N. Masuhara, J. M. Doyle, J.C. Sandberg, D. Kleppner, T.J. Greytak, H.F. Hess and G.P. Kochanski, Phys. Rev. Lett. **61,** 935, (1988).

8. R. Van Roijen, J.J. Berkhout, S Jaakkola, and J.T.M Walraven, Phys. Rev. Lett. **61**, 931 (1988).

9. A. Lagendijk, I.F. Silvera, and B.J. Verhaar, Phys. Rev. B. **33**, 626 (1986).

10. H. Grotch and R.A. Hegstrom, Phys. Rev. A. **4**, 59 (1971).

Atomic Hydrogen in a Magnetic Trapping Field

J.T.M. Walraven, R. van Roijen, and T.W. Hijmans

Natuurkundig Laboratorium, Universiteit van Amsterdam,
Valckenierstraat 65, NL-1018 XE Amsterdam, The Netherlands

We discuss the loading and relaxational decay of spin-up polarized atomic hydrogen (H↑) in a minimum-B-field trap. Our current experiments cover the temperature range from 80 to 225 mK. The maximum obtained density is $n_0 = 3\times10^{14}\text{cm}^{-3}$ at $T \approx 100$mK, corresponding to a total of $N = 4\times10^{13}$ atoms. By covering the walls of the sample cell with either pure ^{4}He or with ^{3}He/^{4}He mixtures it could be demonstrated that the stability of the sample is not sensitive for a variation of the surface adsorption energy by as much as a factor 2.5. Measuring the rate at which H↓ is produced in the trap we can accurately determine the dipolar relaxation rate as a function of temperature. We discuss the possibility of optical detection and the prospects for optical cooling of magnetically trapped hydrogen.

I. Introduction

Substantial progress has been made over the last decade in stabilizing atomic hydrogen at high densities but an important goal of the research, the observation of Bose-Einstein condensation (BEC) still is out of reach.[1][2] To observe BEC density-to-temperature ratios $n^{\frac{2}{3}}/T \approx 6.3\times10^{13}\text{cm}^{-2}\text{K}^{-1}$ have to be reached. The single important factor that seems to frustrate attempts to satisfy this condition by compressing a sample of spin-down polarized hydrogen (H↓) is the fact that absorption of atomic hydrogen on the helium covered surfaces of the sample cell increases dramatically with decreasing temperature, leading to excessive surface recombination. Currently with H↓, $n^{\frac{2}{3}}/T \approx 5\times10^{12}\text{cm}^{-2}\text{K}^{-1}$ has been reached.[3][4][5][6] It was suggested by Hess that a possible way around this problem would be to eliminate the influence of surfaces altogether by trapping a gas of spin-up polarized hydrogen (H↑) in a static magnetic-field minimum.[7]

The absence of surface effects may be very appealing, but a serious disadvantage of trapping in a B-field minimum is that the gas phase of H↑ is much less stable than that of H↓. For densities where three-body effects are negligible the stability of H↓ is limited by nuclear dipolar spin relaxation, opening up a path to the molecular state H_2. The stability of H↑ is determined by electronic dipolar spin relaxation to the high-field seeking hyperfine states, which are ejected from the trap. This dipolar relaxation process limits the attainable densities of H↑ in the trap to about 10^{14}atoms/cm^3. With H↓, densities exceeding 10^{18}cm^{-3} have been reached.[3-6] The real challenge of the trapping experiments is therefore the attainment of ultra-low temperatures in order to satisfy the BEC requirements. For a density $n = 10^{14}\text{cm}^{-3}$ the gas should be cooled to a temperature below 34 μK.

The Hydrogen Atom Editors: G.F. Bassani · M. Inguscio · T.W. Hänsch

The feasibility of surface-free confinement of H↑ was recently demonstrated by Hess et al.(MIT), who reported trapping of up to 5×10^{12} atoms and temperatures as low as $T=40$mK ($n^{\frac{2}{3}}/T \approx 1\times10^{10}\,\mathrm{cm}^{-2}\mathrm{K}^{-1}$).[8] These authors fire a sub-kelvin discharge after which a mixture of four hyperfine states is present in the trapping region. We use the labels *a*, *b*, *c* and *d* for the hyperfine states of H in its electronic ground state in order of increasing energy. Within a second, the field gradients lead to spatial separation of H↑ (a mixture of the *c* and *d* states) and H↓ (a mixture of the *a* and *b* states). The trapping of the H↑ results from interatomic collisions in which potential energy due to the trapping potential is converted into kinetic energy which is carried off by the walls of the sample cell. By allowing hot atoms to escape from the trapping region Hess et al. demonstrated that the gas could be cooled to a temperature well below the wall temperature. This technique is known as evaporative cooling.[7] To study the gas, Hess et al. measured the number of atoms remaining in the trap after a certain holding time by dumping the content of the trap onto a magnetic resonance detector.

In this paper we describe an experiment in which we study the loading and relaxational decay of H↑ in a minimum-*B*-field trap.[9] In our geometry, evaporative cooling is absent. The trap is filled with a pure H↑-flux until a steady state is reached. We monitor the H↑ continuously during (and after) filling by observing the atoms ejected from the trap towards high field after magnetic relaxation. The principle of this experiment and a description of the apparatus is presented in section II. The results are discussed in section III. In section IV we address the prospects of magnetic trapping of H↑. In particular we speculate on the use of an optical method to study and cool a gas of trapped H↑.

II. The experiment

The principle of our experiment is illustrated in Fig.1, where we show a block diagram of the experimental cell as well as the effective trapping potential along the symmetry axis ($r=0$; dashed curve). This curve reflects the magnetic trapping field but also includes the surface adsorption potential due to the wall at the left side of the cell. The solid line corresponds to the field at the wall of the sample cell ($r=6.5$mm). Our trap has a depth equivalent to $\epsilon_{tr}/k_B = 0.92$ K. This depth is somewhat smaller than the binding energy of H on the surface of liquid ^{4}He, $\epsilon_a/k_B \approx 1$ K,[10] and much larger than the binding energy on a ^{3}He/^{4}He mixture, $\epsilon_a/k_B \approx 0.4$ K.[11] The H atoms are produced in high field in a dissociator operated at a temperature $T\approx 600$mK. The low field seeking atoms (H↑) are guided through a capillary to the trapping volume which encloses the *B*-field minimum but also extends to the high field zone. All surfaces are covered with a film of liquid helium. The H↑ atoms, entering the cell thermalize with the walls and fill the trap. These atoms cannot escape the low field region due to the presence of walls

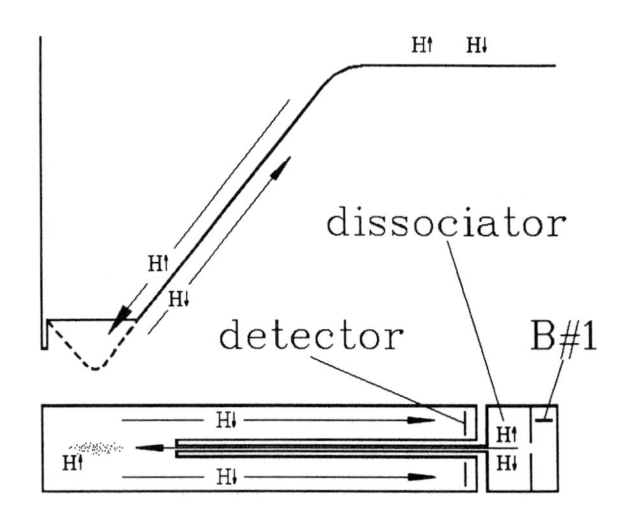

Fig.1. *Principle of the experiment. The low field seeking atoms are driven to the B-field minimum. The gas is studied by observing the atoms that escape the trap after relaxation*

and the high field barrier. The H↓ atoms produced in the discharge tend to stay in the dissociator region where they are sorption pumped by a helium-free bolometer chip (B#1). Similarly, H↓ atoms spuriously entering the cell are continuously sorption pumped by a helium free bolometer plate ('pumping plate'[12][13]) positioned at the high-field end of the sample cell. To study the trapped H↑ we observe the H↓ atoms which are ejected from the trap as a result of magnetic relaxation in the H↑ sample. The dominant relaxation processes are spin exchange and magnetic dipolar relaxation of the electron spins.[14] As spin exchange is very fast but only occurs in *c-c* collisions (*cc→aa*, *ac*, *bd*) it leads to preferential depletion of the *c* states, leaving the gas in the *d* state which is doubly polarized.[14] By properly choosing the dimensions of the experimental cell we could achieve that essentially all of the relaxing atoms arrive at the detector.

The experimental cell is shown in Fig.2. For dissociation we use a compact and rugged $\frac{1}{4}\lambda$ helical cavity with a $Q \approx 300$, resonating at 718 Mhz and driven with $0.1\,W \times 50\,\mu s$ pulses at a 50 Hz repetition rate. The method of RF dissociation at low temperatures was pioneered by Hardy et al.[15] and discussed extensively by Helffrich et al.[16]. The bolometer B#1, used to remove H↓ fraction trapped in the dissociator region, is mounted in a separate volume connected to the dissociator volume by a 1 mm diameter drilled hole. The dissociator operates optimally at $T \approx 600$mK and can produce an H↑ flux of up to 5×10^{12}/s which is guided to the trapping volume through a thin-walled german silver (GS) tube (3.6 mm i.d.). Thermometry is done against a ^{3}He melting line thermometer. The temperatures of cell and dissociator may be varied independently. The 'pumping plate' is mounted in the annular part at the lower end of the cell. This part is positioned in high field ($B \approx 4$ T). The plate

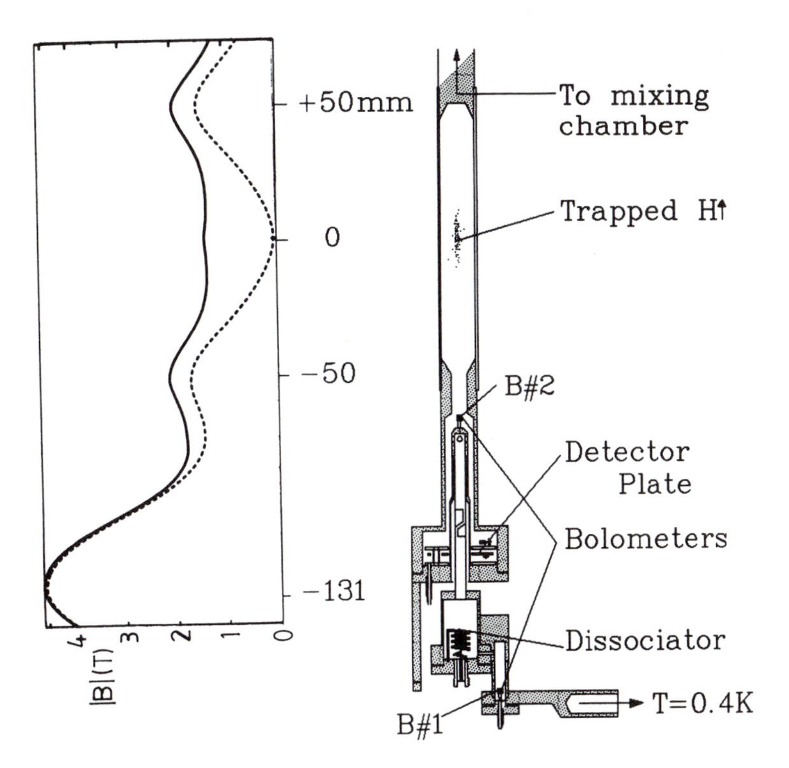

Fig.2. *Drawing of the experimental cell. On the left we show the magnetic field profile for both the symmetry axis (dashed curve) and for a distance of 6.5 mm from the axis*

has an area of 3.5 cm^2 and is suspended by three 16 μm tungsten wires. The assembly requires 20 μW to boil off the helium film and is stabilized at $T = 1.4$ K. The minimum detectable flux is $2{\times}10^{11}$ at/s. The detector efficiency is approximately 60% of the recombination heat.[17] A second bolometer (B#2) is mounted on top of the end cap of the GS filling tube, located in the fringe field of the trap. This bolometer enables us to trigger recombination of the gas in the trap.

The trapping field is similar to that discussed by Pritchard [18] and Hess[7] and is generated by a superconducting coil system operated during the measurements in persistent mode. For radial confinement we use four racetrack shaped coils which provide a quadrupole field. At maximum current (36 A) the quadrupole field reaches 1.4-1.5 Tesla at $r = 6.5$ mm, the surface of the sample cell. Two dipole fields are used for axial confinement. They are located near the ends of the racetracks at $z = +50$ mm and $z = -50$ mm with respect to the center of the trap. These coils produce fields of 1.7 and 1.5 Tesla respectively. The B-field minimum may be adjusted between $B = 0$ to $B = 1$ T with a trim coil incorporated at $z = 0$. At the lower end of the racetracks we mounted a 4.4 T dipolar coil at $z = -131$ mm which serves to separate H↑ and H↓. In Fig.2 we also show the field profile of our trap for both $r = 0$ (dashed curve; cell axis) and $r = 6.5$ mm (solid curve; cell wall).

A typical measurement cycle is shown in Fig.3. During the full cycle both bolometer B#1 and the pumping plate are actively pumping H↓. At $t = -62$ s the dissociator is switched on and the film is removed from B#2. Hence no H↑ density can build up in the trap. The flux observed to appear rapidly is caused by H↓ spuriously entering the cell as was verified by varying the B-field in the dissociator region and changing the temperature of the GS filling tube. At $t = -50$ s B#2 is switched off and the trap starts to fill as witnessed by the slowly growing flux escaping the trap region. When the dissociator is switched off at $t = 0$, H↓ keeps emerging from the cell during 50 seconds after which the remaining H↑ is removed by reactivating B#2. This procedure ensures a proper zero flux base line for the data analysis. We start the analyses at $t = 2$ s when the spurious H↓ signal is known to have disappeared as could be established by keeping B#2 activated during a full cycle.

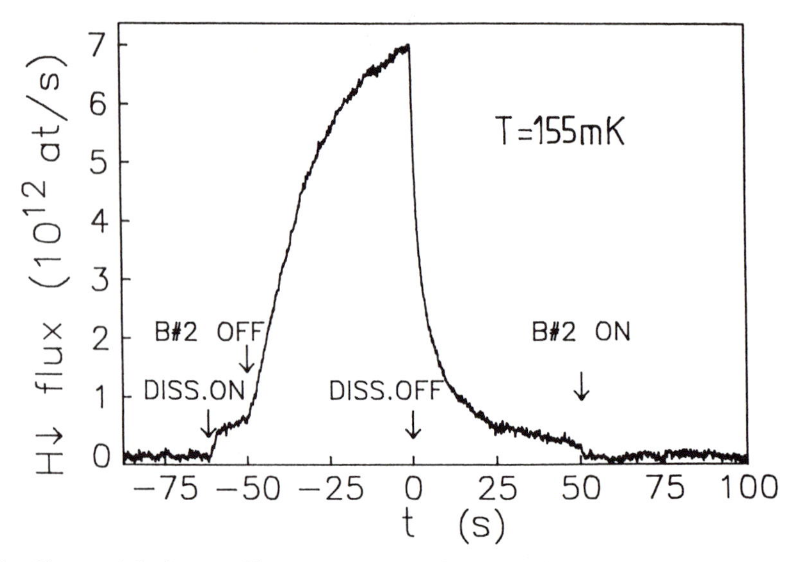

Fig.3. *Observed hydrogen flux versus time (raw data). For discussion see text*

To study the relaxation of H↑ to the high field seeking hyperfine states we plot our data as $V_\gamma \dot{N}/N^2$ versus time as shown in Fig.4. $N(t)$ is the total number of atoms in the trap at a given time t, obtained by integrating the observed flux $\dot{N}$ from t to ∞. V_γ is the effective volume of the sample defined by $V_\gamma = V_{1e}^2/V_{2e}$, where $V_{me} = \int (n(\vec{r})/n_0)^m d\vec{r}$. The density of the sample at the center of the trap is given by n_0. For our trap $V_\gamma T_g^{-\frac{5}{2}} \approx 180\ \mathrm{cm^3 K^{-\frac{5}{2}}}$. In plotting the data we set the gas temperature T_g equal to the wall temperature T_w. For temperatures above 100 mK the

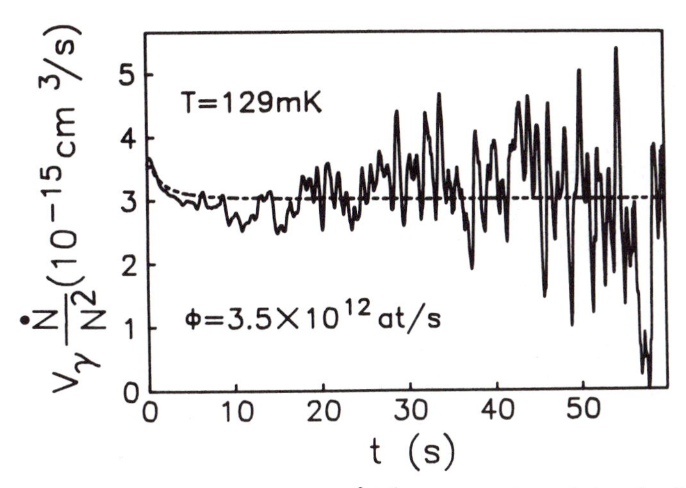

Fig.4. *Typical decay curve plotted as* $V_\gamma \dot{N}/N^2$ versus time; ϕ *is the H↑ filling flux and the dashed line represents a simulation*

initial decay tends to be faster than the decay expected for dipole-dipole relaxation alone. This enhanced initial decay (EID) is observed during up to 8 s after switching off the discharge. As discussed below, this effect is attributed to spin-exchange relaxation and nuclear polarization. For $T < 100$ mK a reduced initial decay (RID) is observed for up to 15 s. The RID is attributed to loss of thermal contact between gas and cell walls.

Using the theoretical results of Stoof et al.[19] we calculate $\dot{N}(t)$ by averaging the field and temperature dependent relaxation rates over a thermal density distribution. The contribution due to an individual process may be written as

$$\dot{N} = (\gamma_0 G_0 / V_\gamma) N_{h1} N_{h2} \qquad (1)$$

where G_0 is the rate constant of the process, evaluated for the conditions at the center of the trap, and $\gamma_0 \equiv \int \{G(\vec{r})/G_0\}(n(\vec{r})/n_0)^2 d\vec{r}/V_{2e}$ includes all effects due the field dependence of the rate. The effective volume V_γ accounts for all effects associated with the spatial distribution of the gas. N_{hi} refers to the total number of atoms in the trap in hyperfine state *hi*. The correction factor γ_0 varies substantially with temperature. For T_g increasing from 80 to 225 mK, γ_0 increases from 2.1 to 3.5 for the dipolar relaxation processes. For spin exchange, γ_0 decreases from 0.28 to 0.08 in this temperature range. Calculating the decay curves we find that the dominance of the spin exchange terms in the rate equations leads to nuclear polarization and to an EID-period after which the polarization reaches a steady state. In Fig.4 the calculated EID denoted by the dashed curve. We find that the asymptotic decay is described to within 3% by $V_\gamma \dot{N}/N^2 \equiv \gamma_{dd} G_{dd}$, where $G_{dd} \equiv 2G^{d}_{ddaa} + G^{d}_{ddac} + G^{d}_{ddad}$ in the notation of ref.[14].

III. Results

In our experiments densities up to $n_0 = 3\times10^{14}\mathrm{cm}^{-3}$ were reached at the center of the trap for $T \approx 100\mathrm{mK}$. This corresponds to a total of $N = 4\times10^{13}$ atoms. The results for the dipole-dipole relaxation rate are given in Fig.5. The data points represent the overall second order decay rate $\gamma_0 G_0 \equiv V_\gamma \dot{N}/N^2$ versus T obtained from plots like Fig.4 by discarding the EID/RID-period of 15 s. The open circles refer to data taken with ^{4}He covered surfaces. The data represented by solid circles were taken after 1% ^{3}He was added to the cell. For ^{4}He surfaces the measurements extend over a temperature range from 135 mK to 225 mK. The lower limit is caused by a lack of filling flux due to surface recombination in the filling tube. Above 225 mK the H↑ density near the detector plate becomes non-negligible and gives rise to systematic errors due to recombination of H↑ on the detector. With ^{3}He in the cell the detector was found to be unreliable for $T > 160$ mK due to heat conduction by the ^{3}He vapor. At the lower end of the temperature range we were limited by the cooling power of our dilution refrigerator.

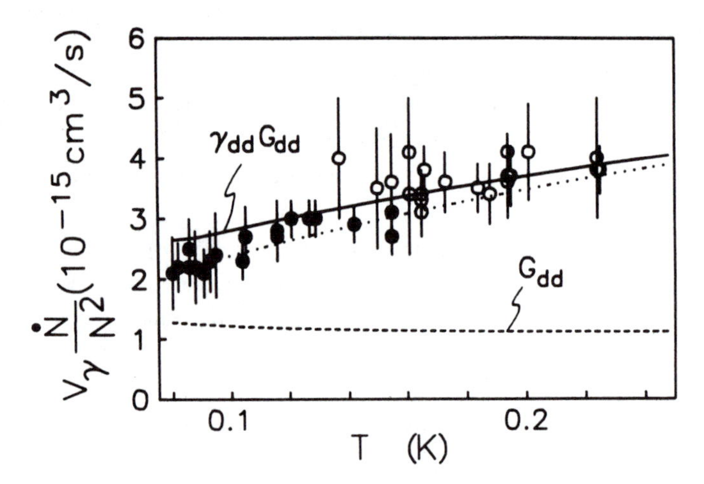

Fig.5. *The second order decay rate plotted versus temperature* (o: *^{4}He wall coverage*; ● *^{3}He/^{4}He coverage*). *Dashed line*: *theoretical result for the bottom of the trap. Solid line*: *including the field average. Dotted line*: *see text*

We find good overall agreement with the theoretical curve for $\gamma_{dd}G_{dd}$ (solid curve in Fig.5). Hence our data reflect the strong temperature dependence of V_γ. From this we infer that the gas in the trap is thermalized to the wall temperature. The slight deviation of the data with respect to the solid curve below 100 mK could in principle be explained by assuming T_g to be 8mK higher than T_w as illustrated by the dotted curve in Fig.5. However, this effect could also be accounted for by a 17% systematic error, not included in the error bars, related to the limited calibration accuracy of the detector. Comparing the open and closed circles one observes that the relaxation rate is insensitive for the surface coverage, as is to be expected for $\epsilon_{tr} \gtrsim \epsilon_a$. The dashed curve represents G_{dd} for $B = 0.05\,T$ and shows the small intrinsic temperature dependence of the dipolar rate. From a comparison of our data with the solid and dashed curves we infer that our results confirm the increase of the dipolar rate with growing field predicted by theory. Also the experimentally observed increase of the relaxation rate with increasing temperature reflects this behavior. Dividing our data by γ_{dd} we find $G_{dd} = 1.1(2)\times10^{-15}cm^3/s$ for $T = 100$ mK and $B = 0.05\,T$. This is more accurate than the result of Hess et al.[8] and in good agreement with the theoretical value $G_{dd} = 1.2\times10^{-15}cm^3/s$.[19]

The observed RID's can be explained if we assume that just after switching off the discharge, the gas is at a slightly higher temperature ($\Delta T < 10$ mK) than the walls and cools down to T_w in 10-15 s. Even a small ΔT should show up as a substantial RID due to the strong temperature dependence of V_γ. However, on the basis of a computer simulation of the decay we expect an exponential temperature dependence of the thermal accommodation time. This is not in line with the measurements, moreover the observed RID's are larger than we expect. This point deserves further study.

IV. Prospects

Even for the relatively high density $n_0 = 3\times10^{14}cm^{-3}$ obtained at the center of our trap at $T \approx 100$ mK we are far from BEC. The density-to-temperature ratio for these conditions is $n^{\frac{2}{3}}/T \approx 4.5\times10^{10}cm^{-2}K^{-1}$, which is still two orders of magnitude lower than the value obtained by compressing H↓. Therefore the temperature has to be reduced by three orders of magnitude to reach BEC. On the basis of our present results we estimate that our current technique is applicable down to approximately 55 mK. To achieve lower temperatures we are developing a Lyman-α (L_α) optical cooling method. Since this leaves the number of particles intrinsically unaffected high densities should be attainable at temperatures below 10 mK. The optical approach is also very attractive from the point of view of detection and thermometry since it is both extremely sensitive and non-destructive.

We briefly describe our optical plans. Fig.6 shows the hyperfine structure of

TABLE 1
Recoil effects.

recoil velocity: $\delta v = (\hbar\omega/mc)$
recoil energy: $E_r = \frac{1}{2}m(\delta v)^2$ (v=0)
frequency shift: $\delta\omega/\omega = (v/c)\cos(\vec{k},\vec{v}) + \frac{1}{2}\delta v/c$
thermal velocity: $\bar{v} = (8kT/\pi m)^{\frac{1}{2}}$

$\bar{v}$ is given for T=50mK

$\delta v/\bar{v} = (\hbar\omega/c)(\pi/8mkT)^{\frac{1}{2}}$
$\delta\omega/\omega = \frac{1}{2}\delta v/c$ (v=0)

	δv (m/s)	$\bar{v}$ (m/s)	$\delta v/\bar{v}$ (%)	$\delta\nu$ (MHz)	E_k/k_B (mK)
H	3.25	32.4	10.0	13.5	0.64
D	1.63	22.9	7.1	6.8	0.32
^{23}Na	0.03	6.7	0.4	25kHz	0.0012

the 1S, 2S and 2P manifolds of the hydrogen atom. In a minimum-*B*-field trap only atoms in the low field seeking hyperfine states are present, primarily the *d* state. For temperatures below 20 mK thermometry is probably best done by measuring the ratio of the intensities of the π_1 and σ_1 lines. At present it is hard to assess the limitations of this method but accurate thermometry in the 10-100 μK regime should be feasible.

Recoil effects, when absorbing or emitting L_α, are extraordinary large in hydrogen as a direct result of the small mass of the atom and the large momentum of the Lyman-α photon. This is illustrated in Table 1 for absorption events with maximum momentum transfer. The large recoil makes it fascinating to analyze the prospects for optical cooling H↑ (or D↑) in the minimum-*B*-field trap. To cool the gas in our trap optically one has to tune to the low frequency wing of the

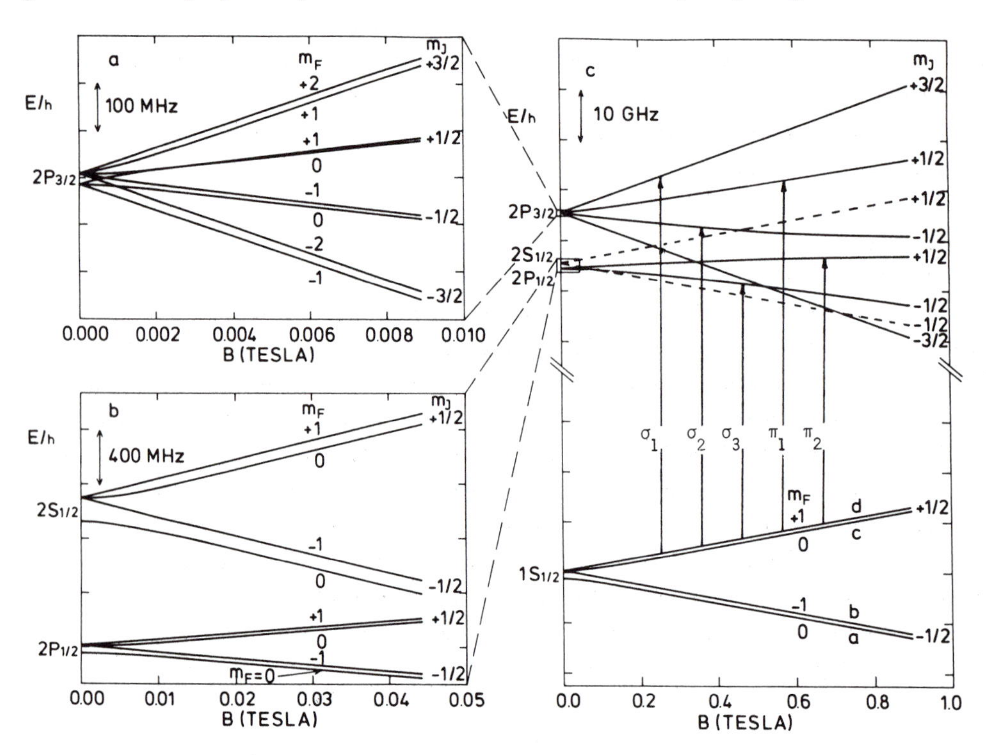

Fig.6. *The hyperfine structure of the 1S, 2S and 2P manifolds of the H-atom. We also show the transitions from the d-state which are allowed within a fine-structure picture*

transition to the $2^2P_{\frac{3}{2}}(m=\frac{3}{2})$ level (σ_1). This state decays back only to the *d* state. To evaluate the cooling efficiency let us assume the trap to be loaded with a total of 10^{10} atoms at 80 mK ($n_0 = 7\times10^{11}\,cm^{-3}$), a condition at which the intrinsic lifetime of the sample is of order hours. From Table 1 one observes that only 10 absorption events with maximum momentum transfer are needed to 'stop' an H-atom.

For our trap, where the atoms are moving in random directions and experience a spatially varying magnetic field, a computer simulation showed that about 80 photons/atom are needed to cool to below 10mK. Assuming a photon flux of 10^{11}-10^{12}photons/s temperatures below 10 mK can thus be reached well within 10 s. The required intensities (5×10^9photons/pulse; 100 Hz repetition rate) and band widths (100 MHz) are state of the art in VUV generation.[20][21]

To reach BEC one has to cool beyond the optical quantum limit $\frac{1}{2}\hbar\Gamma/k_B = 2.2$mK for H. Here Γ is the natural line width of the L_α transition. For this an evaporative cooling scheme seems to be indispensible. Apart from the procedure suggested by Hess [7] in which the trapping fields are lowered, the evaporation may also be induced by optical pumping to one of the high-field-seeking ground state hyperfine levels. For this purpose the σ_2 or π_2 transitions could be used. Also magnetic resonance may be used to induce evaporation. A preliminary analysis has shown that this type of evaporation can be as efficient as the method proposed by Hess.

In the final stage of the preparation of this manuscript we received a preprint from Masuhara et al.[22] reporting $n_0 = 7.6\times10^{12}\text{cm}^{-3}$ at $T = 3$ mK ($n^{\frac{2}{3}}/T \approx 1.3\times10^{11}\text{cm}^{-2}\text{K}^{-1}$), obtained by evaporative cooling.

V. Acknowledgements

The authors benefitted from many useful discussions with the Eindhoven theory group of B.J. Verhaar and the group of A.Lagendijk. The trapping experiments were done in collaboration J.J. Berkhout and S. Jaakkola. This work is part of the research program of the 'Stichting voor Fundamenteel Onderzoek der Materie (FOM)', which is financially supported by the 'Nederlandse Organisatie voor Wetenschappelijk Onderzoek (NWO)'.

References

1. I.F.Silvera and J.T.M.Walraven, Progr.Low Temp.Phys. edited by D.F.Brewer (North-Holland, Amsterdam, 1986), Vol X, 139.
2. T.J. Greytak and D. Kleppner, New Trends in Atomic Physics, edited by G.Grynberg and R.Stora (Elsevier, Amsterdam, 1984), Vol.II, 1125.
3. R.Sprik, J.T.M.Walraven, and I.F.Silvera, Phys.Rev.B 32, 5668 (1985).
4. D.A.Bell, H.F.Hess, G.P.Kochanski, S.Buchman, L.Pollack, Y.M.Xiao, D.Kleppner, and T.J.Greytak Phys.Rev.B 34, 7670 (1986).
5. T.Tommila, E.Tjukanov, M.Krusius, and S.Jaakkola, Phys.Rev.B 36, 6837 (1987).
6. J.D.Gillaspy, I.F.Silvera, and J.S.Brooks (1988), to be published.
7. H.F. Hess, Phys.Rev.B(RC) 34, 3476 (1986).
8. H.F.Hess, G.P. Kochanski, J.M. Doyle, N.Masuhara, D.Kleppner, and T.J.Greytak, Phys.Rev.Lett. 59, 672 (1987).
9. R.van Roijen, J.J.Berkhout, S.Jaakkola, and J.T.M.Walraven, to be published.
10. For a compilation of measurements of the binding energy of H on ^{4}He see W.N. Hardy, M.D. Hürlimann, and R.W. Cline, Jap.J.Appl.Phys. **26**, 2065 (1987) suppl. 26-3 (LT18).
11. G.H. van Yperen, A.P.M. Matthey, J.T.M. Walraven, and I.F. Silvera, Phys.Rev.Lett. 47, 800 (1981).

12. J.J.Berkhout, O.H.Höpfner, E.J.Wolters, and J.T.M.Walraven, Jap.J.Appl.Phys. 26, 231 (1987) suppl. 26-3 (LT18).
13. J.J.Berkhout, E.J.Wolters, R.van Roijen, and J.T.M.Walraven, Phys.Rev.Lett. 57, 2387 (1986).
14. A.Lagendijk, I.F.Silvera en B.J.Verhaar, Phys.Rev.B (RC) 33, 626 (1986).
15. W.N.Hardy, M.Morrow, R.Jochemsen, B.W.Statt, P.R.Kubik, R.M.Marsolais, and A.J.Berlinsky, Phys.Rev.Lett. 45, 453 (1980).
16. J. Helffrich, M. Maley, M. Krusius, and J.C. Wheatley, J.Low Temp.Phys. 66, 277 (1987).
17. J.J.Berkhout, O.H.Höpfner, E.J.Wolters, and J.T.M.Walraven, Jap.J.Appl.Phys. 26, 231 (1987) suppl. 26-3 (LT18).
18. D.E.Pritchard, Phys.Rev.Lett. 51, 1336 (1983). V.S.Bagnato, G.P.Lafyatis, A.G.Martin, E.L.Raab, R.N.Ahmad-Bitar and D.E.Pritchard, Phys.Rev.Lett. 58, 2194 (1987).
19. H.T.C.Stoof, J.M.V.A.Koelman, B.J.Verhaar (1988), to be published and private communication.
20. S.Chu, A.P.Mills, A.G.Yodh, K.Nagamine, Y.Miyake, and T. Kuga, Phys.Rev.Lett. 60, 103 (1988).
21. R.Hilbig and R.Wallenstein, IEEE J.Quantum Electron. 15, 1566 (1981).
22. N. Masuhara, J.M. Doyle, J.C. Sandberg, D. Kleppner, T.J. Greytak, H.F. Hess and G.P. Kochansky, to be published.

Frequency Standards in the Optical Spectrum*

*D.J. Wineland, J.C. Bergquist, W.M. Itano, F. Diedrich**, and C.S. Weimer*
National Bureau of Standards, Boulder, CO 80303, USA

1. Introduction

A long standing problem of spectroscopic measurements in the optical region of the spectrum has been the difficulty of obtaining accurate frequency reference standards to which spectral measurements can be compared. As will be discussed in this symposium on the hydrogen atom, this affects the value of the Rydberg constant as derived from optical measurements in hydrogen. As the experiments on hydrogen improve, the need for better reference points becomes more acute.

In this paper, we address two aspects of this general problem. First, we discuss the problem of frequency standards in the optical spectrum. (An analogue in the microwave region of the spectrum is the cesium beam frequency standard.) If one or a few of these reference frequencies can be accurately calibrated (perhaps by a frequency synthesis chain[1]) then it may be possible to compare optical spectra to these standards. As an example of the precision that might be achieved, we discuss only optical standards based on stored ions. Second, we discuss the problem of frequency comparison of unknown frequencies to the standards. Here we primarily restrict discussion to generation of wideband frequency "combs".

2. Optical Frequency Standards

One way an optical standard could be provided is by harmonic multiplication of a microwave frequency standard in a synthesis chain. By use of this technique, a laser at 88 THz (3.39 μm) has been made phase coherent with a microwave oscillator.[1] The best optical frequency standards may be made by locking a local oscillator (laser) to an atomic or molecular resonance line. State-of-the-art accuracies are characterized by measurements on methane stabilized He-Ne lasers in which reproducibilities in the 10^{-13} range have

---

* Supported by the U.S. Air Force Office of Scientific Research and the U.S. Office of Naval Research. Contribution of the U.S. National Bureau of Standards; not subject to U.S. Copyright.

** Supported by the DFG.

been reported.[1] There are many other schemes for optical frequency standards.[1-3] Here, we will only discuss possibilities with single stored ions in order to give an idea of what stabilities and accuracies might be achieved.

3a. Single Ions

The use of single trapped ions as optical frequency standards was suggested by Dehmelt[4] in 1973. Since that time, demonstrations of laser cooling,[5] 100% detection efficiency,[6-8] and lasers with sub-Hz resolution,[1,9] have increased confidence that a single-ion frequency standard is viable. As an example, we summarize the current status[10] and discuss future prospects using single Hg^+ ions. Similar experiments using different kinds of ions have been reported at Seattle,[4,6] and Hamburg,[7,11] and other labs are initiating experiments.

In the NBS work, the transition of interest for a frequency standard is the Hg^+ $5d^{10}6s\ ^2S_{\frac{1}{2}} \rightarrow 5d^96s^2\ ^2D_{5/2}$ quadrupole transition at 281.5 nm shown in Fig. 1. The $^2D_{5/2}$ level has a lifetime of 86 ms, corresponding to a natural width of 1.8 Hz. Use of the single photon quadrupole transition has an advantage over two-photon Doppler free transitions[12] because ac Stark shifts are negligible.

A mercury atom that was ionized by a weak electron beam was captured in a miniature Paul (radio frequency) trap that has internal dimensions of $r_o \cong 466$ μm and $z_o \cong 330$ μm. The rf trapping frequency was 21.07 MHz with a peak voltage amplitude of about 730 V. The ion was laser cooled by a few microwatts of cw laser radiation that was frequency tuned below the 6s $^2S_{\frac{1}{2}}$ - 6p $^2P_{\frac{1}{2}}$ electric dipole transition near 194 nm. When the Hg^+ ion was cold and the 194 nm radiation had sufficient intensity to saturate the strongly allowed S-P transition, 2×10^8 photons/s were scattered. With our collection efficiency, this corresponded to an observed peak count rate of about $10^5\ s^{-1}$ against a background of less than 50 s^{-1}.

Optical-optical double-resonance utilizing quantum amplification[4,6-8] was used to detect transitions of the ion driven by the 281.5 nm laser to the metastable $^2D_{5/2}$ state. This method makes use of the fact that the 194 nm fluorescence intensity level is bistable; high when the ion is cycling between the S and P states (the "on" state) and nearly zero when it is in the metastable D state (the "off" state). The fluorescence intensity in the on state is high enough that the state of the atom can be determined in a few milliseconds with nearly 100% efficiency. The full measurement cycle was as follows: A series of measurements of the 194 nm fluorescence was made, using a counter with a 10-ms integration period. As soon as the counter reading per measurement period was high enough to indicate that the ion was in the on

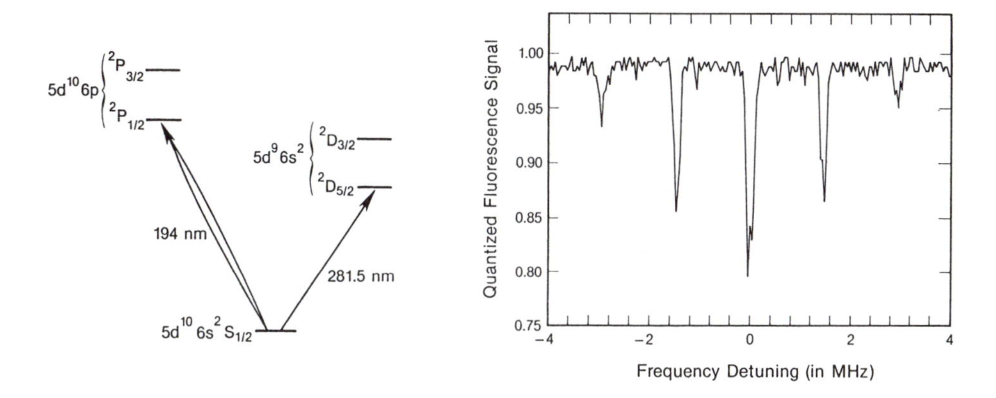

Fig. 1. On the left is a simplified energy-level diagram for $^{198}Hg^+$. The 281.5 nm quadrupole "clock" transition can be observed by monitoring the 194 nm fluorescence. If the ion has made a transition from the $^2S_{\frac{1}{2}}$ to the $^2D_{5/2}$ level the 194 nm flourescence disappears. For the figure on the right, on the horizontal axis is plotted the relative detuning from line center in frequency units at 281.5 nm. On the vertical axis is plotted the probability that the fluorescence from the 6s $^2S_{\frac{1}{2}}$ - 6p $^2P_{\frac{1}{2}}$ first resonance transition, excited by laser radiation at 194 nm, is on immediately after the 281.5 nm pulse. The electric-quadrupole-allowed S-D transition and the first-resonance S-P transition are probed sequentially in order to avoid light shifts and broadening of the narrow S-D transition. The recoilless absorption resonance or carrier (central feature) can provide a reference for an optical frequency standard. (From ref. 11)

state, the 194-nm radiation was shut off and the 281.5 nm radiation was pulsed on for 20 ms. Then, the 194 nm radiation was turned on again, and the counter was read. If the reading was low enough to indicate that the ion had made a transition to the $^2D_{5/2}$ state (the off state), the signal was defined to be 0. Otherwise, it was defined to be 1. The 281.5 nm laser frequency was then stepped, and the cycle was repeated. As the 281.5 nm laser frequency was swept back and forth through the resonance, the quantized measurement of the fluorescence signal at each frequency step was averaged with the previous measurements made at that same frequency. Since we could detect the state of the ion with nearly 100% efficiency, there was essentially no instrumental noise in the measurement process. Occasionally, while the 194 nm radiation was on, the ion decayed from the $^2P_{\frac{1}{2}}$ state to the metastable $^2D_{3/2}$ state rather than directly to the ground state.[13] This process led to a background rate of false transitions which was minimized by the quantized data-collecting method described above and by decreasing the 194 nm fluorescence level thereby decreasing the $^2P_{\frac{1}{2}}$-$^2D_{3/2}$ decay rate. Neglecting decay to the $^2D_{3/2}$ level, the quantized measurement scheme removes any contribution to the signal base line due to intensity variations in the 194-nm source. The 281.5 and 194 nm radiation were chopped so that they were never on at the same time. This eliminated shifts and broadening of the narrow 281.5-nm resonance due to the 194 nm radiation.

Figure 1 shows the signal from an 8 MHz scan of the 281.5 nm laser through a Zeeman component of the ion. The central feature or "carrier" can be the reference for an optical frequency standard. The laser could be servoed to this resonance by probing on both sides of the resonance to develop an error signal which forces the average frequency of the laser to the center of the resonance. Two primary requirements for good performance are: (1) immunity from and ability to characterize systematic frequency shifts (accuracy) and (2) sufficient signal-to-noise ratio and line Q to reach the accuracy level in a reasonable length of time (stability).

3b. Stability

If we make the simplifying assumption that the time-domain Ramsey method is used to interrogate the clock transition, then the frequency stability (two-sample Allan variance[14]) is given as[15]

$$\sigma_y(\tau) \equiv [<(<\omega_k>_\tau - <\omega_{k+1}>_\tau)^2>_k/2\omega_0^2]^{\frac{1}{2}} = (2\omega_0^2 T_R \tau)^{-\frac{1}{2}} . \qquad (1)$$

In this expression, $<\omega_k>_\tau$ is the kth measurement of the frequency of the locked oscillator averaged over time τ, $< >_k$ denotes an average over many measurements, ω_0 is the nominal frequency of the clock transition, and T_R is the interrogation time between pulses in the Ramsey method. Thus $\sigma_y(\tau)$ is a measurement of the rms fluctuations of the average frequency between adjacent measurements of duration τ. Eq. 1 assumes that optical state preparation (cooling and optical pumping) and fluorescence detection takes a time much less than T_R. We have also assumed $T_R \ll \tau(D_{5/2})$ and that the frequency of the 281.5 nm laser changes by much less than $(2T_R)^{-1}$ in time T_R. For our Hg^+ example, assuming T_R = 25 ms, $\sigma_y(\tau) \cong 7 \times 10^{-16}\tau^{-\frac{1}{2}}$. Therefore, at averaging times of 1 s, the measurement imprecision on a single ion should be on the order of 1 part in 10^{15} or 1 Hz. Even better performance is expected on ions with narrower linewidths. To realize such stabilities, very narrow, tunable lasers are required but this technology is now available.[9]

3c. Accuracy

The advantage of single ions over other experimental configurations is the relative immunity from systematic frequency shifts. Laser cooling gives the lowest second order Doppler shifts for single (as opposed to many) trapped ions.[16] In the Doppler cooling limit,[5] the magnitude of the second order Doppler shift for Hg^+ ions in an rf trap is fractionally 2.4×10^{-18}. Laser cooling to the zero point energy, which has recently been achieved,[17] gives even smaller shifts. One must also account for all perturbing influences on the ion's internal structure.[4,18] For example, one must consider the perturbations due to static and time varying multipole interactions for electric, magnetic, and gravitational fields. These include atom trapping field interactions, collisions with neutral atoms, ac Stark shifts due to laser beams, stray electric and magnetic fields, gravitational red shifts, etc.

For the $^2S_{\frac{1}{2}} \rightarrow {}^2D_{5/2}$ quadrupole transition in Hg^+, the limiting accuracy may be due to the uncertainty in the interaction of the $^2D_{5/2}$ atomic quadrupole moment with static electric fields of quadrupole symmetry.[4,18] The interaction with the quadrupolar electric fields of the trap can be calibrated since they affect the ion oscillation frequencies of the trap in a known way. More difficult to control are the effects due to patch fields or stray charge build up. Shifts may be on the order of 0.1 Hz. We can significantly reduce the uncertainty due to these shifts, however, by taking the mean quadrupole transition frequency for three mutually orthogonal orientations of a quantizing magnetic field. In this case, the mean quadrupole shift is zero. For the $^1S_0 \rightarrow {}^3P_0$ transitions of the group IIIA ions, such shifts are absent.[4]

Although experimental verification is still lacking, it would appear that single ion optical frequency standards will eventually yield extremely high performance. Accuracies and stabilities better than 1 part in 10^{15} seem quite feasible; eventually they could exceed 1 part in 10^{18}.

4. Frequency Comparison

To appreciate the problem of frequency comparison in the optical region of the spectrum one needs only to plot the electromagnetic spectrum on a linear frequency scale. The number (and spectral density) of accurate optical frequency standards is very small to begin with; this is to be compared with the situation in the microwave spectrum (and below) where commercially available synthesizers provide state-of-the-art precision at any frequency.

A laser whose frequency is unknown can be compared to a reference laser by heterodyne methods to high precision.[1] Beat frequency measurements up to 2.5 THz in the visible spectrum have already been made.[19] An alternative to simple heterodyne schemes is harmonic mixing by use of synthesis chains.[1] Here, the unknown laser frequency is compared (via heterodyne methods) to a harmonic of some well known reference line such as the methane stabilized He-Ne laser at 3.39 μm.[1]

Comparisons can also be made by wavelength methods. These have the advantage that two lasers of quite different frequency can be compared but are currently limited in accuracy to about the 10^{-10} to 10^{-11} range.[20]

For high accuracy comparison, heterodyne detection and synthesis chains are proven methods but have not been demonstrated in the visible yet. Moreover, since they are somewhat cumbersome and their coverage in the optical spectrum somewhat sparse, it is useful to pursue alternative methods.

5a. Wideband Comb Generation

Precise frequencies relative to a frequency standard (at frequency ω_0) can be provided by comb generation. If we amplitude or frequency modulate (at frequency Ω) a source at the standard frequency, then spectral components at frequencies $\omega_0 \pm n\Omega$ (n an integer) are generated. If n and Ω are chosen so that $\omega_0 + n\Omega \cong \omega(\text{unknown})$ then $\omega(\text{unknown})$ can be precisely determined by heterodyne methods. The challenge is to make $(n\Omega)$ very large.

5b. Amplitude Modulation

As a specific example, let

$$A(t) = \sin(\omega_0 t + \varphi)\, f(t), \tag{2}$$

where f(t) is periodic with period $2\pi/\Omega$. We will assume f(t) is an even function and near t = 0, has the form

$$f(t) \cong \exp[-(t/\tau)^2] \qquad \left(-\frac{\pi}{\Omega} < t < \frac{\pi}{\Omega}\right), \tag{3}$$

where we also assume $\tau \ll 2\pi/\Omega$. (See Fig. 2) Expanding A(t) in a Fourier series, we find

$$A(t) \cong \frac{\Omega\tau}{2\sqrt{\pi}} \Bigg\{ \sin(\omega_0 t+\varphi) \qquad (4)$$

$$+ \sum_{n=1}^{\infty} \exp(-(n\Omega\tau/2)^2) \left(\sin[(\omega_0+n\Omega)t+\varphi_0] + \sin[(\omega_0-n\Omega)t + \varphi_0]\right) \Bigg\}.$$

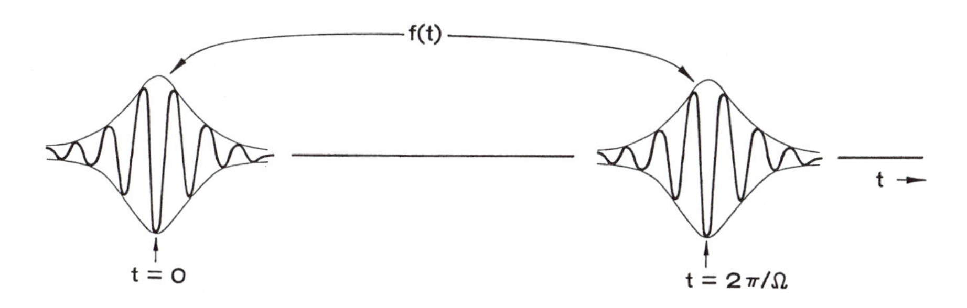

Fig. 2. A sine-wave carrier (frequency ω_0) is assumed to be periodically amplitude modulated at frequency Ω with a Gaussian envelope. Exact frequency division (ω_0/Ω = integer) is obtained if the phase of the carrier is fixed to the phase of the modulation envelope.

To make the spectral component at $\omega_0 \pm n\Omega$ as large as possible we want $n\Omega\tau/2$ close to unity. For harmonic generation, $\omega_0 = \Omega$. For either harmonic generation or comb generation around a high frequency ω_0, the comb bandwidth is approximately equal to $1/\tau$. High order multiplication to the visible or large bandwidth combs in the visible are difficult to achieve for two reasons: (1) it is difficult to make τ small enough and (2) the modulator must be sufficiently broadband to pass the fundamental and its harmonics or sidebands.

We remark that Fig. 2 is characteristic of the output of a mode locked laser[21] or a free electron laser with pulsed input electron beam.[22] Mode-locked lasers have already been suggested as devices for frequency translation in the optical spectrum.[21]

5c. Frequency Modulation

Consider sine wave modulation of a carrier at frequency ω_0:

$$A(t) = A_0 \sin[\omega_0 t + \frac{\Delta\omega}{\Omega}\sin\Omega t + \varphi]$$
$$= A_0 \sum_{n=-\infty}^{\infty} J_n(\frac{\Delta\omega}{\Omega}) \sin[(\omega_0 + n\Omega)t + \varphi], \quad (5)$$

where $\Delta\omega$ is the frequency swing. Assume we are interested in the spectral component at a particular (large) value of n. We make the substitution $\Delta\omega/\Omega = n\beta$, and will limit ourselves to the case where $\beta < 1$. From Meissel's formula[23] for $J_n(n\beta)$, for large n and $\beta < 1$, we can write

$$J_n(n\beta) \cong \left[\frac{\beta e^{(1-\beta^2)^{\frac{1}{2}}}}{1+(1-\beta^2)^{\frac{1}{2}}}\right]^n [2\pi n(1-\beta^2)^{\frac{1}{2}}]^{-\frac{1}{2}} \quad . \quad (6)$$

We can make the spectral component at $\omega_0 + n\Omega$ reasonably large by making β close to unity or the modulation index $\Delta\omega/\Omega$ close to n. This is the primary practical problem for n large.

6a. Frequency Division

Using amplitude or frequency modulation of a carrier at frequency ω_0, we can achieve exact frequency division if we make sidebands of the carrier to such low frequency that we can force the condition $\omega_0 - n\Omega = (n+2)\Omega - \omega_0 = \Omega$ so that $\omega_0/\Omega = n+1$. For example, if we examine Fig. 2, we can achieve exact frequency division by any means which locks the phase of the carrier to the phase of the amplitude modulation; that is, the undulations of the carrier do not "slip" under the envelope of the amplitude modulation. A divider based on these principles would be quite useful if Ω is in the microwave region (or below) where precise frequency synthesis is possible. Since Ω and n could be freely chosen, any value of ω_0 could be measured in a single device.

6b. Single Electron Frequency Divider

The required extreme amplitude or frequency modulation might be accomplished with the interaction of an electromagnetic wave (at frequency ω_0) with a single electron oscillating at frequency Ω. For example, Ω might be the cyclotron frequency of the electron in a magnetic field.

As originally proposed,[24] extreme amplitude modulation of the electromagnetic wave (at the site of the electron) could be accomplished by focussing the wave with a fast lens or slow wave structure (miniature waveguide) onto a portion of the electron's cyclotron orbit. The energy coupled into the electron at frequency $\omega_0 - n\Omega = (n+2)\Omega - \omega_0 = \Omega$ is balanced by synchrotron radiation and phase locking can occur. Lens focussing requires the electron to be relativistic, $\gamma \equiv (1-(v/c)^2)^{-\frac{1}{2}} \gtrsim 2$ to achieve a significant coupling[24] (v = electron speed, c = speed of light). Use of a slow wave structure would allow use of lower energy electrons.

Driving the electron could also be accomplished using an inverse synchrotron radiation geometry.[25] Here, a collimated laser beam is made to coincide with a tangent to the electron's orbit. We require the electron to be relativistic to achieve a significant interaction. If the laser beam waist is made much larger than the electron orbit dimensions, then we have the case of extreme frequency modulation of a plane wave as first analyzed by Kaplan.[26] Here, the electric field at the site of the electron is given by Eqs. 5 and 6 with $\beta = v/c$. Again, we require the electron to be somewhat relativistic to make J_n and J_{n+2} sufficiently large ($\omega_0 - n\Omega = \Omega$).

In all cases above, the center of the electron's orbit (in the plane of the orbit) must be held within $\lambda\!\!\!^{-}(= c/\omega_0)$. A specific scheme for accomplishing this is discussed in Ref. 24. At NBS we have initiated a project along these lines. At first, low order division ($n < 10$) where ω_0 and Ω are both microwave frequencies will be tried. Aside from the practical application discussed here, such an experiment is interesting in terms of the fundamental nonlinear interactions of simple elementary systems.

We thank L. Hollberg and D. Jennings for helpful comments on the manuscript.

1. D.A. Jennings, K.M. Evenson, and D.J. Knight, Proc. IEEE 74, 168 (1986); S.N. Bagaev and V.P. Chebotayev, Usp. Fiz. Nauk 148, 143 (1986) (Sov. Phys. Usp. 29, 82 (1986)), and references therein.
2. Proc. Fourth Symp. Freq. Standards and Metrology, Ed. A. DeMarchi, Ancona, Italy, Sept. 1988 (Springer-Verlag, Heidelberg) to be published.
3. Proc. Third Symp. Freq. Standards and Metrology, J. de Physique, Vol. 42, Colloque C-8, 1981.
4. H.G. Dehmelt, Bull. Am. Phys. Soc. 18, 1521 (1973) and 20, 60 (1975); H.G. Dehmelt, IEEE Trans. Instrum. Meas. IM-31, 83 (1982); H.G. Dehmelt, in Advances in Laser Spectroscopy, Ed. by F.T. Arecchi, F.S. Strumia, and H. Walther, (Plenum, New York, 1983), p. 153.

5. See for example, D.J. Wineland and W.M. Itano, Phys. Today, 40, no. 6, p. 34 (1987), and references therein.
6. W. Nagourney, J. Sandberg, H. Dehmelt, Phys. Rev. Lett. 56, 2797 (1986).
7. T. Sauter, W. Neuhauser, R. Blatt, P.E. Toschek, Phys. Rev. Lett. 57, 1696 (1986).
8. J.C. Bergquist, R.G. Hulet, W.M. Itano, D.J. Wineland, Phys. Rev. Lett., 57, 1699 (1986).
9. J.L. Hall, D. Hils, C. Salomon, and J.M. Chartier, in Laser Spectroscopy VIII, ed. by W. Persson and S. Svanberg, (Springer, Berlin Heidelberg, 1987) p. 376.
10. J.C. Bergquist, W.M. Itano, D.J. Wineland, Phys. Rev. A36, 428 (1987).
11. P.E. Toschek, in Tendances Actuelles en Physique Atomique, Les Houches, Session 38, G. Grynberg and R. Stora, eds., (North-Holland, 1984) p. 381.
12. J.C. Bergquist, D.J. Wineland, W.M. It[illegible], H. Hemmati, H.U. Daniel, G. Leuchs, Phys. Rev. Lett. 55, 1567 (1985).
13. W.M. Itano, J.C. Bergquist, R.G. Hulet and D.J. Wineland, Phys. Rev. Lett. 59, 2732 (1987).
14. J.A. Barnes, A.R. Chi, L.S. Cutler, D.J. Healey, D.B. Leeson, T.E. McGunigal, J.A. Mullen, Jr., W.L. Smith, R.L. Sydnor, R.F.C. Vessot, and G.M.R. Winkler, IEEE Trans. Instrum. Meas. IM-20, 105 (1971).
15. D.J. Wineland, W.M. Itano, J.C. Bergquist and F.L. Walls, in Proceedings of the Thiry-fifth Annual Symposium on Frequency Control (1981), p. 602 (copies available from Electronics Industry Assn., 2001 "Eye" St., N.W. Washington, D.C. 20006).
16. D.J. Wineland, in Precision Measurement and Fundamental Constants II, Natl. Bur. Stand. (U.S.) Special Publ. No. 617, edited by B.N. Taylor and W.D. Phillips, (U.S. GPO, Washington, D.C., 1984), p. 83.
17. F. Diedrich, J.C. Bergquist, W.M. Itano, D.J. Wineland, to be published.
18. D.J. Wineland, Science 226, 395 (1984), and references therein.
19. R.E. Drullinger, K.M. Evenson, D.A. Jennings, F.R. Petersen, J.C. Bergquist, L. Burkins, and H.U. Daniel, Appl. Phys. Lett. 42, 137 (1983).
20. R.G. DeVoe, C. Fabre, K. Jungmann, J. Hoffnagle, and R.G. Brewer, Phys. Rev. A37, 1802 (1988).
21. S.R. Bramwell, D.M. Kane, and A.I. Ferguson, Opt. Commun. 56, 12 (1985); J.A. Valdmanis, R.L. Fork, and J.P. Gordon, Opt. Lett. 10, 131 (1985); see also D.M. Kane, S.R. Bramwell, and A.I. Ferguson, Appl. Phys. B39, 171 (1986).

22. S. Penner et al., in Proc. 9th Int. F.E.L. Conf., Williamsburg, VA, 1987, to be published.
23. G.N. Watson, Theory of Bessel functions (Cambridge Press, 1952), p. 227.
24. D.J. Wineland, J. Appl. Phys. 50, 2528 (1979); J.C. Bergquist, D.J. Wineland, Natl. Bur. Stand. (U.S.), Tech Note 1086, (U.S. G.P.O., 1985) p. TN-30.
25. R.H. Pantell and J.A. Edighoffer, J. Appl. Phys. 51, 1905 (1980).
26. A.E. Kaplan, Optics Lett. 12, 489 (1987).

Precision RF Spectroscopy of Circular Rydberg Atoms

M. Gross, J. Hare, P. Goy, and S. Haroche

Laboratoire de Spectroscopie Hertzienne de l'ENS,
24, rue Lhomond, F-75231 Paris Cedex 05, France

INTRODUCTION

The Rydberg R is the best known fundamental constant in physics (precision of about 3×10^{-10}). It has been measured by various consistent optical spectroscopy experiments performed on atomic hydrogen [1, 2, 3]. The current precision is in fact limited by the reliability of wavelength measurements in the optical domain (1.6×10^{-10}). It would be very desirable to improve further the precision of this measurement.

A natural direction to improve the resolution of R measurements beyond the present level is to attempt the experiments at longer wavelengths, in a domain where the frequency of the radiation can be linked directly to the atomic time standard. A possible way is to measure R on microwave transitions involving what is called circular Rydberg atoms [4, 7].

THE CIRCULAR RYDBERG ATOM

Circular Rydberg atoms are highly excited atoms, i.e. Rydberg atoms (with a large principal quantum number, $n \geqslant 10$) with a maximum value of the angular momentum l and of its projection m along a quantization axis ($l = |m| = n-1$). These atoms are quasi-classical : the wave function of the valence electron has the shape of a thin torus lying around the classical circular orbit of the Bohr theory, whose radius is $a_0 n^2$ (where a_0 is the Bohr radius). These atoms realize exactly Bohr's atom model and are well adapted to the Rydberg constant metrology. The advantages of using these atoms for a Rydberg constant measurement are numerous :

(i) These atoms have a very long natural life-time. Their radiative decay occurs entirely towards the $n-1$, $l=m=n-2$ lower circular level within a time proportional to n^5, typically of the order of 10^{-3}s for $n=30$.

The Hydrogen Atom Editors: G.F. Bassani · M. Inguscio · T.W. Hänsch

(ii) Circular Rydberg atoms have non-permanent electric dipole moment and hence no linear Stark effect. Stray electric field shifts and broadenings, which are usually very important in the spectroscopy of the low angular momentum Rydberg levels, can thus be minimized and much better controlled.

(iii) In circular atoms, the Rydberg electron remains always very far from the nucleus. Hence, all the contact terms, which become significant corrections at the 10^{-10} level in the optical experiments and which depend upon the not-so-well known proton form factor, are in circular states completely negligible. Lamb-shift corrections are also very small for these states. From the point of view of Q. E. D. corrections, circular atoms are, by far, the best candidate for R metrology.

(iv) At the 10^{-11} precision level, it is not even required to perform the Rydberg measurement on a Hydrogen circular state : in alkali atoms, the circular electron sees the core made of the nucleus and the Z-1 remaining electron mainly as a point-like charge. The small correction due to the core polarizability can be expressed as a very small quantum defect in Lithium [5]. Actual knowledge of this polarizability corresponds to uncertainties in the 10^{-11} range.

We present the preliminary results of an experiment performed on quasi-hydrogenic circular Rydberg states of Lithium atoms whose microwave transitions are induced in a Fabry-Perot cavity [6]. We have observed transit time width-limited resonances between adjacent circular Rydberg levels at frequencies around 447 GHz and 270 GHz. Our results can presently be interpreted as a measurement of R with an accuracy of 6×10^{-9}. It is still one order of magnitude off the present accuracy of the optical experiments and two orders of magnitude away from the planned goal (10^{-11}). However this experiment is interesting since it has allowed us to study in detail the properties of circular Rydberg atoms and to analyze their behavior in electric and magnetic fields. This is an essential task to design carefully the final experiment and to control the various systematic effects susceptible to cause errors in the measurement. The improvement of the present experiment mainly involves the lengthening of the atom-microwave interaction time. This can be done either by slowing down the atoms or by using a long cavity geometry such as a Ramsey design split cavity.

THE EXPERIMENT

The circular atom microwave spectroscopy experimental set-up is sketched on Fig. 1-a. A thermal beam of Li atoms crosses three sections of the apparatus : the excitation, the microwave interaction region and the detection zone. The whole set-up is protected from room temperature thermal radiation by a liquid nitrogen cooled shield (which can be replaced in a later stage of the experiment by a liquid helium cooled one).

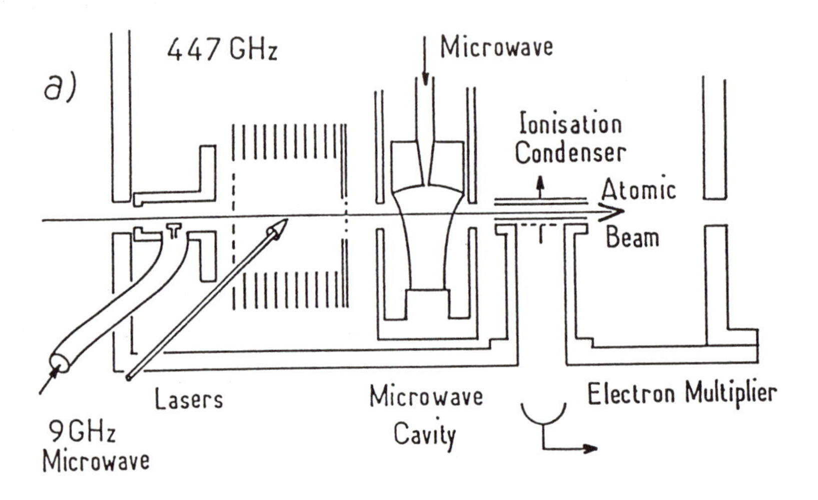

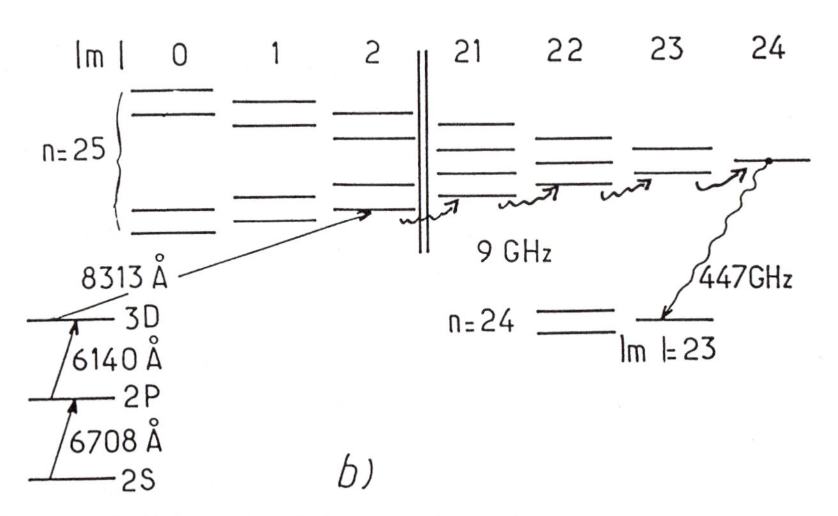

Fig. 1: a) Schematic of the experimental set-up

b) Diagram of the Li^7 levels involved in the experiment

The preparation of the Rydberg circular states is performed by pulsed laser excitation followed by an adiabatic microwave transfer method (A.M.T.M.) already described in ref. [8]. This method requires that the atom interacts with an homogeneous electric field F_1, produced by the stack of equally spaced metallic plates shown on Fig. 1-a. This field removes the degeneracy of the various n-manifolds.

The laser excitation prepares the atoms in the lowest energy Stark state with $|m|=2$ in a chosen n-manifold (n=24 for example) (see Fig. 1-b). This excitation is itself a stepwise process involving three pulsed dye laser beams in resonance with the 2S-2P, 2P-3D and 3D-n, $|m|=2$, $n_1=0$ transitions respectively (wavelengths 6708, 6104 and about 8300 Å). From this Stark sublevel, the atoms undergo a succession of fast adiabatic passage transitions bringing them into the circular state of the manifold ($|m|$ = n-1, n_1 = 0). These transitions, shown by the arrows on Fig. 1-b, are induced by a 9 GHz microwave field produced by an X-band klystron and the successive $m \to m+1$ steps are tuned in resonance one after the other by sweeping down the electric field F_1 and taking advantage of the slight frequency difference of the successive transitions (due to second order Stark effect). The microwave field is sent on the atom through a 25 mm diameter hole pierced in the stack of the Stark plates.

The circular atoms then cross a Fabry-Perot semi-cofocal cavity in which a resonant millimeter wave is fed through a coupling hole. The cavity, tuned by slight mirror translation, sustains a Gaussian mode, corresponding to a time-varying Gaussian envelope of the field seen by each atom as it crosses the cavity. The microwave around 447 GHz (for the one-photon 24-25 transition) is produced by a phase-locked backwave oscillator (Thomson-CSF) driven by a microcomputer, whose frequency can be stabilized, swept and measured continuously with reference to a time standard. After crossing the microwave cavity, the atoms enter into the Rydberg ionization zone, where they are selectively detected by the field ionization method [7]. A ramp of electric field $F_2(t)$ is applied after an appropriate delay between two metallic plates surrounding the atomic beam. This field reaches at different times the ionization threshold for the various n, m, n_1 states involved. It turns out in particular that the circular state of the n-manifold ionizes in a higher field than the lowest m levels. The time resolved ionization signals thus provide an unambiguous signature of the circular Rydberg atoms. As the microwave frequency is scanned across a resonance line, the time-dependent Rydberg ionization signals change, since the circular n and n+1 or n−1 states have different ionization thresholds and appear at different times in the field ramp $F_2(t)$. The resonance signal can be conveniently plotted as a function of the microwave frequency. A computer is used to compute the circular state to circular state transfer rate (ratio of the peak intensities corresponding to the final and initial states populations) and this rate is plotted versus measured microwave frequency.

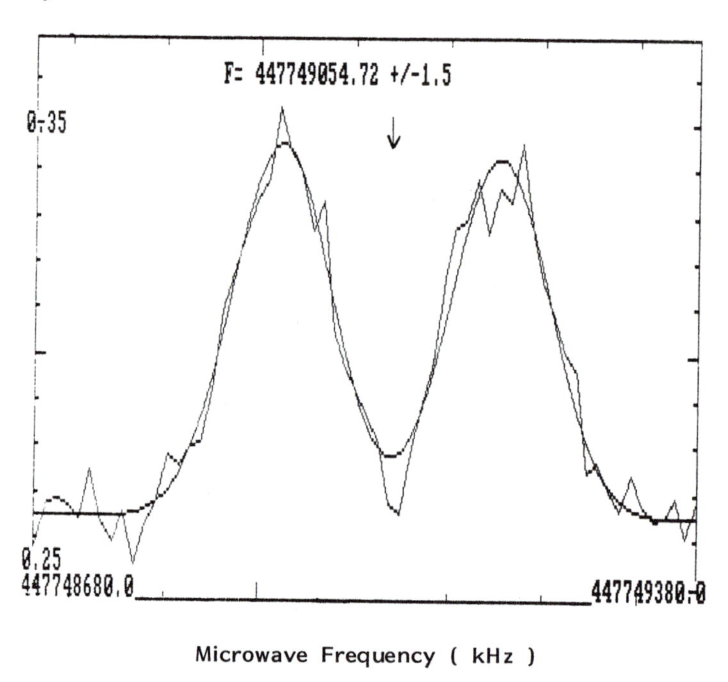

Fig. 2: n=24 to n=25 circular-circular transition microwave spectrum. The population transfer rate is plotted as a function of the microwave frequency. The smooth curve corresponds to the Gaussian fit. F is the fitted central frequency (arrow)

Fig. 2 shows such a typical resonance signal. It has been recorded with the atoms interacting with an electric E field (about 1V/cm) and a magnetic B field (about 0.1 Gauss) both oriented in the cavity axis vertical direction. The B field splits the line in two Zeeman components, corresponding to both circular orbits' helicities. The E field shifts both Zeeman components by a small amount (about 7 kHz).

The Fig. 2 experimental points result from the averaging of 20 pulses recorded for each frequency. The frequency scan is performed randomly to avoid systematic drift effects. The width at half maximum of the resonance : 120 kHz, corresponds to the theoretical limit due to the finite transit time across the waist of the microwave Gaussian beam. The observed Zeeman components have not the same weight. This is just related to the resonant 447 GHz microwave source whose polarization is partially elliptical.

THE ANALYSIS OF THE RESULTS

We have performed a precise measurement of the n=24 to n=25 circular-circular transition in Lithium. This measurement is based on 63 atomic resonance spectra recorded in about the same conditions as the spectrum seen in Fig. 2.

Using a least squares iterative method, each spectrum is fitted with two Gaussian lines. The smooth curve seen on Fig. 2 corresponds to such a Gaussian fit. This procedure provides a rather accurate value of F_c, the frequency of the center of the two Zeeman lines. It provides also the F_c uncertainty interval. This interval is related to the difference between the best fit and the experimental spectrum, which is here mainly due to noise. Typically, for a sweep such as the one presented on Fig. 2, F_c is determined to within ±2 kHz (variance).

The about 1V/cm E electric field is measured, and F_{exp} the unperturbed transition frequency is then deduced from F_c by substracting to F_c the calculated Stark shift (the other corrective effects are negligible at our scale of accuracy). The E uncertainty, lower than 5%, corresponds to ±50 Hz on the frequency, and can be neglected.

Fig. 3 summarizes the results. F_{exp} is given for each spectrum with its error bars. Data are seen in time order in three groups separated by a small vertical blank space, corresponding to three different days. The first group of data has been recorded with a first cavity, where the interaction time is shorter, making the linewidth typically twice as large. In this first group, the error bars are typically twice as large. The two following groups have been recorded with a second cavity.

The averaged value of F_{exp} (weighted by the inverse square of the uncertainties) is :

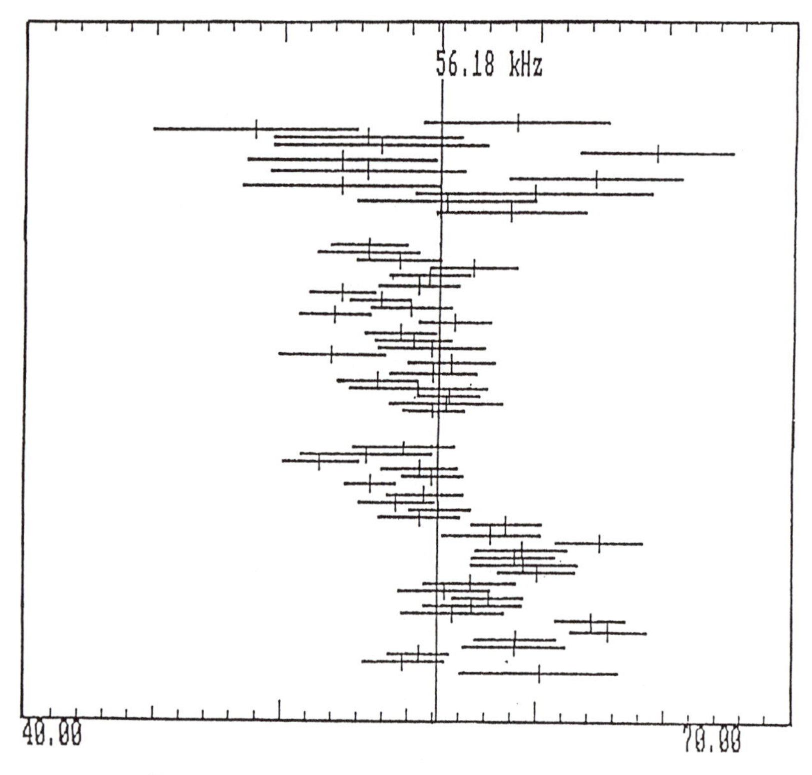

Resonance Frequency Last Digit (kHz)

Fig. 3: Plot of the resonance frequency (447 749 0XX kHz) of the n=24 to n=25 circular-circular transition of Li^7. The vertical straight line corresponds to the data weighted average

$$F_{exp} = 447\ 749\ 056.18 \text{ kHz}$$

The theoretical predicted value is :

$$F_0 = 447\ 749\ 056.43 \text{ kHz}$$

In the F_0 calculation, we have not taken into account the following interactions :

- the hyperfine coupling (about 5 Hz on F_0),
- all the exchange or correlation effects between the core electrons and the Rydberg one (below 1 Hz),
- the Q.E.D. corrections, i.e. Lamb-shift (50 Hz),
- the B and E fields higher order terms (less than a few Hz).

The theoretical relative uncertainty on F_0, except the 1.7×10^{-10} one due to the Rydberg itself [9], is about 4×10^{-11}. It is related to the actual one percent precision on the static dipolar polarizability of the LiII (1s)2 ground state, needed to evaluate the electron-core interaction : the uncertainty on the atomic mass (about 1 10^{-8}) step is only at the 4 10^{-12} level.

At first, we can note that the F_{exp} mean value is very close to F_0 (250 Hz), and therefore compatible with the very low uncertainty about 250 Hz (accuracy of 6 10^{-10}), that is precisely obtained if we assume that the noise on the spectra is the only cause of dispersion. However, as it appears clearly on Fig. 3, this dispersion, characterized by the standard deviation of 2.7 kHz, is too large in comparison with the noise's amplitude. The effective distribution of F_{exp} suggests a dependance versus time, probably related to fluctuating systematic shifts.

In our case, the possible fluctuating systematics are the following :

* The horizontal components of the magnetic field (in particular, the one parallel to the laser axis) cannot be exactly controlled. It results in a time-dependent drift within ±300 Hz.
* The cavity is tuned by maximizing the atomic transfer rate. By this way, the cavity center frequency equals Fc within about 25% of the cavity resonance half-width. The slow variation of the MW power through the linewidth shifts the line in direction of the cavity center frequency. Since the cavity has a rather low quality factor, this shift is at the worst ±500 Hz. Due to thermal drift, the cavity tuning may vary at the time scale of an hour, and yield a time-dependent ±500 Hz systematic.
* For large atomic populations (about 1000 atoms detected), we have observed shifts of a few kHz, attributed to collective effects. Although all the reported measurements have been performed with less than 300 atoms, we are actually unable to exclude the possibility of a residual shift of about ±500 Hz.

These three fluctuating effects, with their respective orders of magnitude, account for the observed standard deviation $\pm 6 \times 10^{-9}$ (±2.5 kHz), which therefore represents the effective uncertainty on our measurement. Residual first order Doppler effect and relativistic corrections are negligible at this scale.

Finally, this preliminary experiment, where the microwave interaction time is two orders of magnitude shorter than the atomic lifetime, can be reinterpreted as a R Rydberg constant measurement with a 6×10^{-9}

accuracy. It demonstrates the feasibility of a competitive R measurement in frequency unit on microwave transition involving circular states.

REFERENCES

[1] P. ZHAO, W. LICHTEN, H.P. LAYER, J.C. BERGQUIST, Phys. Rev. A34, 5138 (1986)

[2] E.A. HILDUM, U. BOESL, D.H. McINTYRE, R.G. BEAUSOLEIL and T.W. HÄNSCH, Phys. Rev. Lett. 56, 576 (1986)

[3] F. BIRABEN, J.C. GARREAU and L. JULIEN, Europhys. Lett. 2, 925 (1986)

[4] J. LIANG, M. GROSS, P. GOY and S. HAROCHE, Phys. Rev. A33, 4437 (1986)

[5] R.R. FREEMAN, D. KLEPPNER, Phys. Rev. A14, 1614 (1976)

[6] J. HARE, M. GROSS, P. GOY, S. HAROCHE, Post deadline paper in XVth International Quantum Electronics Conference, Baltimore (1987)

[7] T. DUCAS, M. LITTMAN, R. FREEMAN, D. KLEPPNER, Phys. Rev. Lett. 35, 366 (1975)

[8] R.G. HULET, D. KLEPPNER, Phys. Rev. Lett. 51, 1430 (1983).

[9] F. BIRABEN, L. JULIEN, private communication.

Part II

Positronium, Muonium, and Other Hydrogen-Like Systems

Laser Spectroscopy of Positronium and Muonium

S. Chu

Physics Department, Stanford University,
Stanford, CA 94305, USA

I. Introduction

The laser spectroscopy of leptonic atoms is of interest because of the simplicity of the atoms. Leptons possess no known structure. Unlike the hydrogen atom with its complicated and poorly understood proton, the behavior of leptonic atoms should in principle be calculable to much greater precision. In the case of muonium, the theoretical uncertainty will be limited by our knowledge of fundamental constants such as m_{μ}/m_{p} and α. Given the 72 kHz linewidth of the 1S state due to the 2.2μsec lifetime of the muon, one can ultimately expect a measurement of the 1S-2S splitting better than $\Delta\nu/\nu < 10^{3}$ Hz/10^{15} Hz = 10^{-12}. In the case of positronium, the 140 nsec lifetime should lead to an ultimate precision of $\Delta\nu/\nu = 10^{-11}$. Thus, this system provides a unique opportunity for the precise study of a purely leptonic two-body system, and QED corrections to that system. In addition to QED tests, the laser excitation of these atoms could provide thermal, sub-thermal, and eventually cryogenic sources of positrons and muons that can be used in a variety of applications.

This article will outline the experimental techniques we have used in the laser spectroscopy of these atoms and briefly indicate current plans for the refinement of these measurements. As Fig. 1 shows, the laser spectroscopy of positronium and muonium is not competitive with comparable measurements in hydrogen, largely due to the low density sources of these atoms. In the case of positronium, the first measurements were done at peak densities of a few atoms/cm^{3} during a laser pulse. The muonium work was limited by atom densities ~ 10^{-2} atom/cm^{3} per laser pulse. As feeble as these sources might seem to spectroscopists of less exotic atoms, one must remember that these instantaneous densities represent many orders of magnitude improvement of above pre-existing sources of thermal positronium and muonium. Clearly, improved sources will lead to more precise measurements.

II. Positronium

Figure 2 shows the lowest lying energy states of positronium and the energy intervals that have been measured to date. The optical spectroscopy of the positronium $1^{3}S_{1}$-$2^{3}S_{1}$ resonance has been previously described. [1] Briefly, the long lived triplet positronium was excited by two counterpropagating 486

The Hydrogen Atom Editors: G.F. Bassani · M. Inguscio · T.W. Hänsch

nm laser beams via the two-photon Doppler-free technique. Atoms in the n=2 state were then ionized with high probability and the detected positrons were recorded as the laser frequency was tuned through the resonance.

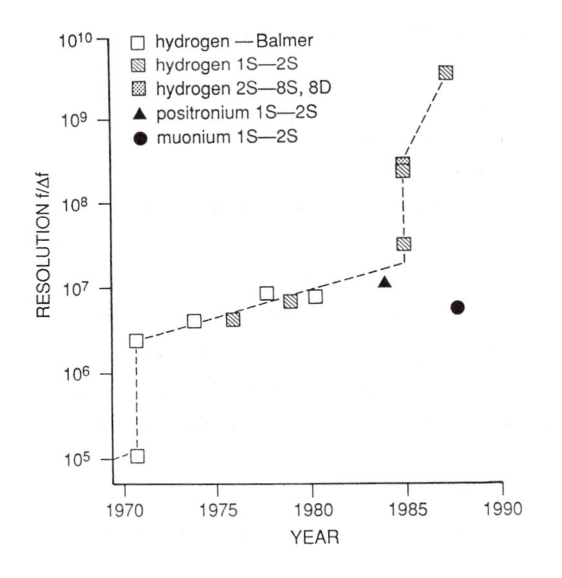

Fig. 1 Laser spectroscopy of hydrogen, positronium and muonium as a function of time.

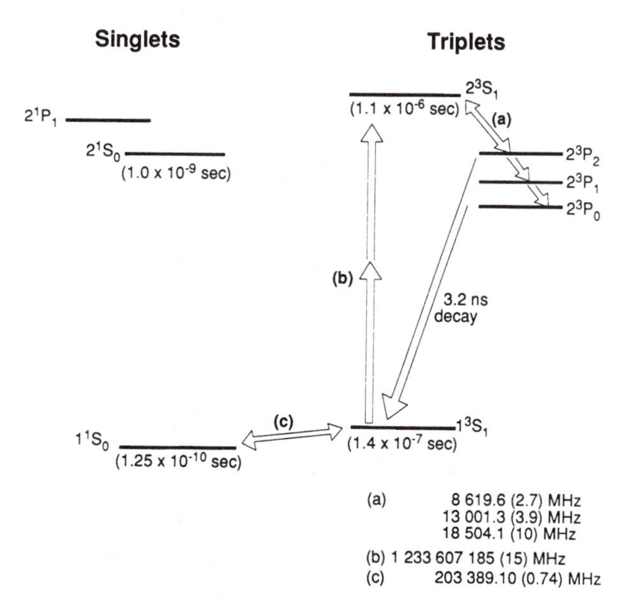

Fig. 2 Lowest energy level structure of positronium showing the measured energy level splitting.

Figure 3 shows a block diagram of the experimental apparatus. Starting with greater than 10^9 positrons/sec emitted from a ^{58}Co β^+ source, on the order of 10^6 positrons/sec are extracted from a cooled, single crystal tungsten moderator and directed into an e^+ magnetic bottle for storage. The bottle allows us to match the e^+ source to the duty cycle of the pulsed laser used to excite the atoms. A harmonic buncher in the magnetic bottle could be pulsed at a kilohertz rate, producing sub-keV positrons in 10 nsec pulses. The approximately 80 e^+/pulse are directed to an Al(111) target heated to 300°C. By heating the metal surface, positrons that would normally annihilate in surface states are thermally desorbed as free thermal Ps in vacuum. During the experiment, roughly 20 thermal Ps would be emitted from the 1 cm dia Al target per bunching pulse.

The laser system consisted of a cw dye laser oscillator amplified by a XeCl excimer pumped 4-stage dye laser amplifier. Nominally 50 mw of cw power was amplified to 20-25 mJ, 10-nsec laser pulses with a frequency bandwidth of 70 MHz. The cw laser frequency was measured relative to the deuterium $2S_{2/3}$-$4P_{3/2}$ line in a three-step process. We first used a theoretical value for the deuterium line from Erickson's calculation and measured the shift between that line and a particular molecular line (labeled 1326 in tellurium atlas [5]) in the tellurium spectra. Next, the Ps resonance was measured relative to the tellurium standard. Finally, the frequency of the amplified pulsed laser beam is measured relative to the cw laser. Since the Doppler free counter-propagating beams used to excite the Ps atoms are produced in a flat-flat Fabry Perot interferometer inside the vacuum chamber (free spectral range = 450 MHz, finesse ~ 35) the beam pointing instability of the pulsed laser translated into a time averaged laser bandwidth of 30-35 MHz. A 300 MHz free spectral range confocal interferometer was used to measure the frequency shift of the pulsed laser light transmitted through the vacuum Fabry-Perot cavity relative to the cw light.

The dominant uncertainties in the measurement were the dc Stark shift of the deuterium line in the discharge tube, the ac Stark shift of the Ps line, the second-order Doppler shift of the thermal Ps and the amplified laser frequency shift relative to the cw laser. Counting statistics show up in the measurement of the Ps - tellurium frequency shifts as one extrapolates to zero laser power and zero atomic velocity.

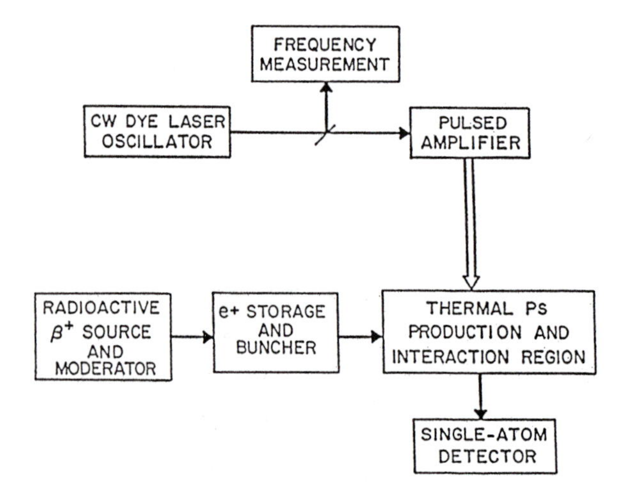

Fig. 3 Block diagram of the positronium experiment.

The quoted uncertainty in the measurement of the Te_2 line relative to the hydrogen reference was ± 10.5 MHz while the uncertainty of the Te_2 line relative to the Ps resonance was ± 10.6 MHz. The final result of the measurement was $\Delta\nu$ (1^3S_1 - 2^3S_1) = 1 233 607 185 ± 15 MHz. [1] Since that measurement, the tellurium line was recalibrated by McIntyre and Hansch [6], causing the experimental value to be shifted significantly. The new value is $\Delta\nu$ (Ps) = 1 233 607 142.9 (10.7) MHz.

The theoretical value of the energy interval is given as an expansion in powers of α: $\Delta\nu$ (1^3S_1 - 2^3S_1) = $3/8\ R_\infty C\ \{1 + K_2\alpha^2 + K_3\ \alpha^3 + K_4\ \alpha^4 + ...\}$ where the contributions are listed in Table I. The α^4 and $\alpha^4 \ln\alpha$ terms have not been calculated, although the $\alpha\ \ln\alpha$ term is estimated by scaling $1/n^3$ behavior of the hfs splitting.[7] The α^4 term would be 3.5 MHz if the uncalculated K_4 coefficient turned out to be 1. Thus, the uncertainty due to the uncalculated terms could easily be ± 30 MHz. If we ignore the uncertainties due to the uncalculated terms, $\Delta\nu_{theory} - \Delta\nu_{expt}$ = 56.4 (11) MHz. Clearly, more experimental and theoretical work needs to be done.

Table I Positronium Theory

$3/8\ R_\infty c$	1 233 690 730 (1.3) MHz
$K_2\alpha^2$	- 82 005.616 (18)
$K_3\alpha^3$	- 1 527.440 3 (5)
$K_4\alpha^4$	(3.5 MHz if K_4 =1)
$\alpha^4 \ln\alpha$	(-4.2 MHz ± ?)

$\Delta\nu$ (1^3S_1 - 2^3S_1) = 1 233 607 197.(1.3)

In addition to tests of QED, precision measurements of the energy levels of Ps can be used to establish upper limits in other areas of physics. For example, we have implicitly assumed that the Ps Rydberg R_{Ps} is exactly 1/2 R_∞. Agreement between experiment and theory can be used to establish an upper limit on $m_{e+} - m_{e-}$. [1] Precision Ps spectroscopy can also be used to set upper limits on electron coupling to light scalar particles. The unexpected observation of correlated back-to-back emission of e^+e^- pairs in heavy ion collisions [8] has led to the suggestion of a new neutral particle of mass M ~1.7 MeV. If the new particle has a scalar coupling $g\phi\bar{\psi}\psi$, measurement of the electron g-factor limits the coupling constant to $g^2/4\pi < 10^{-8}$. [9] Given this upper limit the 1S-2S splitting in Ps can be off from the QED value by 100 KHz for a point-like particle. If the mystery particle has structure, the form factor would tend to suppress its contribution to the electron g-2 value relative to Ps energy level shifts. Schafer, *et al.* [10] have pointed out that the agreement with theory in the ground state Ps sets good limits on possible vector, axial vector, and pseudo-scalar couplings, but does not say anything about a scalar coupling.

Work is currently underway to re-measure the 1S-2S energy level splitting. An electron accelerator beam dump at AT&T Bell Laboratories has been shown to produce 4 x 10^4 slow positrons/pulse at 30 Hz, and further improvements are expected.[11] We anticipate a 10^3 increase in the number of available thermal Ps. In collaboration with M. Fee and K. Danzmann at Stanford, we are also working on an improved laser system. Fig. 4 shows a block diagram of some of the essential features of the new laser system. The cw dye laser will be stabilized to < 100 KHz with an improved servo system, and an electro-optic rf side-band will be locked to the Te_2 line. Frequency scanning will be done by using an rf frequency synthesizer to shift the rf sideband locked to the Te_2 line. The shift between the amplified dye laser and the cw oscillator will be measured by beating the two laser beams together after the pulsed laser is passively filtered to ~ 6 MHz bandwidth.

Precision measurements with high powered pulsed lasers are difficult because of the incomplete knowledge of the optical electric field. A common problem with amplified laser pulses is that the beam is never Fourier transform limited because of the rapid index of refraction modulation of the amplifying medium. Filtering with a Fabry-Perot cavity after amplification does not eliminate the problem. As an example, we assumed a linear frequency-chirped gaussian beam a factor of two worse than the Fourier transform limit, computer modeled the transmission through a Fabry Perot cavity, and then calculated numerically the two-photon transition rate. A 6 MHz filtered laser beam showed a 1.5 MHz red shift relative to the peak of the frequency spectral distribution. If the time varying electric field were known for each laser pulse (by simultaneously measuring the beat frequency of the pulsed laser with a stable oscillator and the quadrature beat freqency) this systematic effect could be accounted for. However, such a measurement may not be practical, and ultimately, we plan to create cryognic Ps atoms, either by collisional cooling or by laser cooling. [12] Assuming that the technical difficulties can be overcome, cooling the atoms to few Kelvin temperatures will allow us to measure the interval using cw excitation without excessive transit time broadening. Under these conditions, precision approaching the $\Delta\nu/\nu = 10^{-11}$ level should be achievable.

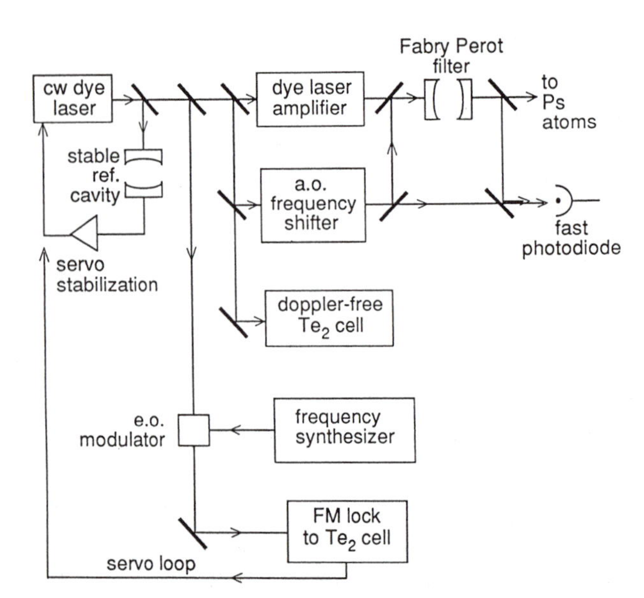

Fig. 4 Present plans for an improved laser system for positronium and muonium spectroscopy.

III. Muonium

The laser spectroscopy of muonium was recently made possible by the development of thermal muonium sources. As in the positronium work, μ^+ stopping in gas targets is considered unacceptable because of the perturbing collisions of the gas molecules. Extensive efforts to create thermal muonium in vacuum have given us two methods: (1) μ^+ stopping near the surface of a hot tungsten foil has been shown to desorb in the form of muonium analogous to the production of hydrogen and thermal Ps. [13] (2) Thermal muonium has also been produced by emission from silica-powder target. [14] In that case, it has been shown that incident μ^+ convert to muonium inside the SiO_2 balls, and then diffuse into the vacuum space between the SiO_2 grains. [15] If the powder remains uncompressed, enough interstitial space is left to allow a reasonable fraction of the muonium to escape into the vacuum.

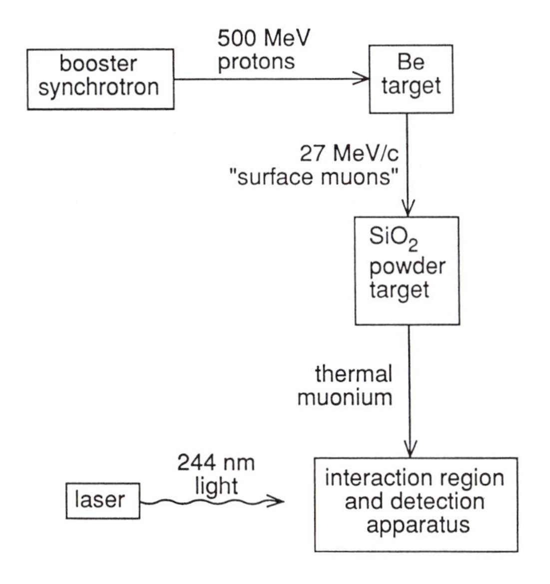

Fig. 5 Block diagram of the muonium experiment.

The first experiment on optically excited muonium [16] was performed at the Booster Meson Facility, Meson Science Laboratory, University of Tokyo, located at the National Laboratory for High Energy Physics (KEK) at Tsukuba, Japan. [17] Figure 5 shows a schematic of the experiment. A pulsed beam of 500-MeV protons incident on a Be target was used to create a low-energy muon beam resulting from the decay of π^+ stopped near the surface of the target. The bursts of 27.5 MeV/c "surface" muons were directed to a SiO_2 powder target. Typically, $100\mu^+$ per pulse yielded one thermal muonium in the space in front of the target. Our 1% yield is lower but not inconsistent with the 17% yield reported in Ref. 14, given the different target geometry and beam parameters.

The muonium atoms were excited and ionized by counter-propagating light beams at 244nm. The brightness of the dye laser system was increased to 80 mJ/pulse in a 30 MHz bandwidth. Using that source, we were able to generate 10-15 mJ/pulse at 244nm in a Ba_2BO_4 doubling crystal. The laser fluence needed to obtain a 50% excitation probability is 0.25J/cm² for a Fourier-transform limited pulse. Since our laser has about twice the bandwidth of an ideal pulse, we tried to operate at a fluence level of ~0.4 J/cm².

Thermal muonium that is resonantly ionized by the pulsed laser is collected by an electrostatic immersion lens, accelerated to 4kV, and directed in a 2.5m long zig-zag path to a microchannel plate detector. (CEMA) Because of the low density of muonium atoms and the small volume irradiated by the uv light, it was necessary to reduce the background counting rate to less than one count per hour in order to see a clear signal. Accordingly, the long path between the SiO_2 target and the CEMA detector was painted with colloidal graphite (Aqua-dag), and baffles and vanes were added around the electrostatic mirrors to reduce the background due to scattered light. Lead shielding 15 cm thick surrounded the CEMA.

We also found it necessary to set a very narrow time window (100 nsec) for counted events that may be due to muonium ionization. The identification of the various time-of-fight peaks in a histogram of the CEMA counts, particularly the H^+ ions emitted from the grid and target allowed us to calculate that the expected arrival time of the μ^+ is 1.43± 0.03μsec after the laser pulse. A retangular region defined by the intersection of the appropriate time cut and a 150 MHz frequency cut (75 MHz per Fabry-Perot fringe) shows six counts. The average number of counts in a rectangle of the same area arbitrarily placed in the scatter plot is 0.39 ± 0.05. Assuming a Poisson distribution, the probability that six or more counts will appear in such a rectangle is 4 x 10^{-6}. We therefore interpret the data as evidence for the laser excitation of muonium.

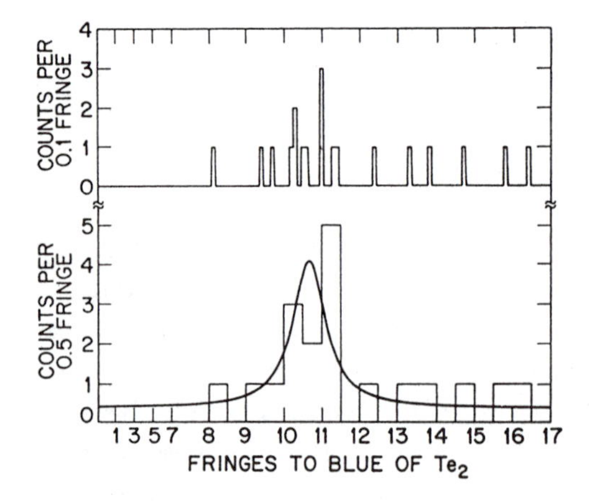

Fig. 6 Frequency spectrum for all runs taken where the laser frequency and μ^+ arrival time were recorded. The Lorentzian curve is fitted to the individual events.

In order to avoid biasing our data selection, we also include the data taken under conditions where the intense uv light was known to have partially destroyed the optics inside the vacuum chamber, the statistical significance of our result diminishes. Figure 6 shows the inclusion of 16 hours of integration time with the same time cut in the CEMA counts. A Lorentzian curve fitted to the nineteen counts given in Fig. 6 gives a line center at 10.7 ± 0.4 fringes with a fwhm of 1.1 ± 0.7 fringes. The peak signal is 7.4 ± 5.0 counts per fringe. The probability that ten or more counts between fringes 10 and 12 would occur by chance decreases to 6×10^{-4}. Thus, the effect shown in our best 2h run survives the inclusion of all the data. Comparison of our observation with the QED calculation was previously reported. [16] Since the resonance line is inherently narrow, our reported observation of 1/4 of the frequency of the $F = 1 \rightarrow 1$ transition 613 881 924 ± 46 MHz represents a resolution $\Delta\nu/\nu = 10^{-8}$. The result is in reasonable agreement with the QED prediction of 613 881 989.90 (90). We stress that this first work should not be taken as a "measurement" of the 1S-2S splitting in muonium.

Clearly, a more intense source of thermal muonium is necessary before precise laser spectroscopy can be done on this atom. Active work to develop efficient muon moderators analogous to the positronium moderators is under way in several laboratories. Conversion of the 30 MeV surface muons into few eV muons with a moderator would allow one to generate thermal muonium very efficiently in convenient target geometries. Brighter pulsed muon sources are also under development. The next generation experiment would also use improved uv optics and possibly a brighter laser source. With these improvements one might be able to achieve a 10^4 increase in the number of counts for the same integration time. Another improvement would be the calibration of the experiment with a 1S-2S measurement in hydrogen in the same apparatus. Many of the systematic effects associated with a pulsed measurement of muonium would be similar to the effects in hydrogen where the 1S-2S splitting is measured with greater accuracy. The ionization signal can also be used to better determine the timing windows for the muonium signal.

References

1. S. Chu, A.P. Mills, Jr. and J.L. Hall: Phys. Rev. Lett. *52*, 1689 (1984); S. Chu: In *Positron Studies of Solids, Surfaces and Atoms*, A.P. Mills, W.S. Crane, and K.F. Canter, eds., World Scientific, Singapore (1986); and references contained within. For the most recent radio frequency measurements, see:
   (a) S. Hatamian, R.S. Cont., and A. Rich, Phys. Rev. Lett. *55*, 1833 (1987)
   (b) M.W. Ritter, *et al.* Phys. Rev. *A30*, 1331 (1984)
2. A.P. Mills, Jr., S. Berko and K.F. Canter: Phys. Rev. Lett. *34*, 1541 (1975)
3. M.W. Ritter, *et al.* Phys. Rev. *A30*, 1331 (1984)

4. G.W. Erickson: J. Chem. Phys. Ref. Data *6*, 831 (1977) as corrected by the new value of R_∞ from S.R. Amin, C.D. Caldwell and W. Lichten: Phys. Rev. Lett. *47*, 1234 (1981)
5. J. Carion and P. Luc: *Atlas du Spectre d'Absorption de la Molecule de Tellure*, Lab. Aime-Cohon C.N.R.S., Orsay, France (1980)
6. D.H. McIntyre and T.W. Hansch: Phys. Rev. *A34*, 4504 (1986)
7. T. Kinoshita and J. Sapirstein: In *Atomic Physics* 9, eds., R.S. Van Dyke and E.N. Forston, World Scientific, Singapore (1984)
8. T. Cowan, *et al.*: Phys. Rev. Lett. *56*, 446 (1986)
9. S.J. Brodsky, E. Mottola, I. Muzinich, and M. Soldate: Phys. Rev. Lett. *56*, 1763 (1986)
10. A. Schafer, J. Reinhardt, W. Greiner and B. Muller: Mod. Phys. Lett. *A1*, 1 (1986)
11. A.P. Mills: private communication
12. S. Chu, L. Hollberg, J.E. Bjorkholm, A. Cable and A. Ashkin: Phys. Rev. Lett. 55, 48 (1985) see E.P. Liang and C.D. Dermer, Lawrence Livermore Lab preprint for a discussion of possible laser cooling of Ps.
13. A.P. Mills, Jr., *et al.*: Phys. Rev. Lett. *56*, 1463 (1986)
14. G.A. Beer, *et al.*: Phys. Rev. Lett. *57*, 671 (1986)
15. G.M. Marshall, *et al.*: Phys. Lett. *65A*, 351 (1978)
16. S. Chu, *et al.*: Phys. Rev. Lett. *60*, 101 (1988)
17. K. Nagamine: Hyperfine Interact *8*, 787 (1981)

Positronium Decay Rates

A. Rich[1], *R.S. Conti*[1], *D.W. Gidley*[1], *and P.W. Zitzewitz*[2]

[1]Department of Physics, The University of Michigan, Ann Arbor, MI48109, USA

[2]Department of Natural Sciences, The University of Michigan, Dearborn, MI48128, USA

1. Introduction

In this paper we briefly review all positronium (Ps) decay rate measurements (including those of the excited states and the Ps negative ion) that have been completed to date. The results are compared with theoretical values. The Ps system represents the most rigorous confrontation with theoretical decay rate calculations for any QED system.

2. Measurement of the 1^3S_1 decay rate of Ps

In a recent publication [1], a new 200 ppm measurement of the vacuum decay rate, λ_T of triplet Ps formed in a gas was presented. The result, $\lambda_T = 7.0516 \pm 0.0013 \mu s^{-1}$, represents a factor of four improvement over previous measurements and is in substantial agreement with existing experimental results, the most recent of which are (see bibliography in reference 1) $7.056 \pm 0.007 \mu s^{-1}$, $7.045 \pm 0.006 \mu s^{-1}$, $7.051 \pm 0.005 \mu s^{-1}$, and $7.050 \pm 0.013 \mu s^{-1}$. These latter values are all 1-2.5 standard deviations above the present theoretical value and the new measurement exceeds theory by 10 experimental standard deviations.

The theoretical value of λ_T may be expressed as the sum of decay rates into three gammas (λ_3), five gammas (λ_5), *etc.*: $\lambda_T = \lambda_3 + \lambda_5 + \ldots$. The contribution of λ_5 has been calculated [2], [3] to be $\lambda_5 \sim 10^{-6}\lambda_3$, and is thus negligible. The leading term is:

$$\lambda_3 = \frac{\alpha^6 mc^2}{\hbar}\frac{2(\pi^2-9)}{9\pi}\left[1 + A_3\left(\frac{\alpha}{\pi}\right) - \frac{1}{3}\alpha^2 \ln \alpha^{-1} + B_3\left(\frac{\alpha}{\pi}\right)^2 + \ldots\right]. \quad (1)$$

The two most recent calculations give $A_3 = -10.266 \pm 0.011$ [4] and $A_3 = -10.282 \pm 0.003$ [5]. The coefficient B_3 is still uncalculated, and one obtains through order $\alpha^2 \ln \alpha$, $\lambda_3 = 7.03830 \pm 0.00005 \mu s^{-1}$ [$1 \times (\alpha/\pi)^2$ adds only 0.00005 μs^{-1} to λ_3]. If one assumes that the disagreement between the theoretical and experimental values of λ_T is due

The Hydrogen Atom Editors: G.F. Bassani · M. Inguscio · T.W. Hänsch

to the $(\alpha/\pi)^2$ term, then $B_3 = 340 \pm 33$ is required to bring theory and experiment into agreement. We note here that it may be more appropriate [6] to write the second order term as a coefficient times α^2 rather than $(\alpha/\pi)^2$, so that the (still rather large) coefficient of 34 would explain the difference. Exotic, non-QED decay modes of o-Ps have been considered [7], [8] to account for the discrepancy but recent axion searches place severe restrictions on such decays [9].

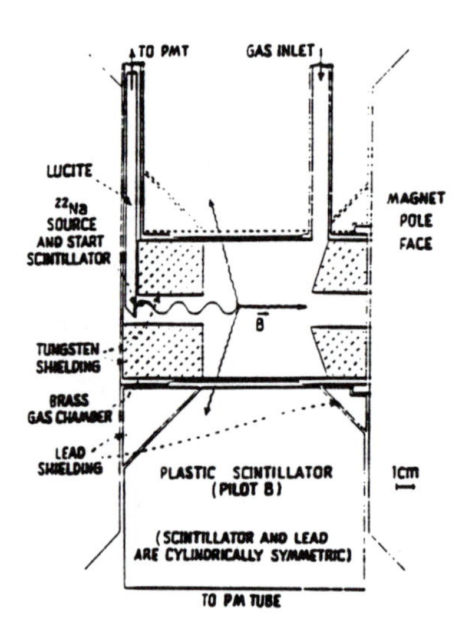

Fig. 1. The gas-filled Ps formation chamber and detector arrangement.

In the experimental technique used in our new measurement [1] Ps is formed in a gas in a magnetic field of 6.8 kG (Fig. 1). The field confines positrons to the axis of the chamber and reduces the $1^3S_1(m = 0)$ lifetime to 13 ns. The $1^3S_1(m = \pm 1)$ states are unperturbed and continue to decay in the field with a rate, λ, which depends on the particular gas and its density. The decay rate is determined by fitting of the annihilation lifetime spectrum and λ_T is then determined by extrapolating λ to zero gas density. The results of this extrapolation are: isobutane 7.0524 ± 0.0013 μs^{-1}, neopentane 7.0551 ± 0.0026 μs^{-1}, nitrogen 7.0487 ± 0.0018 μs^{-1}, and neon 7.0501 ± 0.0023 μs^{-1} with the weighted average value being 7.0514 ± 0.0013 μs^{-1}. The uncertainty is obtained from the isobutane result with the other gases treated as systematic tests.

The measurement of λ_T has included extensive systematic tests [1]. These include: i) measurement of λ_T to at least ± 0.0026 μs^{-1} (350 ppm) in four different gases, ii) use of two independent digital timing systems with, and without, electronic noise rejection, iii) use of different positron source strengths to search for effects related to the signal-to-noise ratio. A detailed discussion of possible systematic errors related to physical effects in the gases has been published [10]. The most important of such effects include: 1) any non-linearities in λ vs. gas density, 2) formation of long-lived excited states of Ps, and, 3) incomplete thermalization of Ps that could result in time dependent collisional quenching ("pickoff") of Ps. With these systematics in mind we have recently performed an extrapolation of λ vs. gas density using only high density (pressure greater than one atmosphere) N_2 and Ne gases where Ps thermalization and excited state quenching should be rapid. Using only our data with $P > 1$ atmosphere and published [11] high pressure data for N_2 (7-36 atmospheres) and Ne (7-39 atmospheres) we extrapolate to zero density to find $\lambda_T = 7.0491 \pm 0.0021\ \mu s^{-1}$ and $7.0483 \pm 0.0029\ \mu s^{-1}$ respectively. The close agreement of these values with the values measured entirely below two atmospheres is a strong systematic check on the gas related effects mentioned above.

All of the systematic tests [1, 10] to date support the results of reference [1]. The 1900 ppm difference with theory remains unresolved at this time. A new and systematically very different experiment designed to reach $\sim$ 100 ppm accuracy [12] using a slow positron beam with Ps formation in an evacuated cavity is now underway. Results are expected within one year.

3.<u>Measurement of the 1^1S_0 Ps Decay Rate</u> There is only one precision measurement of the singlet ground state (parapositronium) decay rate [13] with sufficient accuracy to test the first order radiative connections to λ_S. The singlet decay rate may be expressed as $\lambda_S = \lambda_2 + \lambda_4 + \lambda_6 \ldots$. Since $\lambda_4(\sim 1.5 \times 10^{-6}\lambda_2)$ is small, [3, 14] we need only concentrate on λ_2 in the present discussion. The expression [4] for λ_2 is:

$$\lambda_2 = \frac{mc^2}{2\hbar}\alpha^5\left[1 + A_2(\frac{\alpha}{\pi}) + \frac{2}{3}\alpha^2 \ln \alpha^{-1} + B_2(\frac{\alpha}{\pi})^2 \ldots\right] \quad (2)$$

where $A_2 = -(5 - \pi^2/4) = -2.532$ and B_2 is, as yet, uncalculated. Through order $\alpha^2 \ln \alpha^{-1}$ the decay rate is $\lambda_2 = 7.9866\ ns^{-1}$ $[1 \times (\alpha/\pi)^2$ would add $0.0004\ ns^{-1}]$.

In the experiment, Ps is formed in isobutane gas in a uniform magnetic field of about 4 kG (the experimental arrangement is almost identical to that shown for λ_T in

Fig. 1). The magnetic field mixes the m= 0 triplet and singlet states and, as a result, the m= 0 triplet decay rate is increased to

$$\lambda'_T = \frac{1}{1+y^2}\lambda_T + \frac{y^2}{1+y^2}\lambda_S \quad ,$$

$$\text{where} \qquad y = \frac{x}{1+\sqrt{1+x^2}}, \text{ and } x = \frac{2g'\mu_B B}{h\Delta\nu} \approx \frac{B(kG)}{36.287} \quad . \tag{3}$$

Thus the annihilation lifetime spectrum has two exponential components: the unperturbed decay from the m= ±1 states; and the "quenched" decay from the m= 0 state (at 4 kG the lifetime is about 30 nsec). Measurement of these decay rates, λ_T and λ'_T, at gas pressures ranging from 200 − 1400 torr allows one, after extrapolation to zero gas density, to solve (3) for λ_S. Measurements were made at three different magnetic fields and the average result is $\lambda_S = 7.994 \pm 0.011$ nsec^{-1}, in agreement with the λ_2 calculation at the 1400 ppm level.

Considering the 1900 ppm difference between the measured value of λ_T and λ_3, it would be interesting to measure λ_S at comparable precision ($\sim$ 200 ppm). The current $\lambda_\mathrm{S} - \lambda_2$ difference is (1100 ± 1400) ppm and cannot distinguish such an effect. A measurement at the 200 ppm level would be of immediate interest since the computationally simpler B_2 will probably be calculated before B_3. We will shortly begin construction of a λ_S experiment that is designed to reach the 200 ppm level of precision.

4. <u>Measurement of the Ps$^-$ Decay Rate</u> The Ps negative ion, consisting of two electrons and one positron, is a relatively simple system for testing many-electron calculational schemes [15]. There is recently additional interest in measuring the ground state decay rate of Ps$^-$, Γ, [16] to see if the 1900 ppm discrepancy surrounding λ_T enters into Γ by way of λ_S. Approximately 98% of Γ is given simply by the spin average, $\Gamma \approx 0.25\lambda_\mathrm{S} + 0.75\lambda_\mathrm{T}$. Thus, if there is a sizable discrepancy in λ_S (see previous discussion) it would show up at the same level in Γ.

The best theoretical value [17, 18] of Γ is 2.0861 ns^{-1}. This two photon, Hylleras-type calculation also includes order-α corrections for 3-photon annihilation and 2-photon radiative corrections [17].

The Ps$^-$ ion was first observed by Mills [19], who has reported the only measurement of Γ to date [20]. In this experiment, Ps$^-$ is formed on a thin carbon film and accelerated by applying a constant potential, V, to two grids separated by a distance, d. Measurements of the number of ions reaching the second grid as a function of d (and hence the proper time since Ps^- emission) yields $\Gamma = 2.09 \pm 0.09$ ns^{-1}. A remeasurement [16] of Γ using an improved variation of this time-of-flight technique is presently underway with the goal of achieving 1000 ppm accuracy.

5. Measurement of the Radiative Decay Rate of Ps in the 2^3P_J (J = 0, 1, 2) States

The radiative decay rate γ of 2^3P_J Ps was measured as a byproduct of an experiment [21] to determine the frequency intervals ν_J between the 2^3S_1 and 2^3P_J states (J=0,1,2). The 2^3P_J states decay to the 1^3S_1 state by emission of a Lyman-α photon at 243 nm. To lowest order in α, the expected decay rate

$$\gamma = 1/\tau = \frac{5\pi}{64}\alpha^3\frac{Ry}{h} = 313.8\mu s^{-1} \tag{4}$$

is half that for the corresponding transition in hydrogen.

In this experiment Ps in the 2^3S_1 state is irradiated with microwaves of varying frequency and a resonant increase in 2^3P_J states is observed as the result of stimulated emission. The signature for the 2^3P_J state is its emission of a Lyman-α photon leaving the Ps in the 1^3S_1 state which subsequently annihilates to three gammas with a lifetime $\lambda_T^{-1} = 142$ ns.

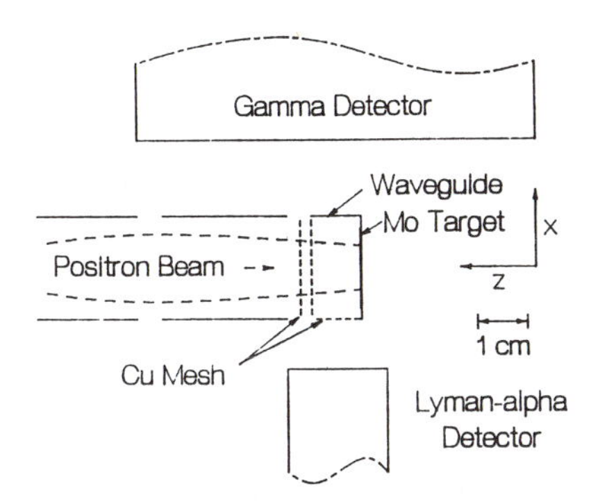

Fig. 2 Experimental Apparatus. (See text for a detailed description.)

A schematic representation of the apparatus [21] is shown in Fig. 2. An electrostatically focused beam (10^5 e$^+$/s at 65 eV) enters a section of waveguide (2.3cm × 1.0cm) and strikes a polycrystalline molybdenum target attached to the opposite inner wall of the waveguide. A fraction (3×10^{-4}) of the incident positrons is emitted from the target as $n = 2$ Ps. Assuming equal distribution in all of the $n = 2$ magnetic substates, 3/16 of them will be in the 2^3S_1 state, and 9/16 will be in the 2^3P_J ($J = 0, 1, 2$) states. Lyman-α photons are detected in a solar-blind photomultiplier with an overall detection efficiency of 1%. One or more of the gamma rays from the subsequent annihilation of the 1^3S_1 states are detected in two Pilot-B plastic scintillators with a combined detection efficiency of 15%.

In the absence of rf in the waveguide, only those 2^3P_J states originally formed contribute to the signal rate $R(0, \nu)$ of a Lyman-α photon with a delayed γ ray. As the rf frequency is scanned the signal rate $R(I, \nu)$ will increase resonantly at the transition frequencies, ν_J. The theoretical expression [21] for the ratio $r \equiv [R(I, \nu) - R(0, \nu)]/R(0, \nu)$ is fitted to data taken in the vicinity of all three transition frequencies and at a variety of rf intensities I. The results are displayed in Table 1. In addition to the results for the ν_J, this fit provides the first measurement of the natural width γ of the 2^3P_J states. The value of γ is taken to be the same for all three 2^3P_J states (in theory the largest difference among the decay rates for different J states due to their energy splittings is 25 ppm) resulting in an averaged value for γ.

Table 1. Results of fit to r. T is the 2^3S_1 transit time across the waveguide and γ is the radiative decay rate of 2^3P_J Ps.

Parameter	Value	σ_{stat}	σ_{syst}	Theory
ν_o	18504.1 MHz	10.0 MHz	1.7 MHz	18496.1 MHz
ν_1	13001.3 MHz	3.9 MHz	0.9 MHz	13010.9 MHz
ν_2	8619.6 MHz	2.7 MHz	0.9 MHz	8625.3 MHz
T	17.3 ns	1.2 ns	4.0 ns	—
γ	284 μs^{-1}	24 μs^{-1}	63 μs^{-1}	313.8 μs^{-1}

The systematic error assigned to γ is due to a simplification in the fitting function used. The fitting function was generated with the assumption that a single time T described the transit of 2^3S_1 Ps across the waveguide, when, in fact, there is an unknown distribution of transit times. Two fits were made with multiple-value distributions (one with two equally likely transit times and the other with a Gaussian distribution of transit times), and the 5 MHz error was accordingly assigned as a conservative estimate of this error.

The result is in good agreement with the theory. The presence of the above systematic error makes the use of this technique to measure γ to better than the 10 μs^{-1} level dependent on detailed knowledge or control of the 2^3S_1 velocity spectrum.

6. Summary

The decay rates of Ps discussed in this paper are summarized in Table 2 below.

Table 2. Positronium Decay Rates (Units - μs^{-1}, Ex=Experiment, Th=Theory)

n = 1	
$\lambda_{Ex}(1^1S_0 \rightarrow 2\gamma) \equiv \lambda_S = 7994(11)$	Gidley, Rich, Sweetman, West (1982)
$\lambda_{Th}(1^1S_0 \rightarrow 2\gamma) \equiv \lambda_S = 7986.6(0.1)$	Caswell, Lepage (1979), Adkins (1983)
$\lambda_{Th}(1^1S_0 \rightarrow 4\gamma) \sim 1.5 \times 10^{-6} \lambda_{Th}(1^1S_0 \rightarrow 2\gamma)$	Adkins, Brown (1983)
$\lambda_{Ex}(1^3S_1 \rightarrow 3\gamma) \equiv \lambda_T = 7.0514(13)$	Gidley, Westbrook, Conti, Rich (1987)
$\lambda_{Th}(1^3S_1 \rightarrow 3\gamma) \equiv \lambda_T = 7.03830(5)$	Caswell, Lepage (1979), Adkins (1983)
$\lambda_{Th}(1^3S_1 \rightarrow 5\gamma) = 10^{-6} \lambda_{Th}(1^3S_1 \rightarrow 3\gamma)$	Lepage et al. (1983), Adkins (1983)
n = 2	
$\lambda_{Ex}(2^3P \rightarrow 1^3S_1 + hv) \equiv \gamma = 284(72)$	Hatamian, Conti, Rich (1987)
$\lambda_{Th}(2^3P \rightarrow 1^3S_1 + hv) \equiv \gamma = 313.8$	
Ps^-	
$\lambda_{Ex}(1^1S^e \rightarrow 2\gamma) = 2090(90)$	Mills (1983)
$\lambda_{Th}(1^1S^e \rightarrow 2\gamma) = 2086.1(0.2)$	Bhatia, Drachman (1983), Ho (1983)

7. Acknowledgements

This work has been supported by National Science Foundation Grants No. PHY-8403817 and PHY-8605544 and by grants from the office of the Vice President for Research of the University of Michigan.

REFERENCES

1. Westbrook, C.I., Gidley, D.W., Conti, R.S., and Rich, A., Phys. Rev. Lett. 58, 1328 (1987).
2. Adkins, G.S., Annals of Physics 146, 78 (1983).
3. Lepage, G.P., Mackenzie, P.B., Streng K.H., and Zerwas, P.M., Phys. Rev. A 28, 3090 (1983).
4. Caswell, W. and Lepage, G.P., Phys. Rev. A 20, 36 (1979).

5. Adkins, G.S. and Brown, F.R., Phys. Rev. A 28, 1164 (1983).
6. Lepage, G.P. and Adkins, G.S., private communication (1987). This modification was suggested since bound state corrections of order α^2 will occur as well as radiative corrections of order $(\alpha/\pi)^2$.
7. Cleymans, J. and Ray, P.S., Nuovo Cimento 37, 569 (1983).
8. Samuel, A.L., submitted for publication (1987).
9. Wahl, W., Proceedings of the Symposium "Production of Low-Energy Positrons with Accelerators and Applications", Giessen, FRG (1986) (to be published in Appl. Phys. A).
10. Gidley, D.W., Westbrook, C.I., Conti, R.S., and Rich, A., in "Atomic Physics with Positrons, NATO ASI Series B, Volume 169", edited by J.W. Humberston and E.A.G. Armour, (Plenum Press, NY, 1988), p 277.
11. Coleman, P.G., Griffith, T.C., Heyland, G.R. and Killeen, T.L., J. Phys. B 8, 1734 (1975).
12. Nico, J.S., Gidley, D.W., Zitzewitz, P.W., and Rich, A., Bull. Am. Phys. Soc. 32, 1051 (1987).
13. Gidley, D.W., Rich, A., Sweetman, E. and West, D., Phys. Rev. Lett. 49, 525 (1982).
14. Muta, T. and Niuya, T., Prog. Theor. Phys. 68, 1735 (1982).
15. Bhatia, A.K. and Drachman, R.J., Phys. Rev. A 35, 4051 (1987).
16. Friedman, P.G., Mills, A.P., Jr. and Zuckerman, D., Bull. Am. Phys. Soc. 33, 953 (1988).
17. Bhatia, A.K. and Drachman, R.J., Phys. Rev. A 28, 2523 (1983).
18. Ho, Y.K., J. Phys. B 16, 1503 (1983).
19. Mills, A.P., Jr., Phys. Rev. Lett. 46, 717 (1981).
20. Mills, A.P., Jr., Phys. Rev. Lett. 50, 671 (1983).
21. Hatamian, S., Conti, R.S., and Rich, A., Phys. Rev. Lett. 58, 1833 (1987).

New Schemes for the Production and Spectroscopy of Positronium

G. Werth

Institut für Physik, Universität Mainz,
D-6500 Mainz, Fed. Rep. of Germany

The rate of positronium formation has been increased by 2-3 orders of magnitude using recently developed accelerator based slow positron sources. This opens the possibility of improvements of precision experiments on the Ps atom as well as new experiments on excited states. First evidence for enhanced metastable Ps formation is presented and future possibilities are discussed.

1. Introduction

Thirty-six years after the first production of positronium by Deutsch it seems somewhat surprising that our experimental knowledge of this system is restricted to the states of principal quantum numbers n=1 and n=2. Table I summarizes the present most precise data.

Table I Presently known values of energy differences and decay rates in positronium

	Experiment			Theory	
1^1S_0 decay rate	7.944(11)	ns^{-1}	/1/	7.9842	ns^{-1}
1^3S_1 decay rate	7.0516(13)	μs^{-1}	/2/	7.0383	μs^{-1}
Ground state Hfs	203 388.7(7)	MHz	/3/	203 404.0	MHz
1^3S_0-2^3S_1	1 233 607 185(15)	MHz	/4/	1 233 607 197	MHz
2^3S_1-2^3P_2	8 619.6(3.6)	MHz	/5/	8625.2	MHz
	8 628.4(2.8)	MHz	/6/		
2^3S_1-2^3P_1	13 001.3(4.8)	MHz	/5/	13 010.9	MHz
2^3S_1-2^3P_0	18 504.1(11.7)	MHz	/5/	18 496.1	MHz

The Hydrogen Atom Editors: G.F. Bassani · M. Inguscio · T.W. Hänsch

There exist no experiments at all for states above n=2, which would be an interesting test case for a relativistic two-body system. In view of the fundamental nature of the system more extensive and more precise data would be highly desirable as a test for bound state QED calculations. With the exception of the 1S-2S energy difference measurement, which is limited in precision by the laser wavelength measurement, all present experiments suffer from low statistics due to the small number of Ps atoms. It seems therefore worthwile discussing recent advances of increased Ps production and in particular methods to populate the metastable 2^3S_1 state, which could serve as a basis for excited state spectroscopy. While laser excitation of states above n=2 from the ground state requires wavelengths between 205 nm (L_β) and 182 nm (ionization limit), which are very difficult to obtain with sufficient power, the corresponding wavelengths range from 2^3S_1 is 1310 nm (H_α) to 728 nm (ionisation limit). Lasers with such wavelengths are commercially available.

2. Slow Positron Production

Positronium atoms are formed in collision of low energy positrons in gases or on surfaces. The most obvious way to enhance the Ps formation rate is to increase the number of low energy positrons. Conventional sources use radioactive e^+-emitters like ^{22}Na or ^{58}Co. The high energy positrons from these sources are stopped in thin metal foils. A small number of them diffuses to the surface before annihilation and is released into the vacuum with an energy of about 1 eV due to the negative work function of metals for positron emission. Extensive work has been done by different groups to improve on the moderator efficiency. Using carefully cleaned single crystal surfaces, about 1 out of 10^3 fast positrons is re-emitted at low energies having an energy spread of about 0.2 eV. The total amount of slow positrons depends on the radioactive source strength. Taking up to 100 mCi, which can be handled decently in the laboratory, low energy positron rates up to 10^6 s^{-1} are obtained routinely. Apart from safety requirements the source strength is limited by self-absorption inside the source material unless one uses extended surfaces, which in turn reduces the beam brightness. The main characteristics of such beams are that they are continuous in time and have a spin polarization due to the natural helicity of the particles. Higher intensities of slow positron beams require the use of high flux reactors or of electron accelerators. The first approach has been made by LYNN et al. /7/, who used a neutron activated ^{64}Cu source with half-life of 12.8h to obtain a high specific activity of 600 Ci/g and produced a slow positron beam of 4×10^7 e^+/s. The moderation of high energy positrons appearing in the bremsstrahlung of electron accelerators has been investigated at first in Mainz /8/ and Livermore /9/. A picture of the present setup at Mainz is shown in Fig. 1.

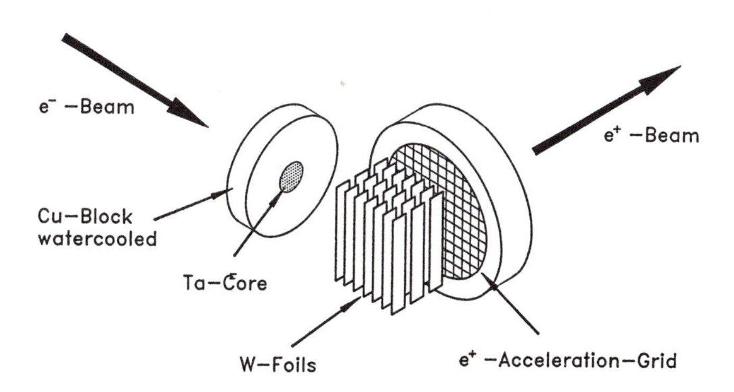

Fig. 1 Target and moderator setup for slow positron production at the Mainz Linear Accelerator

While the moderation process does not differ from that of radioactive sources, the number of particles as well as the time structure of the beam depends on the accelerator. Since the intensity of the bremsstrahlung is approximately proportional to the atomic number Z and the cross-section for pair production scales as Z^2, high Z materials of high melting points like tantalum are suited best as targets. The maximum efficiency of a target-moderator combination, defined as the number of slow positrons per fast electron, depends on the electron energy and the target's thickness: if for a given energy the target is too thin, only a fraction of the beam is used for positron production. On the other hand in thick targets many of the positrons annihilate before they reach the surface. The optimum thickness is roughly two radiation lengths, which is the distance in which a relativistic electron beam energy is reduced to 1/e of its initial value. Data from different groups (Mainz, Livermore, Tsukuba, Gießen) indicate that total slow positron fluxes of more than 10^8 e^+/s are obtained almost routinely.

Equally important as the source strength is the time structure. Accelerators may be continuous in time or may have a pulse structure, which sometimes can be varied according to the requirements of the experiment. Figure 2 shows as an example the time structure of the 400 MeV LINAC at Mainz.

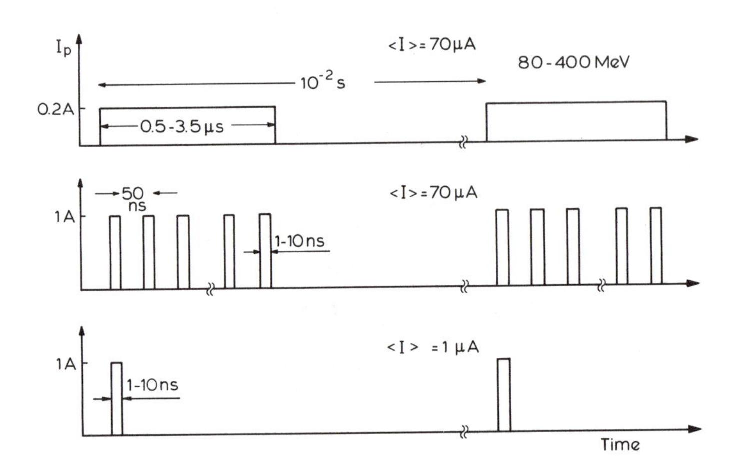

Fig. 2 Possible modes of operation of the Mainz LINAC

Long pulses of up to 3.2 μs as well as single shot operations with pulse duration ranging from 1 to 10 ns are possible. In particular the single pulse mode with 100 Hz repetition rate is very similar to the time structure of commercially available high power lasers. This can be of advantage in laser excitation of positronium atoms as discussed below. It should be pointed out that the time structure is only partially preserved in slow positron beams. Slow positron beam transport over distances of several meters is a necessary requirement because of the high radiation and background level at accelerator targets. Due to the initial energy spread, short pulses are spread out in time, in particular if the transport system uses magnetic guidance fields and contains bends. In practice we have obtained 40 ns e^+-pulses, starting from 10 nsec accelerator pulses after a beam transport over 20 meters.

3. Ground State Positronium Production

Slow positrons impinging on metal or metal-oxide surfaces at energies ranging up to a few hundred eV may form positronium atoms. The probability depends strongly on the surface temperature. While in experiments using radioactive sources as positron emitters, in general clean and well defined crystal surfaces have been used to obtain a maximum positronium yield, the high flux of slow positrons at electron accelerators facilitate substantially the experimental setup. Untreated metal surfaces show the same temperature dependence of positronium formation as previously observed with single crystals (Fig. 3), and even with somewhat reduced efficiency compared to single crystals positronium intensities of several $10^7\ s^{-1}$ are obtained.

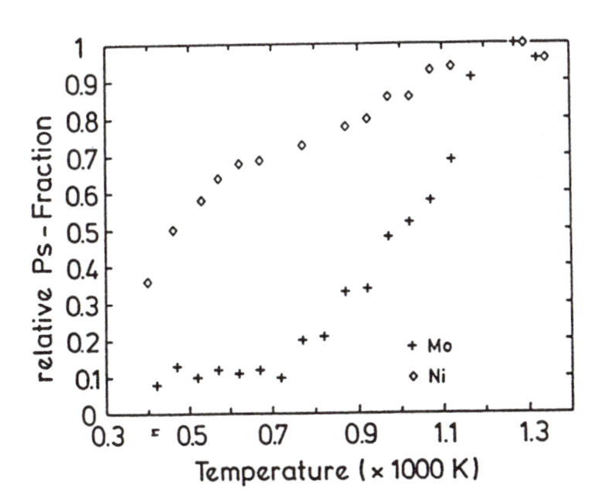

Fig. 3 Temperature dependence of Ps formation on untreated metal surfaces (from ref. 16)

The energy distribution of positronium atoms emitted from metal surfaces has been investigated by MILLS /10/. The main part has a mean kinetic energy of about one eV and is formed when a positron diffuses from the inside of the target material to the surface, is emitted into the vacuum by the negative work function of the metal, and catches a surface electron. A second part of the Ps atom has lower energies and arises from thermal desorption of positrons bound in a potential well at the surface by the image force. The activation energy E_a for this process is a delicate balance between the binding energy E_b of the positron at the surface, the work function ϕ_- of the target material and the binding energy 1/2 Ry of the Ps atom /11/:

$$E_a = E_b + \phi_- - 1/2\ Ry.$$

Typical values for E_b are 2-4 eV, while E_a ranges from 0.2 eV to 1 eV depending on the crystal face and the surface contamination.

4. Excited State Positronium Formation

In 1975 CANTER et al. /12/ discovered that about 1 out of 10^4 Ps atoms emitted from solid surfaces is in the metastable 2^3S_1 state. Although this number is small it was sufficient to allow first experiments on the fine structure of the n=2 state /5,6/. The stronger sources now available at accelerators may lead to intensities of metastable Ps atoms which are comparable to the number of ground state Ps in earlier experiments. In a first attempt our group has tried to observe fine structure transitions on n=2 Ps in a setup at an electron accelerator and found a good signal-to-noise ratio in the observed L_α count rate following a microwave induced fine structure transition /13/ (Fig. 4,5).

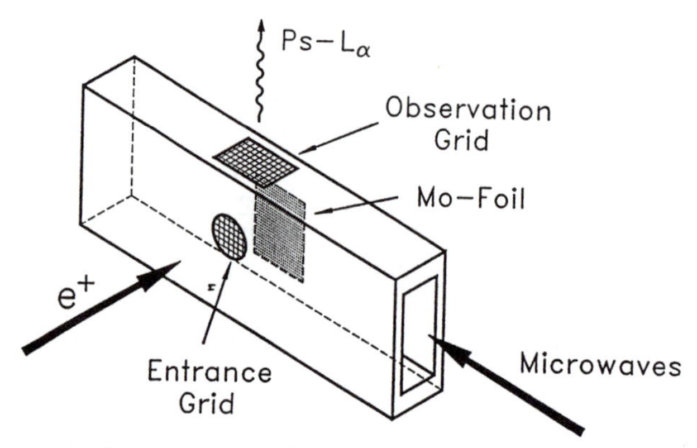

Fig. 4 Experimental setup to observe L_α photons after microwave induced fine structure transitions of n=2 Ps

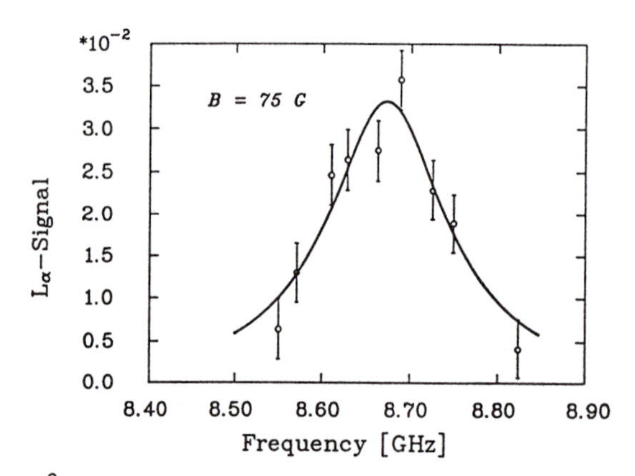

Fig. 5 2^3S_1 - 2^3P_2 resonance. The L_α-detector count rate difference with and without microwaves, normalized to the total count rate, is plotted against the microwave frequency. Running time was 5 min per data point

Although the resonance is shifted by Zeeman- and motional Stark-effects due to a residual magnetic field of about 75 G in the transition region, the results are promising and lead to the expectation of improved precision in future excited state experiments.

A disadvantage of this simple approach to create n=2 Ps is the fact that thermal activation from surface states, as in the case of ground state Ps, is energetically not possible because of the small binding energy of n=2 Ps. Thus only the fast component (about 1 eV) of n=2 Ps exists, which may lead to transient time problems in spectroscopy.

A different approach, which could further increase the number of n=2 Ps and at the same time preserve the slow velocity component is the optical excitation from the ground state into 2^3S_1. A direct two-photon 1^3S_1 - 2^3S_1 transition at 243 nm does not lead to n=2 population since the strong required laser power immediately leads to ionisation of n=2 Ps by a third photon. It should be possible, however, to excite the allowed transitions to the n=3 and n=4 levels from the ground state at 205 nm and 194 nm, respectively. From the branching ratio of the excited states one finds that 12% of the excited atoms end up in the 2^3S_1 state in each case. The required wavelengths are commercially available with the kW output power from Raman shifted excimer-pumped dye lasers, e.g. The time structure of many accelerators of several ns pulse width and 100-1000 Hz repetition rate fits particularly well to the time structure of those lasers (see Fig. 2). Although the short pulse width of accelerators is not completely preserved in the positron time structure, as mentioned above; one can anticipate intensities of about 10^5 n=2 Ps atoms per second in the near future.

5. Possible Experiments

Accelerator based slow positron sources, which produce 2-3 orders of magnitude higher intensities of Ps atoms than radioactive sources, could be used to improve earlier experiments on Ps and could make possible the spectroscopy of higher excited states. Most obvious would be a remeasurement of the fine structure splitting in n=2 Ps. The good signal-to-noise ratio should allow to reduce the microwave power to avoid power broadening and to obtain the natural linewidth of 50 MHz. Although it is difficult to estimate the final precision of those experiments, the need for higher-order corrections above the presently calculated $\alpha^3 R$ correction is indicated.

Increased metastable n=2 Ps rates should allow laser excitation to higher states, in particular highly excited Rydberg states. The signature of such a transition could be the reduction of L_α counts, if a n=2 fine structure transition is driven simultaneously.

The specific feature of short pulses of slow positrons could be used to make a new determination of the Ortho-Positronium lifetime. At present the most accurate experimental determination of this quantity differs by 10 standard deviations from the theoretical value, which represents the only serious discrepancy in low energy QED (see Table I). A new experiment, which uses short pulses of many Ps-atoms and observes the decay of the ensemble, would differ substantially in Ps rate and consequently in the electronic setup from earlier single-event experiments and could help to decide whether the discrepancy is due to so far unobserved systematic effects in the experiments or whether the theory needs improvement.

Finally the high Ps rate could be used to improve on the search for rare or forbidden decays of ground state Ps. Table II summarizes the present values of such decay modes and theoretical predictions.

A serious drawback in precision spectroscopy on Ps is the high velocity of $2 \cdot 10^6$ cm/s at initial energies of about 1 eV, which easily leads to transient time broadening. Thus finally laser cooling of Ps may be required to obtain smaller uncertainties. A rough estimate indicates that such experiments are in fact possible: The scattering of a 243 nm photon from the Ps atom reduces its velocity by an amount of $\Delta v = h\ k/m = 10^5$ cm/s. Thus about 20 scattering processes are needed to cool a 1 eV Ps atom to very low temperatures. The minimum time required for this process is 20 times the average lifetime of the first excited 2^3P state, which is 3.2 ns. The total cooling time of 64 ns takes about half the lifetime of the Ps ground state (142 ns). Although the recoil limited final temperature will be by a factor of $(M(\text{atom})/m\ (\text{Ps}))^{1/2} \sim 200$ higher than that obtained in heavy atom cooling, it would be a substantial progress to high precision experiments.

Table II Experimental and theoretical values of rare decay modes of Singlet (λ_S) and Triplet (λ_T) positronium

Decay	Experiment	Theory
$\frac{\lambda_T(3_\gamma)}{\lambda_S(2_\gamma)}$	$\frac{1}{1133(20)}$	$\frac{\alpha(\pi^2-9)\cdot 4}{9\pi} = \frac{1}{1115}$
$\frac{\lambda_S(4_\gamma)}{\lambda_S(2_\gamma)}$		$0.274(\alpha/\pi)^2 = 1.5 \cdot 10^{-6}$
$\frac{\lambda_T(5_\gamma)}{\lambda_T(3_\gamma)}$		$0.177(\alpha/\pi)^2 = 0.96 \cdot 10^{-6}$
$\frac{\lambda_S(3_\gamma)}{\lambda_S(2_\gamma)}$	$<2.8\cdot 10^{-6}$ /14/	0, if C conserved
$\frac{\lambda_T(4_\gamma)}{\lambda_T(3_\gamma)}$	$<8\cdot 10^{-6}$ /15/	0, if C conserved

6. Acknowledgements

The experiments of the Mainz group were performed by R. Ley, K.D. Niebling and A. Schwarz and were supported by the Deutsche Forschungsgemeinschaft. The author enjoyed conversations about the subject with A. Rich, V. Telegdi and C. Westbrook during the symposium on "The Hydrogen Atom" at Pisa, 1988.

References

/1/ D.W. Gidley, A. Rich, E. Sweetman and D. West: Phys.Rev.Lett. 49, 525 (1982)

/2/ C.I. Westbrook, D.W. Gidley, R.S. Conti and A. Rich: Phys.Rev. Lett. 58, 1328 (1987)

/3/ W.M. Ritter, P.O. Egan, V.W. Hughes and K.A. Woodle: Phys.Rev. A30, 1331 (1984)
A.P. Mills: Phys.Rev. A27, 262 (1983)

/4/ S. Chu, A.P. Mills and J.L. Hall: Phys.Rev.Lett. 52, 1689 (1984)
/5/ S. Hatamian, R.S. Conti and A. Rich: Phys.Rev.Lett. 58, 1833 (1987)
/6/ A.P. Mills, S. Berko and K.F. Canter: Phys.Rev.Lett. 34, 1541 (1975)
/7/ M. Weber, K.G. Lynn et al.: In Electron and Atomic Collisions, ed. by D.C. Lorents, W.E. Meyerhof and J.C. Peterson (North-Holland, Amsterdam 1986), p. 227
/8/ M. Begemann, G. Gräff, H. Herminghaus, H. Kalinowski and R. Ley: Nucl.Instr.Meth. 201, 287 (1982)
/9/ R.H. Howell, R.A. Alvarez and M. Stanek: Appl.Phys.Lett. 40, 751 (1982)
/10/ A.P. Mills and L. Pfeiffer: Phys.Rev.Lett. 43, 1961 (1979)
/11/ A.P. Mills: Solid State Comm. 31, 623 (1979)
/12/ K.F. Canter, A.P. Mills and S. Berko: Phys.Rev.Lett. 33, 7 (1975)
/13/ K.D. Niebling: Thesis, Mainz 1988 (unpublished)
/14/ A.P. Mills and S. Berko: Phys.Rev.Lett. 18, 420 (1967)
/15/ K. Marco and A. Rich: Phys.Rev.Lett. 33, 980 (1974)
/16/ J. Dahm, K.D. Niebling, R. Ley and G. Werth: Appl.Phys. A44, 105 (1987)

Some Recent Advances in Muonium

V.W. Hughes

Gibbs Laboratory, Physics Department, Yale University,
New Haven, CT06520, USA

I. INTRODUCTION

Muonium (M) is the bound atomic state of a positive muon (μ^+) and of an electron (e^-) and hence it is a hydrogenic atom. Muonium was discovered[1] in 1960 through observation of its characteristic Larmor precession in a magnetic field. Since then research on the fundamental properties of M has been actively pursued[2,3], as has also the study of muonium collisions in gases, muonium chemistry and muonium in solids.[4]

The principal reason that muonium continues to be important to fundamental physics is that it is the simplest atom composed of two different leptons. The muon retains a central role as one of the elementary particles in the modern standard theory, but we still have no understanding as to "why the muon weighs" and in all respects behaves simply as a heavy electron. Muonium is an ideal system for determining the properties of the muon, for testing modern quantum electrodynamics, and for searching for effects of weak, strong, or unknown interactions in the electron-muon bound state. Basically muonium is a much simpler atom than hydrogen because the proton is a hadron and, unlike a lepton, has a structure that is determined by the strong interactions. Thus muonium provides a cleaner system to study than hydrogen for testing QED and the electroweak interaction.

II. GROUND STATE HYPERFINE STRUCTURE AND ZEEMAN EFFECT

After the discovery of muonium, measurements of its energy levels could be undertaken by microwave magnetic resonance spectroscopy utilizing the facts that the incident μ^+ are polarized so that polarized muonium is formed and that the decay positrons have an asymmetric angular distribution with respect to the muon spin direction.[5] The Breit-Rabi energy level diagram for the ground state of muonium is shown in Fig. 1. With the aim of determining the hyperfine structure interval $\Delta\nu$ and the muon magnetic moment μ_μ, transitions at both weak and strong magnetic fields have been measured as indicated. Starting in 1962 a series of increasingly accurate measurements were undertaken[2,3] by both the Yale-Heidelberg and Chicago groups. The latest experiment at the Los Alamos Meson Physics Facility (LAMPF) was a strong field measurement.[6] A schematic

The Hydrogen Atom Editors: G.F. Bassani · M. Inguscio · T.W. Hänsch

diagram of the experimental arrangement is shown in Fig. 2 and a photograph of the precision solenoid electromagnet in Fig. 3. Typical resonance curves are shown in Fig. 4.

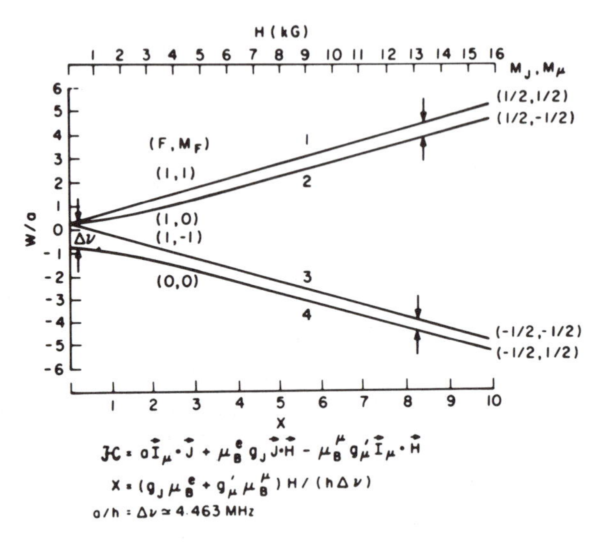

$$\mathcal{H} = a\vec{I}_\mu \cdot \vec{J} + \mu_B^e g_J \vec{J}\cdot\vec{H} - \mu_B^\mu g'_\mu \vec{I}_\mu \cdot \vec{H}$$

$$X = (g_J \mu_B^e + g'_\mu \mu_B^\mu) H/(h\Delta\nu)$$

$a/h = \Delta\nu \simeq 4.463$ MHz

Fig. 1: Breit-Rabi energy level diagram for the ground state of muonium.

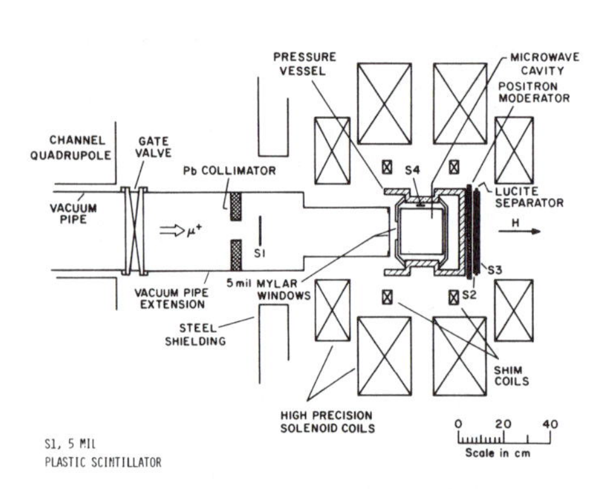

Fig. 2: Experiment at LAMPF in which the latest precision measurement of the hyperfine structure interval $\Delta\nu$ in muonium was made.

Fig. 3: Photograph of the large solenoid electromagnet. The outer diameter of the iron shield is 80 in.

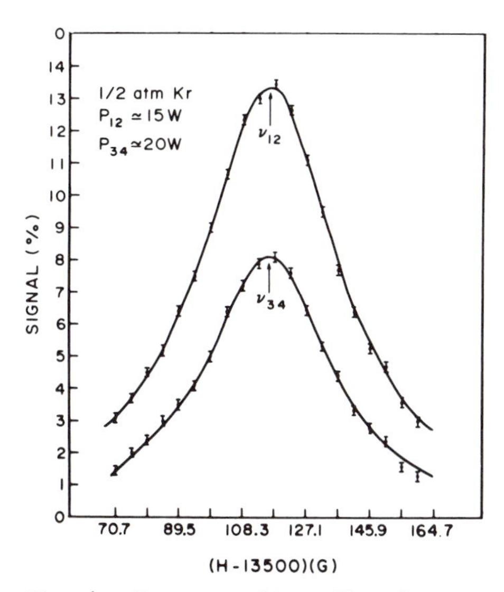

Fig. 4: Resonance lines fitted to the data from the experiment shown in Fig. 2. P_{12} and P_{34} are input powers to the microwave cavity.

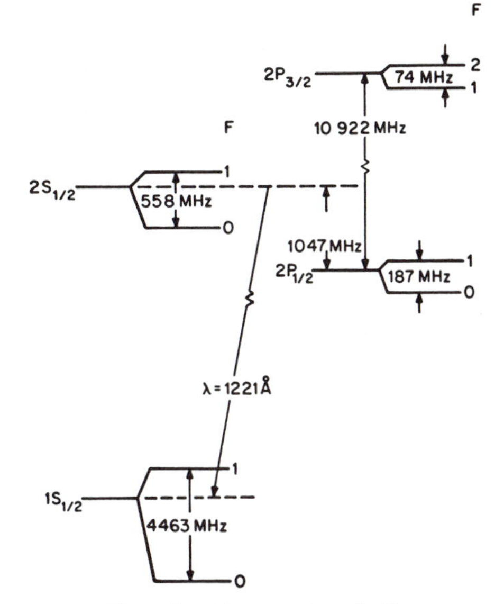

Fig. 5: Energy level diagram of the n=1 and n=2 states of muonium.

The experimental results for $\Delta\nu$ and μ_μ and the current theoretical value for $\Delta\nu$ are given in Table I. The radiative and recoil corrections to the leading Fermi value for $\Delta\nu$ have been computed to high order.[7] The principal error of 1.5 kHz or 0.34 ppm is due mostly to uncertainty in the value of the constant μ_μ/μ_p appearing in the Fermi term E_F. (The condensed matter value[8] of α is accurate to 0.09 ppm.) The error of 0.2 kHz is an estimate of numerical uncertainties in the calculated terms in ε_{QED}, and the 1.0 kHz error is an estimate of the size of uncalculated higher order radiative and recoil terms. A small hadronic vacuum polarization contribution of 0.22(4) kHz is included in the factor with 18.18 in the last term of δ'_μ. The experimental value for $\Delta\nu$ is known to 36 ppb, and the experimental and theoretical values agree well within the theoretical error of 0.4 ppm. This agreement constitutes one of the important, sensitive tests of quantum electrodynamics and of the behaviour of the muon as a heavy electron.

Improvement by a factor of about 10 in the sensitivity of the comparison of theory and experiment for $\Delta\nu$ appears possible at this time.[9] With the use of a chopped muon beam now available at Los Alamos, line narrowing techniques can be employed.[10] Use of a higher magnetic field value will improve the accuracy in determining μ_μ/μ_p. Finally, the intensity and quality of the muon beam has been improved since the last measurement. Considering all these factors, a measurement of $\Delta\nu$ to about 5 ppb and of μ_μ/μ_p to 30 ppb appears possible.

Theoretical computation to about 0.1 kHz or 20 ppb appears realistic. We might remark that weak interaction contributions to $\Delta\nu$ are predicted[11] at the level of 16 ppb.

Table I. Theoretical value of muonium $\Delta\nu$ and comparison with experiment.

$$\Delta\nu_{th} = \left[\tfrac{16}{3}\alpha^2 c R_\infty(\mu_\mu/\mu_B^e)\right]\left[m_\mu/(m_e+m_\mu)\right]^3\left[1+\epsilon_{\mathrm{QED}}\right]$$

$$\Delta\nu_{th} = E_F(1+\epsilon_{\mathrm{QED}})$$

$$\epsilon_{\mathrm{QED}} = \tfrac{3}{2}\alpha^2 + a_e + \epsilon_1 + \epsilon_2 + \epsilon_3 - \delta'_\mu$$

$$a_e = (g_e-2)/2;\ \epsilon_1 = \alpha^2(\ln 2 - 5/2)$$

$$\epsilon_2 = -\tfrac{8\alpha^3}{3\pi}\ln\alpha\left(\ln\alpha - \ln 4 + \tfrac{281}{480}\right);\ \epsilon_3 = \tfrac{\alpha^3}{\pi}(15.38 \pm 0.29)$$

$$\delta'_\mu = \Big\{\tfrac{3\alpha}{\pi}\tfrac{m_R}{m_\mu - m_e}\ln m_\mu/m_e + \alpha^2\tfrac{m_R}{m_\mu+m_e}\left[2\ln\alpha + 8\ln 2 - 3\tfrac{11}{18}\right]$$
$$+(\alpha/\pi)^2 m_e/m_\mu \times \left[2\ln^2(m_\mu/m_e) - \tfrac{13}{12}\ln(m_\mu/m_e) - 18.18(63)\right]\Big\}\tfrac{1}{1+a_\mu}$$

where $m_R = m_e m_\mu/(m_e + m_\mu)$

$R_\infty = 1.097\ 373\ 152\ 1\ (11) \times 10^5 \mathrm{cm}^{-1}$ (0.001 ppm)

$c = 2.997\ 924\ 580 \times 10^{10}$ cm/sec

$\alpha^{-1} = 137.035\ 981\ (12)$ (0.09 ppm)

$a_e = 1\ 159\ 652\ 193\ (4) \times 10^{-12}$ (3.4 ppb)

$\mu_\mu/\mu_B^e = (\mu_\mu/\mu_p)(\mu_p/\mu_B^e);\ \mu_p/\mu_B^e = 1.521\ 032\ 209\ (16)$ (0.01 ppm)

$m_\mu/m_e = 206.768\ 259\ (62)$ (0.3 ppm)

$\mu_\mu/\mu_p = 3.183\ 345\ 47\ (95)$ (0.3 ppm)

$[\mu_\mu/\mu_p = 3.183\,346\,1\ (11)$ (0.36 ppm)] from muonium Zeeman effect

$a_\mu = (g_\mu - 2)/2 = 0.001\ 165\ 923\ 0(84)$ (7.2 ppm)

$E_F = 4\ 459\ 033.4\ (1.5)$ kHz

$\Delta\nu_{th} = 4\ 463\ 303.6(1.5)(0.2)(1.0)$ kHz (0.4 ppm)

$\Delta\nu_{exp} = 4\ 463\ 302.88(0.16)$ kHz (0.036 ppm)

$\Delta\nu_{th} - \Delta\nu_{exp} = (+0.7 \pm 1.8)$ kHz

Determination of α : α^{-1} = 137.035 988 (20) (0.15 ppm), μ^+e^-

137.035 994 (5) (0.038 ppm), $g_e - 2$

137.035 981 (12) (0.09 ppm), condensed matter

III. LAMB SHIFT IN MUONIUM

With the development of a muonium beam in vacuum,[12] measurement of the Lamb shift in muonium became possible. The energy level diagram of the n=1 and n=2 states of muonium are shown in Fig. 5. Two measurements[13,14] have determined the Lamb shift $\mathcal{S}_L$ by observing the $2^2S_{1/2}$ to $2^2P_{1/2}$ transition in a radiofrequency spectroscopy experiment. Table II gives the current theoretical value of $\mathcal{S}_L$ and the two experimental values. The theoretical value for the muonium Lamb shift differs from that in hydrogen by the absence of a proton structure term and by the relatively greater importance of recoil terms. The experimental values agree with the theoretical value within the limited experimental accuracy of about 1%.

Recently, at LAMPF, the fine structure transition $2^2S_{1/2}$ to $2^3P_{3/2}$ in muonium has been studied.[15] Figure 6 indicates the experimental method. Muonium is formed in the metastable $2^2S_{1/2}$ state at an Al foil just downstream of a low gas pressure MWPC. After collimation, the M(2S) beam enters a microwave cavity operating at a frequency of about 10 GHz which drives the transition $2^2S_{1/2} \rightarrow 2^2P_{3/2}$. From the $2^2P_{3/2}$ state M decays to the ground 1S state with a mean life of 1.6 ns, and the Lyman-α 1221 A photon is detected by a UV photomultiplier tube, while the resulting M(1S) atom travels to a microchannel plate where it is detected. The signal due to the microwave field, defined as a delayed triple coincidence between a μ^+ count in the MWPC detector, a Lyman-α photon, and a microchannel plate count, is shown in Fig. 7 as a function of the microwave frequency, together with the predicted line centers. The analysis is in progress so no fitted line shape or result can yet be quoted. Further development of

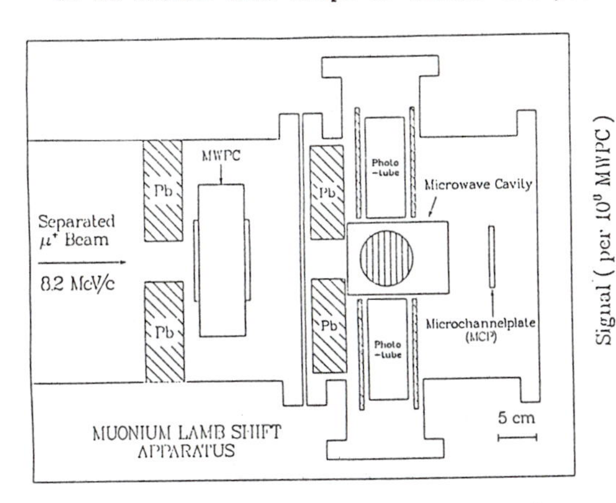

Fig 6: Diagram of the apparatus used in observation of the muonium fine structure transition $2^2S_{1/2} \rightarrow 2^2P_{3/2}$.

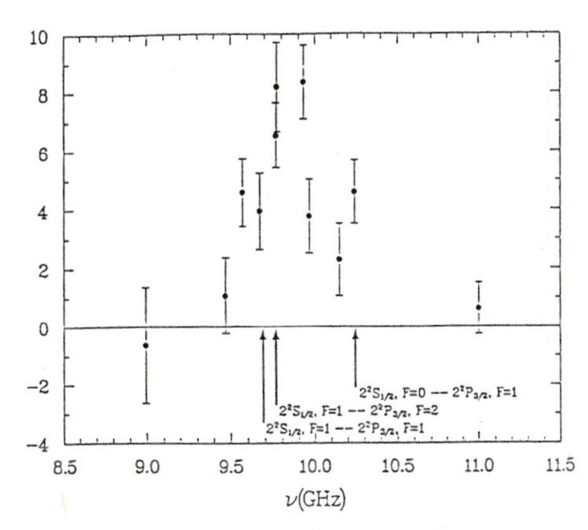

Fig. 7: Muonium $2S^2_{1/2} - 2P^2_{2/3}$ transition showing raw data. (No correction is made for variation of microwave power with frequency).

Table II. Muonium Lamb Shift Theoretical Value

CORRECTION	ORDER (mc^2)	VALUE (MHz)
Self energy	$\alpha(Z\alpha)^4[\ln(Z\alpha)^{-2},1,Z\alpha,...]$	1 085.812
Vacuum polarization	$\alpha(Z\alpha)^4[1,Z\alpha,...]$	-26.897
Fourth order	$\alpha^2(Z\alpha)^4$	0.101
Reduced mass	$\alpha(Z\alpha)^4 m/M_\mu[\ln(Z\alpha)^{-2},1]$	-14.626
Relativistic recoil	$(Z\alpha)^5 m/M_\mu[\ln(Z\alpha)^{-2},1]$	+3.188
Total		1 047.578(300)MHz

The estimated uncertainty in the theoretical value is due to uncalculated terms of higher order in m/M_μ and in $\alpha m/M_\mu$, that is the terms (m/M_μ)(reduced mass term) and α (reduced mass term). An estimate of the size of these terms is 0.3 MHz.

Experimental Value

$\mathcal{S}_{expt}$ = 1054 ± 22 MHz LAMPF

1070^{+12}_{-15} MHz TRIUMPF

this experiment involving the suppression of Doppler effect by the choice of cavity mode relative to the M velocity direction, the use of M(2S) production from the foil at about 30° to the incident μ^+ direction to reduce the background associated with energetic μ^+, and photon detectors with improved efficiency should make possible a precision of about 0.1% in the determination of $\mathcal{S}_L$. This experiment is severely limited by signal rate which at present amounts to about 10 counts/hr, and additional future progress would require higher M(2S) beam intensities.

IV. GROUND STATE $\Delta\nu$ AND THE 1S-2S TRANSITION WITH THERMAL MUONIUM

At SIN, the Heidelberg-Yale group established that thermal muonium emitted from SiO_2 powder is polarized.[16] Figure 8 shows the experimental arrangement and Figure 9 shows the characteristic muonium precession, not only from the region of the powder but also from the free-space region beyond the powder occupied by thermal muonium.

Using a microwave cavity in the free magnetic field region downstream of the SiO_2 target, the hfs transition $\Delta\nu$ = 4 463 MHz in n=1 state muonium was observed.[17] Because of the Doppler effect it does not appear that this type of observation can lead to a high precision determination of $\Delta\nu$, competitive with the existing value.

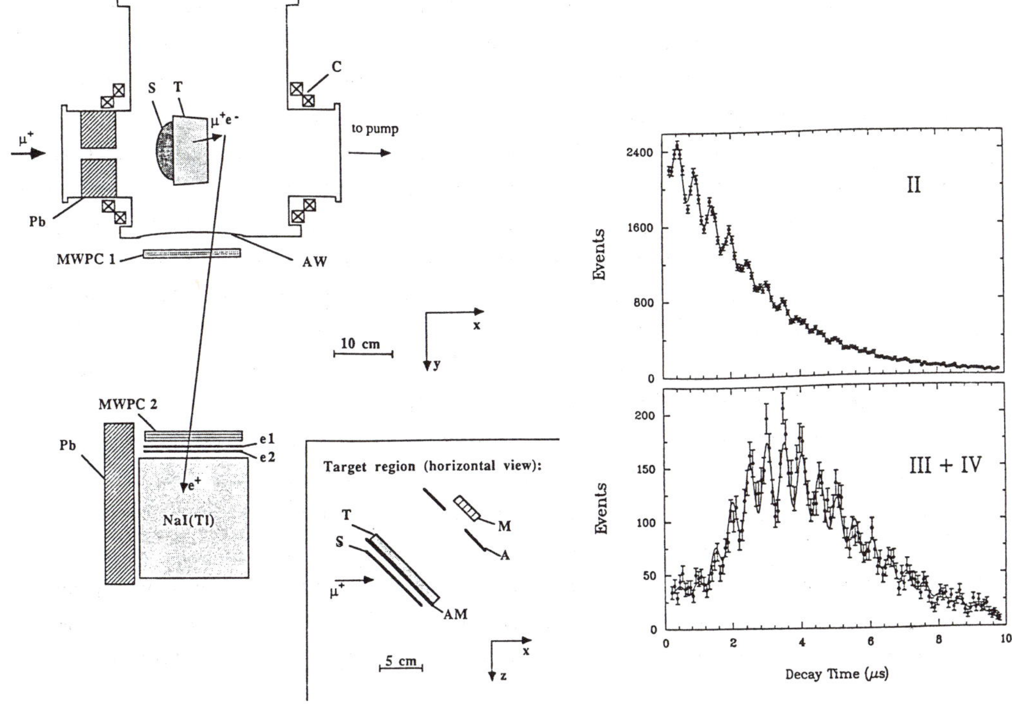

Fig. 8: Schematic of the experimental apparatus showing the μ^+ scintillator (S), SiO_2 target (T), magnetic field coils (C), Al window (AW), and e^+ scintillators (e_1 and e_2). The aperture electrode, the microchannel plate, and one pair of field coils have been omitted in order to simplify the picture. The inset shows a horizontal view of the target region with the μ^+ scintillator (S), aluminized mylar foil (AM), SiO_2 target (T), aperture electrode (A), and microchannel plate (M).

Fig. 9: Observed numbers of positrons as a function of time with an applied magnetic field of 1.4G. The upper histogram corresponds to decays from Region II, which includes the target and the immediate downstream region, and the lower histogram to decays from Regions III and IV, which is the free-space region downstream of the target.

Using the high intensity pulsed muon beam from the Rutherford Laboratory 1 GeV proton synchrotron, the Heidelberg group[18] has initiated a new experiment to measure the 1S-2S transition in muonium by two-photon laser spectroscopy with the objective of determining this interval to about 1 part in 10^9.

V. MUONIUM → ANTIMUONIUM CONVERSION

The muon and the electron may be considered to belong to two different generations of leptons, which thus far appear to remain separate because of the independent conservation laws of muon number and of electron number. Any connection between the muon and the electron, such as a process which would violate muon number conservation, would be an important clue to the relationship between the two generations.[19] Speculative modern theories which seek a more unified theory of particles and their interactions predict muon number violating processes.[20] As yet no such rare decay process has been observed,[21] and with our present knowledge theory has little useful predictive power.

The conversion of muonium (μ^+e^-) to its antiatom antimuonium (μ^-e^+) would be an example of a muon number violating process,[2] and like neutrinoless double beta decay would involve $\Delta L_e=2$. The M-$\bar{M}$ system also bears some relation to the K°-$\overline{K^\circ}$ system, since the neutral atoms M and $\bar{M}$ are degenerate in the absence of an interaction which couples them. In Table III a four-Fermion Hamiltonian term coupling M and $\bar{M}$ is postulated, and the probability that M formed at time t=0 will decay from the $\bar{M}$ mode is given. Present experimental limits[22,23] for the coupling constant G are indicated and are larger than the Fermi constant G_F.

Two new experiments are in progress to search for the $M\to\bar{M}$ transition using thermal muonium from SiO_2. At TRIUMF[24] the signal in the experiment would be an induced radioactivity due to a μ^- nucleus capture. At LAMPF[23,25], the signal would be a coincident high energy e^-($\gtrsim$30 MeV) and a low energy e^+(~10 eV). Figure 10 shows a schematic diagram of the apparatus and Figure 11 is a photograph of the apparatus. A successful checkout of this experiment took place in the summer of 1988 and with the data-taking planned a sensitivity $G_{M\bar{M}} \sim 10^{-2}\ G_F$ might be achieved.

Table III. Muonium-antimuonium conversion including present experimental limits.

$\mu^+e^- \to \mu^-e^+$

Muon number, L_μ, +1(-1) for $\mu^-(\mu^+),\nu_\mu(\bar{\nu}_\mu)$, 0 for other particles.

Violates additive conservation law for muon number, ΣL_μ = constant.

Allowed by multiplicative conservation law, $(-1)^{\Sigma L_\mu}$ = constant.

$$\mathcal{H}_{M\to\bar{M}} = \frac{G_{M\bar{M}}}{\sqrt{2}}\ \bar{\psi}_\mu\ \gamma_\lambda(1+\gamma_5)\ \psi_e\bar{\psi}_\mu\gamma^\lambda(1+\gamma_5)\ \psi_e + \text{H.C.}$$

$$P(\bar{M}) = \int_0^\infty \gamma\ {}^{-\gamma t}\left|\langle\bar{M}|\psi(t)\rangle\right|^2 dt = 2.5\times10^{-5}\left(\frac{G_{M\bar{M}}}{G_F}\right)^2$$

Probability of decay of M from $\bar{M}$ state reduced by collisions in a gas by factor of (number of collisions during muon lifetime)$^{-1}$.

$G_{M\bar{M}} < 5800\ G_F$ ($M\to\bar{M}$ in gas; Nevis: 1968)

$G_{M\bar{M}} < 600\ G_F$ ($e^-e^-\to\mu^-\mu^-$; HEPL; 1969)

$G_{M\bar{M}} < 20\ G_F$ ($M\to\bar{M}$ in powers; TRIUMF; 1982,1986)

$G_{M\bar{M}} < 7.5\ G_F$ (M with keV kinetic energies in vacuum; LAMPF; 1987).

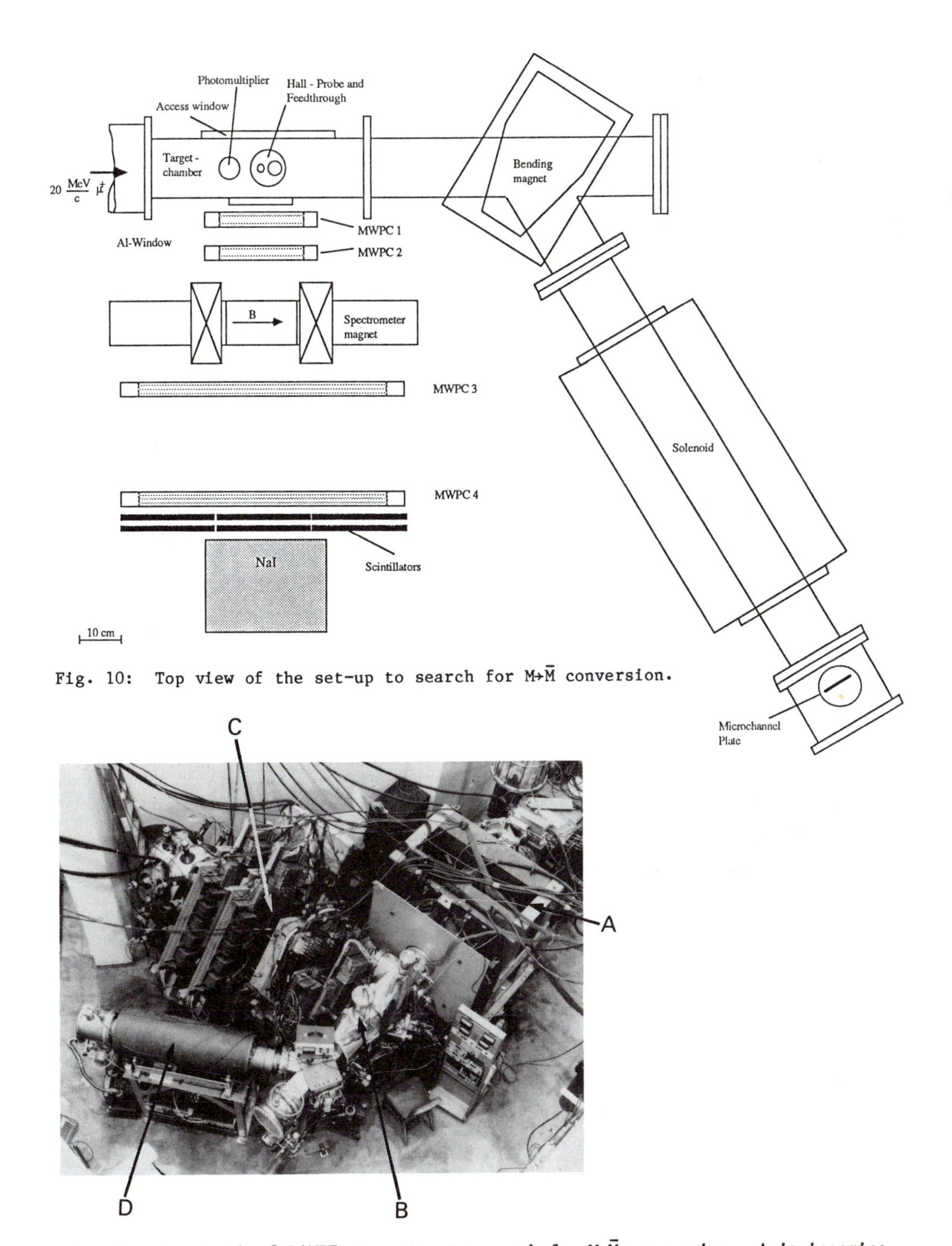

Fig. 10: Top view of the set-up to search for $M \to \bar{M}$ conversion.

Fig. 11: Photograph of LAMPF apparatus to search for $M \to \bar{M}$ conversion. A is incoming μ^+ beam line; B is region where thermal muonium is formed and decays; C is high energy $e^{\pm}$ detector with magnet and MWPC; D is low energy $e^{\pm}$ detector with low energy keV spectrometer and microchannel plate.

VI. SUMMARY

As a concluding remark we emphasize that research on the fundamental properties of muonium is flourishing with many important recent advances and with bright prospects for the future. As this exciting Symposium has shown, fascinating and fundamental research on many aspects of the hydrogen atom, which we have known for some 200 yrs., is still most active, so it is not surprising that the same is true of muonium, a most fundamental isotope of hydrogen discovered less than 30 years ago.

Research supported in part by DOE under contract DE-AC02-76ER03075.

REFERENCES

1. V.W. Hughes, et al., Phys. Rev. Lett. 5, 63 (1960); Phys. Rev. A1, 595 (1970).
2. V.W. Hughes and T. Kinoshita, in Muon Physics I, ed. by V.W. Hughes and C.S. Wu, (Academic Press, New York, 1977), p. 11.
3. V.W. Hughes and G. zu Putlitz, Comments in Nucl. Part. Phys. 12, 259 (1984); V.W. Hughes, Ann. Phys. Fr. 10, 955 (1985).
4. J.H. Brewer et al., in Muon Physics, ed. by V.W. Hughes and C.S. Wu,(Academic Press, New York, 1975), Vol. III p. 3; J.H. Brewer and K.M. Crowe, Ann. Rev. Nucl. Part. Sci. 28, 239 (1978).
5. V.W. Hughes, Ann. Rev. Nucl. Sci. 16, 445 (1966).
6. F.G. Mariam, et al., Phys. Rev. Lett. 49, 993 (1982).
7. J.R. Sapirstein, et al., Phys. Rev. Lett. 51, 982 (1983); J.R. Sapirstein, Phys. Rev. Lett. 51, 985 (1983); T. Kinoshita and J. Sapirstein, in Atomic Physics 9, ed. by R.S. Van Dyck, Jr. and E.N. Fortson, (World Scientific, Singapore, 1985) p. 38; E.A. Terray and D.R. Yennie, Phys. Rev. Lett. 48, 1803 (1982); J. Sapirstein, in Intersections Between Particle and Nuclear Physics, ed. by D.F. Geesaman, (AIP, N.Y., 1986) p. 567.
8. B.N. Taylor, J. Res. Natl. Bur. Stand. 90, 91 (1985).
9. "Ultrahigh Precision Measurements on Muonium Ground State: Hyperfine Structure and Muon Magnetic Moment" LAMPF Proposal, November 1986, V.W. Hughes, G zu Putlitz, P.A. Souder, Spokesmen.
10. V.W. Hughes, in Quantum Electronics, ed. C.H. Townes (Columbia Univ. Press. N.Y., 1960), p. 582; D.E. Casperson, et al., Phys. Lett. 59B, 397 (1975); D. Favart et al., Phys. Rev. Lett. 27, 1336 (1971); Phys. Rev. A8, 1195 (1973).
11. M.A.B. Bég and G. Feinberg, Phys. Rev. Lett. 33, 606 (1974); 35, 130 (1975).
12. P.R. Bolton, et al., Phys. Rev. Lett. 20, 1441 (1981).
13. C.J. Oram, et al., Phys Rev. Lett. 52, 910 (1984); in Atomic Physics 9, ed. by R.S. Van Dyck, Jr. and E.N. Fortson, (World Scientific, Singapore, 1985) p. 75.

14. A. Badertscher, et al., Phys. Rev. Lett. 52, 914 (1984); A. Badertscher, et al., in Atomic Physics 9, ed. by R.S. Van Dyck, Jr. and E.N. Fortson, (World Scientific, Singapore, 1985) p. 83.
15. Yale, Heidelberg, William and Mary Collaboration; Y. Ahn, K.P. Arnold, F. Chmely, M. Eckhause, M. Gladisch, P. Guss, V.W. Hughes, J. Kane, S. Kettell, Y. Kuang, B. Matthias, H.J. Mundinger, B. Ni, H. Orth, G. zu Putlitz, J. Reidy, R. Schaefer, and K. Woodle.
16. K.A. Woodle, et. al., Z. Phys. D9, 59 (1988).
17. Private Communication from K. Jungmann and G. zu Putlitz, University of Heidelberg.
18. Rutherford Proposal, "High Resolution Two Photon Spectroscopy of the Muonium 1S-2S Transition", K. Jungmann, University of Heidelberg, Spokesman.
19. G. Feinberg and L.M. Lederman, Ann. Rev. Nucl. Sci. 13, 431 (1963); T.D. Lee and C.S. Wu, Ann. Rev. Nucl. Sci. 15, 381 (1965); E.D. Commins and P.H. Bucksbaum, Weak Interactions of Leptons and Quarks, (Cambridge Univ. Press, Cambridge, 1983); L.B. Okun, Leptons and Quarks, (North Holland, Amsterdam, 1982); Muon Physics II, ed. by V.W. Hughes and C.S. Wu, (Academic Press, New York, 1975).
20. H. Primakoff and S.P. Rosen, Ann. Rev. Nucl. Part. Sci. 31, 145 (1981); A. Halprin, Phys. Rev. Lett. 48, 1313 (1982); R.N. Mohapatra, in Quarks, Leptons, and Beyond, ed. H. Fritzsch, (Plenum Press, N.Y., 1985), p. 219.
21. H.K. Walter, in Particles and Nuclei, ed. by B. Povh and G. zu Putlitz, (North-Holland, Amsterdam, 1985), p. 409c.
22. B. Ni, et al., Phys. Rev. Lett. 59, 2716 (1987).
23. "Search for Muonium to Antimuonium Conversion: Overview and Recent Experimental Results," H.R. Schaefer, to be published in Intersections Between Particle and Nuclear Physics, Rockport Maine, (1988)
24. "Search for Mixing of (μ^+e^-) with Fermi Coupling Strength," T.M. Huber, G.A. Beer, T. Bowen, C.A. Fry, Z. Gelbart, P.G. Halverson, A.C. Janissen, K.R. Kendall, A.R. Kunsekman, G.M. Marshall, G.R. Mason, A. Olin, and J.B. Warren, to be published in Intersections Between Particle and Nuclear Physics, Rockport, Maine, (1988).
25. Yale, Heidelberg, William and Mary Collaboration, Y. Ahn, F. Chmely, S. Dhawan, M. Eckhause, V.W. Hughes, K. Jungmann, J. Kane, S. Kettell, Y. Kuang, B. Matthias, H. Mundinger, B. Ni, G. zu Putlitz, R. Schaefer, and K. Woodle.

Transitions in Muonic Helium Ions Induced by Laser Radiation

E. Zavattini

CERN, Geneva, Switzerland and
Department of Physics, University of Trieste, Italy

An apparatus to perform a laser-stimulated 3D-3P transition in muonic helium ions (at the Single Burst Extraction Beam of the Brookhaven National Laboratory Alternating Gradient Synchrotron) is presented together with some results.

Some years ago, at the CERN Synchro-cyclotron, a CERN-Pisa Collaboration [1] devised an experimental scheme to perform a laser-induced transition experiment on a muonic helium ion $(\mu^{-4}He)_{2S}$. At that time a rather complex set-up, intimately connected with the operation of the Synchro-cyclotron itself, was installed and made to work, and the energy level differences D_i = 2S - 2P (see Table 1) were measured with an accuracy of 1.5 x 10^{-4}. In Table 1 are presented the experimental results as well as the expected theoretical values as given by the quantum electrodynamics (QED) calculations [2]. As can be seen from the table, the quantities D_i (quantities directly measured) are essentially due to the polarization of the vacuum of the electron-positron field and demonstrate a well-known fact, that the situation for muonic atoms is quite different from that for 'normal atoms' where the vertex corrections to the differences D_i are usually dominant; one concludes therefore that the muonic atoms are the ideal systems to 'check', with significant accuracy, the QED vacuum polarization corrections to the energy levels of bound systems.

However, it is important to note from the table that about 20% of D is, in this case, due to corrections induced by the nuclear electric form factor $\langle r^2 \rangle^{1/2}$: consequently, one can look at the experimental results of Table 1 from two points of view.

The Hydrogen Atom Editors: G.F. Bassani · M. Inguscio · T.W. Hänsch

Table 1

Contributions to the n = 2 energy splittings in the $(\mu^{-4}He)^{+}_{2S}$ system

Contributions	Transition energies (meV)	
	$2P_{3/2}-2S_{1/2}$	$2P_{1/2}-2S_{1/2}$
Dirac contribution with Coulomb potential and point-like charges	145.70	0
Nuclear polarizability	3.1 ± 0.6	3.1 ± 0.6
Finite size	-289.5 ± 2.8	-289.5 ± 2.8
Electronic vacuum polarization		
Uehling term: first iteration	1664.44	1664.17
higher iteration	1.70	1.70
Kallen-Sabry term $(\alpha^2 Z\alpha)$	11.55	11.55
$\alpha(Z\alpha)^n$, n > 3	-0.02	-0.02
$\alpha^2 (Z\alpha)^2$	0.02	0.02
Muon vacuum polarization	0.33	0.33
μ-e vacuum polarization	0.02	0.02
Hadron vacuum polarization	0.15	0.15
Vertex corrections and (g-2)		
$\alpha(Z\alpha)$	-10.52	-10.85
$\alpha(Z\alpha)^n$, n > 1	-0.16	-0.16
$\alpha^2\ Z\alpha$	-0.03	-0.03
Recoil terms		
Breit	0.28	0.28
Two photons	-0.44	-0.44
Weak contribution	0.00002	0.00002
Sum theory	1526.6 ± 2.8	1380.3 ± 2.8
Experiment	1527.5 ± 0.3	1381.3 ± 0.5

Firstly, one can assume the QED computations to be valid (at least to experimental precision) and deduce with great accuracy from the data the helium nuclear electric form factor, as seen from a muon probe. One obtains:

$$\langle r^2 \rangle^{1/2} = \langle r_\mu^2 \rangle^{1/2} = (1.673 \pm 0.003) \text{ fm} .$$

Comparison of this value with the one obtained from electron-helium nucleus elastic scattering experiments [3] can define the limits of a possible muon-hadron anomalous interaction [4,5].

The value for $\langle r^2 \rangle^{1/2}$ given above represents the most accurate form factor of a nucleus.

Secondly, one can assume for the form factor the existing experimental value $\langle r_e^2 \rangle^{1/2} = (1.676 \pm 0.008)$ fm [3] (as seen by an electron probe), assume muon-electron universality, and then give a limit within which the QED vacuum polarization contribution is 'tested' by these measurements. In doing this, one can see that such a QED correction is tested (at the momentum transfer implied by the experiment) to the level of 0.17%; the result of this experiment represents to my knowledge one of the best direct tests, so far performed, of a vacuum polarization correction.

It is interesting to note that in this last case the limitation in the test is not given by the experimental uncertainties on the energy level differences shown in Table 1, but rather by the experimental uncertainty of the assumed helium form factor.

A CERN-Columbia Collaboration [6] [assisted by staff of the Alternating Gradient Synchrotron (AGS) Brookhaven National Laboratory] have looked into the possibility of performing an experiment of the same type (i.e. via laser transition in muonic ions), but with transitions between levels not belonging to S states in order to avoid the consequences of the electromagnetic form-factor uncertainties.

In particular, this Collaboration has installed a set-up to measure the 3D-3P energy level differences in the muonic helium ions and here I wish to describe the apparatus, its performance, and give some of the experimental results obtained with it.

In Fig. 1 are shown the first energy levels of the muonic helium ion together with some of their characteristics [6]; in the same figure, the transition $3D_{5/2}-3P_{3/2}$, which is the first one to be measured, is marked with an arrow.

The principle of the experiment is to see an increase in the K_β intensity due to laser-stimulated transitions when negative muons are stopped in the helium gas present in a multipass optical cavity, where a high-density electromagnetic radiation of the correct wavelength is stored; it is crucial here that the lifetimes of the levels in which one is interested are around 10^{-12} s (we recall that the negative muon cascade time is about 1 ns or less).

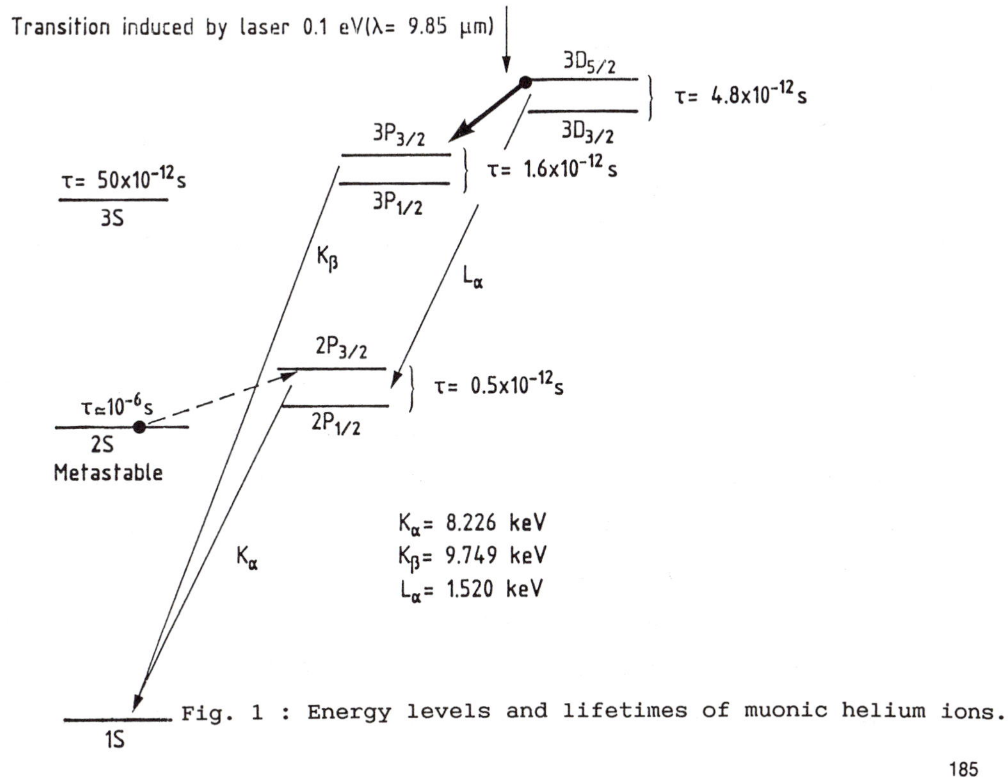

Fig. 1 : Energy levels and lifetimes of muonic helium ions.

From a simple cascade calculation model it can be shown that about 60% of the muons stopped in the target will pass through the D levels: of course, the natural emission of K_β X-rays, during the cascade of the μ^-, represents our main physical background, from which we have to sort out the small increase due to the laser-stimulated emission.

In Table 2 are presented the results of a calculation [6] for the different energy differences 3D-3P; one sees that the wavelength of the radiation to match the resonance conditions is around that of a CO_2 laser and that the width Γ of the lines is 0.437 μm (as deduced from Fig. 1).

It can easily be shown that if N_d is the number of negative muons present in a D level of a helium muonic ion $(\mu^{-4}\text{He})^+_D$ and E/V is the energy density of the radiation at the site of the stopping muon, then the fraction of transitions is given by ($\nu = \omega/2\pi$ is the radiation frequency):

$$\varepsilon = \frac{N_{stim}}{N} = \frac{\Gamma/2}{\hbar^2(\omega-\omega_0)^2 + \Gamma^2/4}\,\frac{1}{\gamma_{3D}}\,|F_{3D-3P}|^2 , \qquad (1)$$

where $\Gamma = (\gamma_{3D} + \gamma_{3P})$ is the width of the transition, γ_{3D} and γ_{3P} are the radiative decay rate of the D and P levels (see Fig. 1; $\gamma = h/\tau$), and $|F_{3D-3P}|^2 = 2.43\,(ea_\mu)^2(E/V)$ is the square of the electric dipole transition matrix element.

Taking V = 0.4 l and E = 4 J in Eq. (1), at the resonance frequency we get : ε = 0.6%. (One of our problems has been to make the volume V as small as possible without losing μ^--stops in the illuminated region and also to limit the background events due to μ^--stopping in regions seen by the X-ray detectors but not illuminated by the radiation of the cavity.)

Figure 2 is a sketch of the target set-up: the effective target is represented by a multipass optical cavity [7] (the mirror surfaces are polished copper), where a CO_2 laser burst is stored (for about

Table 2

Contributions to the 3D-3P energy level differences

Transition	Vacuum polarization α (Uehling-Serber) (meV)	α^2 (Kallen-Sabry) (meV)	Fine structure (Dirac) (meV)	Hyperfine structure (meV)	Total (meV)	λ (μm)
$3D_{3/2}-3P_{1/2}$	110.458	0.905	43.164	0	154.528	8.0235
$3D_{5/2}-3P_{3/2}$	110.458	0.905	14.388	0	125.751	9.8595
$3D_{3/2}-3P_{3/2}$	110.458	0.905	0	0	111.363	11.1334

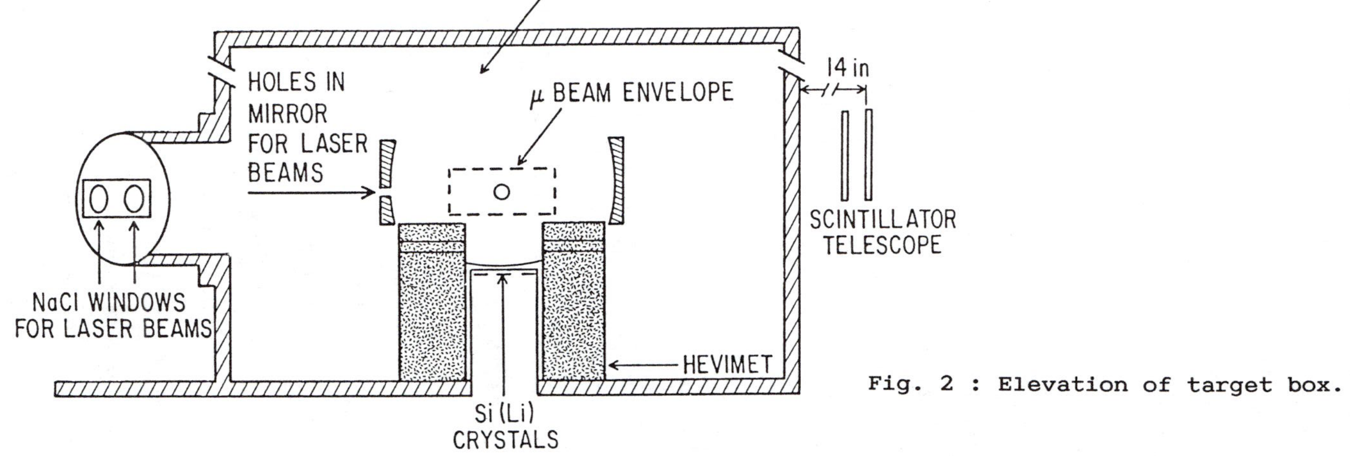

Fig. 2 : Elevation of target box.

100 ns); the cavity is assembled in 3 atm of helium, and in the small-cavity region a burst of negative muons is stopped during the presence of the radiation. At the bottom the K X-ray detecting system is visible; it is composed of 3 Si-Li detectors (2 cm^2 each) situated at about 3.5 in. below the axis of the target cavity.

The cavity assembly is mounted inside a large box filled with 3 atm of helium gas (at room temperature) and the laser beam, split into two parts of about 2 J each (because of the breakdown threshold of the helium gas), enters the box through the NaI-salt windows shown on the left side of the figure. Each beam stays in the cavity for 32 reflections and then exits: the outcoming radiation beams are then used for monitoring and controlling purposes.

To obtain the desired total of 4 J at the correct wavelength, an isotopic $^{13}C^{18}O_2$ laser had to be employed. The laser scheme used is shown in Fig. 3. The oscillator consists of a Transverse Excitation Atmospheric (TEA) $^{13}C^{18}O_2$ laser with a diffracting grating to select a particular frequency (Fig. 4 shows the P branch lines of a $^{13}C^{18}O_2$ laser compared with the expected muonic transition line) in order to explore the entire region of the $3D_{5/2}$-$3P_{3/2}$ transition: one should be able to locate the centre of this resonance to better than 10 Å. Afterwards, the beam from the oscillator passes twice through another $^{13}C^{18}O_2$ TEA laser that functions as an amplifier (laser mixture: 81% He, 12% $^{13}C^{18}O_2$, 8% N).

The 28 GeV/c AGS proton accelerator has been adapted to give to our area (D line) one of the 12 internal proton bursts [Single Burst Extraction (SBE) operation]; this isolated burst contained about 10^{12} protons and it was about 20 ns wide. These protons directed onto our pion-production target were used to generate the burst of low-energy negative muons (about 50 ns wide) to be sent into our helium gas box. Naturally the SBE operation had to be synchronized with our $^{13}C^{18}O_2$ laser firing so that the negative muons and the laser

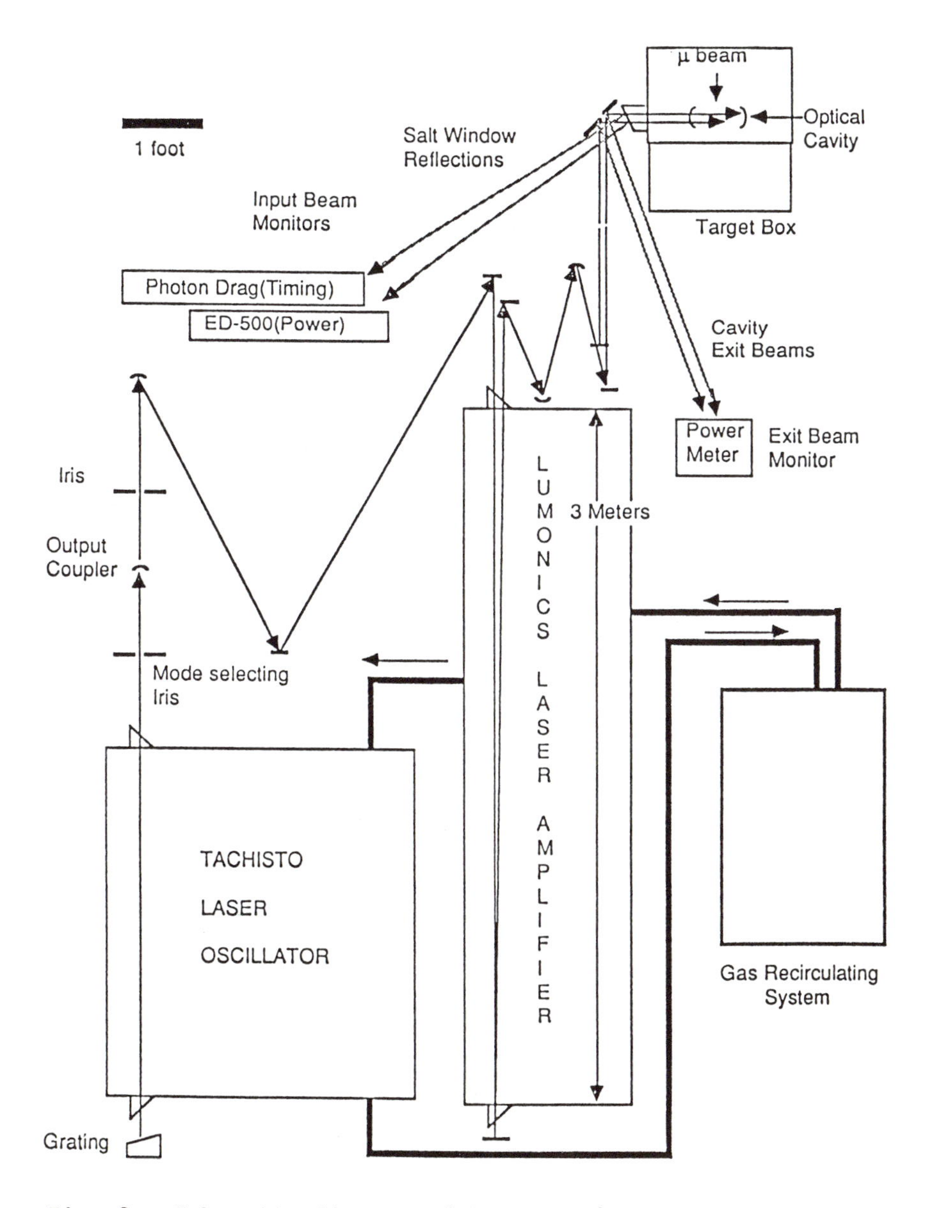

Fig. 3 : Schematic diagram of laser system

radiation, would be present together in the multipass cavity target; Fig. 5 shows the scheme of the different timings.

Table 3 gives the relevant figures for the muon beam (AGS operation: normally 1 burst each 1.4 seconds).

The number of μ^--stops in the table refers to the volume $\bar{V}$ 'seen' by the X-ray detectors: this volume is bigger than the volume

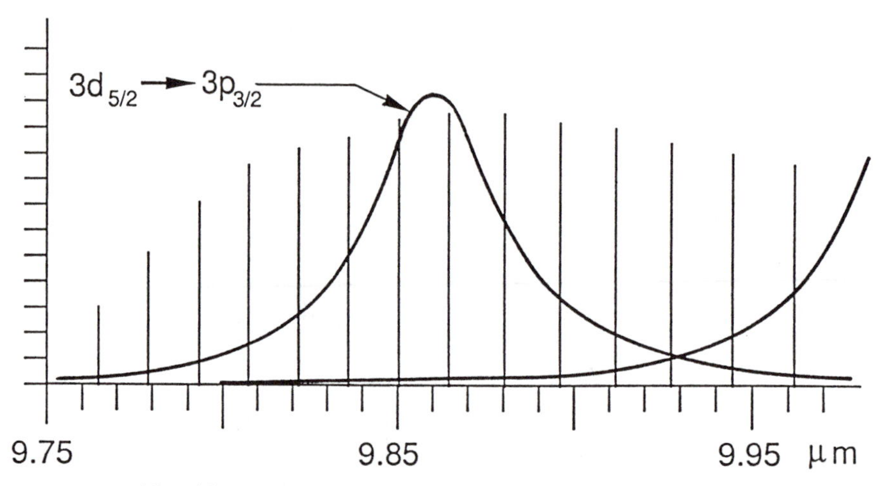

Fig. 4 : $^{13}C\ ^{18}O_2$ laser transitions: 001-020 [II] P branch.

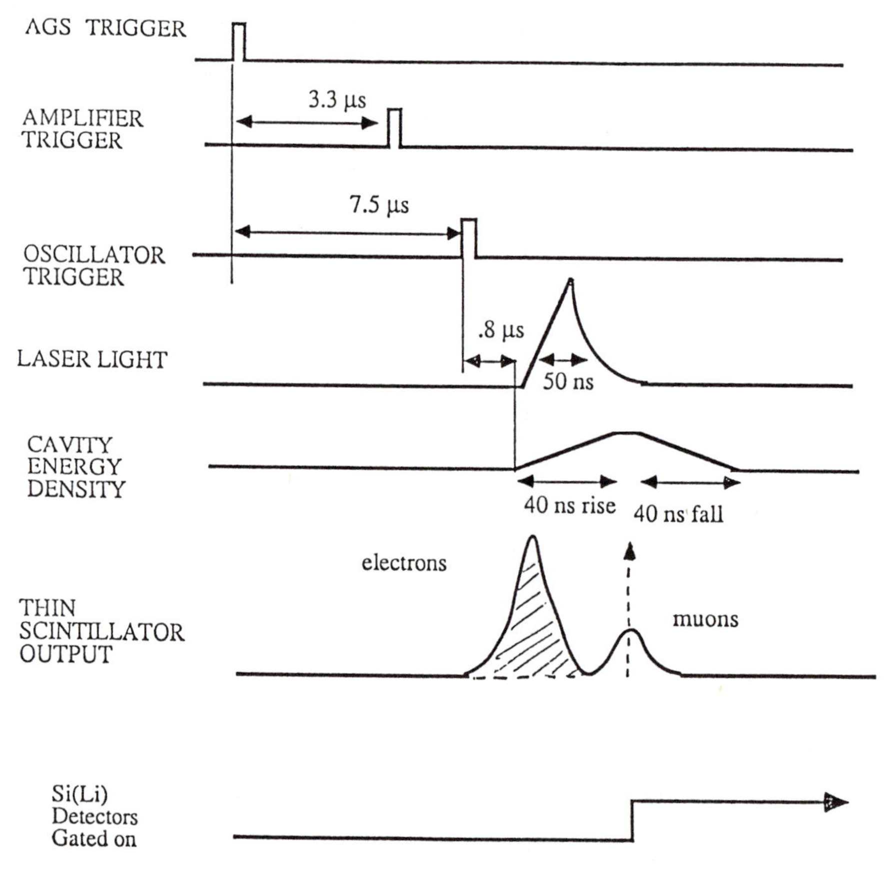

Fig. 5 : Timing of laser system.

Table 3
Muon beam parameters per burst (10^{12} protons on target)

Momentum (MeV/c)	μ^--stops in cavity	μ^--total over 5 cm x 15 cm	Electrons/muons	Si-Li counts
25	300	1000	8	0.65

V illuminated by the reflecting laser beam. Let us put $f = V/\bar{V}$ (f is between 0.7 and 0.5).

Since the pulsed muon beam is essentially instantaneous compared to the integration time of the X-ray detectors, only one count can be accepted by each detector during a single AGS burst. The selection of an optimum integration time is a compromise between the improved resolution of a long time and the possible reduction in background interference using a shorter integration time. The figure in the last column of Table 3 (which represents the sum of all pulses in one Si-Li detector) is the result of such a compromise: in that figure (obviously obtained without laser firing) there is the contribution of the accidental background (for example, neutrons from the pion-production target) as well as the contribution of the physical background (given, for example, by the muon's decay electrons or cascade X-rays from muons stopped in the helium within the cavity volume V and directly seen by the Si-Li detectors). Clearly, here we have the typical difficulty of working with a pulsed beam; to arrive at the given figure of 0.65 we had to spend a great deal of time in trying various shielding arrangements.

Many runs have been done without the firing of the laser, in order to establish the yield of K_β X-rays (our natural physical background) emitted in the prompt de-excitation processes of the helium muonic system; in this region of pressure (1-3 atm) no measurements exist. For us it is essential to know this rate in order to compute the number of events we have to take with the laser on, so as to resolve the peak of the stimulated K_β X-rays over the background.

In addition, this region of pressure seems to be rather important for an understanding of possible mechanisms [8, 9] through which the 2S muonic helium ion, at higher pressures, is so abundantly present after many hundreds of nanoseconds.

The results of the measurements [6] of the K_β yield when negative muons are stopped in a helium gas target at pressures of 1 and 3 atm are presented in Fig. 6 together with all measurements done by other groups on the K_β yield at different pressures. This figure shows that the K_β yield, after about 5 atm, becomes almost proportional to the gas density. This figure also shows our results

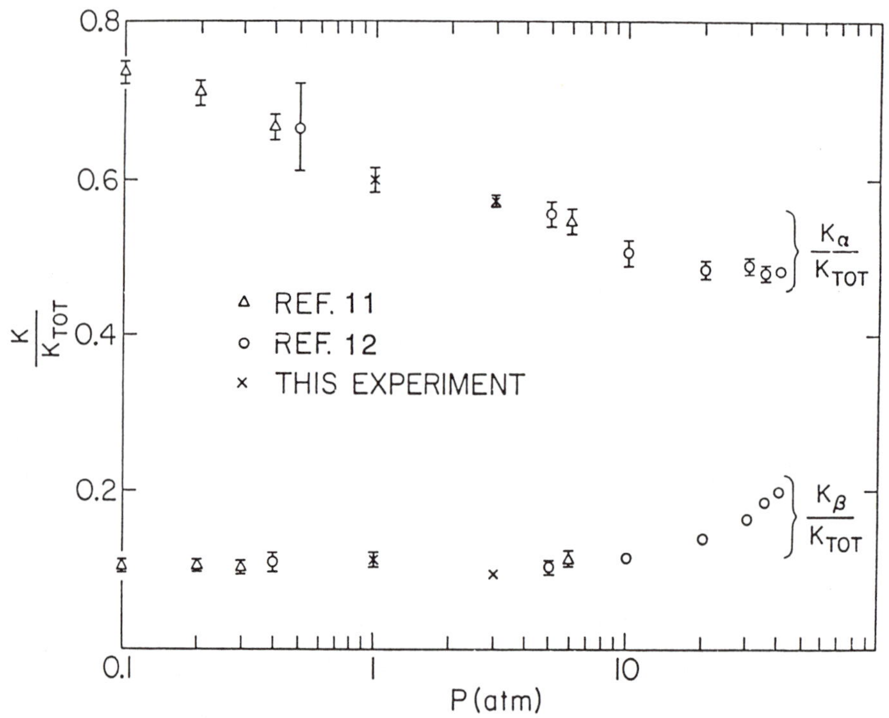

Fig. 6 : Experimental values for K_α/K_{tot} and K_β/K_{tot} as a function of target pressure.

concerning the K_α intensity together with the results of the other groups. At present, there are no theoretical calculations able to reproduce these results.

The practical conclusion we have drawn from Fig. 6 is that, since the K_β yield is a great nuisance, we have chosen to work at 3 atm.

The first runs were done with normal $^{13}C^{18}O_2$ gas in the laser; this was just to study the effect of the laser discharges on our electronics. Obviously it took some time, in these conditions, to learn to shield all the sensitive parts of the apparatus (data taking and monitoring electronics).

Our last runs were with the (expensive) isotopic gas $^{13}C^{18}O_2$ in our TEA laser system. We ran for some days with the diffracting grid set at λ = 9.8595 μm (presumed resonance value), firing the laser with every other beam burst. During this period, the repetition rate of the AGS was 1.4 s. It is easy to see that a 'good' quantity to look at to observe stimulated laser transitions is the ratio K_β/K_α. From the analysis of the spectra we obtained for the following quantity (o = laser on; f = laser off)

$$Y = \frac{(K_\beta/K_\alpha)_o - (K_\beta/K_\alpha)_f}{(K_\beta/K_\alpha)_f} ,$$

the experimental value

$$Y_{exp} = (0.012 \pm 0.014) .$$

The expected value for Y (assuming that we centred the wavelength) is

$$Y_{th} = f \times 0.034 = 0.017 ,$$

putting the f parameter earlier mentioned equal to 0.5.

From these results we concluded that in order to do a significant measurement, i.e. comparable or competitive with the 2S-2P difference measurements [1], it is necessary to collect a total of many thousands of K_β: this is possible, in principle, but with the X-ray detecting system at our disposal we need to do a run too long to be realistic.

A large array of X-ray detectors (total surface 60 cm^2 at least), is needed with high segmentation and prompt enough to reject events from the earlier electrons contained in the burst (see Fig. 5). At present we are investigating various options to solve this problem.

REFERENCES

[1] G. Carboni et al.: Nucl. Phys. A278, 381 (1977),
G. Carboni et al.: Phys. Lett. B73, 229 (1978),
G. Carboni et al.: Nuovo Cimento 34A, 493 (1976).

[2] E. Borie and G.A. Rinker: Phys. Rev A18, 324 (1978).

[3] I. Sick: Phys. Lett. B116, 212 (1982).

[4] E. Zavattini: Proc. First Course of the Int. School of Physics of Exotic Atoms, Erice 1977, eds. G. Fiorentini and G. Torelli (Lab. Naz. di Frascati, 1977), p. 43.

[5] W.G. Bauer and H. Salecker: Found. Phys. 13, 115 (1983).

[6] A.M. Sachs, A. Blaer, J. French, M. May and E. Zavattini: An experiment to measure vacuum polarization in 3D-3P transitions in muonic helium atoms. Proposal Exp. 745, Brookhaven National Laboratory.
See also J. French: Thesis Nevis 263 (1987) R/1379, CU/369; and A. Blaer, J. French, M. May, A.M. Sachs and E. Zavattini: Measurement of K X-rays from muonic helium formed in a low density target in an intense pulsed muon beam, to be published.

[7] D. Herriot et al.: Appl. Opt. 3, 523 (1964).

[8] J.S. Cohen and N.J. Bardsley: Phys. Rev. A23, 46 (1981).
J.S. Cohen: Phys. Rev. A25, 1791 (1982).

[9] L.I. Men'shikov, L.I. Ponomarev and L.P. Sukhanov: preprint IAE-4508/12, I.V. Kurchatov Atomic Energy Institute, Moscow, 1987.

[10] A. Placci et al.: Nuovo Cimento 1A, 445 (1971).
A. Bertin et al.: Nuovo Cimento 26B, 433 (1975).
A. Bertin et al.: Phys. Rev. Lett. 33, 253 (1974).

[11] F.B. Dittus: Experimentelle Untersuchung uber Bildung und Zerfall Myonicher Heliumionen im Metestabilen 2S-Zustand, Thesis ETH Zurich, No. 7877, 1985, see p. 104 (footnote).

[12] M. Eckhause et al.: Phys. Rev. A33, 1743 (1986).

Positrons for Low Energy Antihydrogen Production

G. Gabrielse and B.L. Brown

Department of Physics, Harvard University,
Cambridge, MA 02138, USA

Based upon the capture of low energy antiprotons in an ion trap, we are investigating the possibility of producing antihydrogen by merging extremely cold trapped plasmas of antiprotons and positrons. In principle, the calculated rate for antihydrogen is very high compared to other techniques. Here we survey the possibility of accumulating a high density of trapped positrons.

I. INTRODUCTION

Antiprotons were recently captured within the small volume ($\approx 10\ cm^3$) of an ion trap by the TRAP Collaboration.[1] They were decelerated from GeV production energies down to 21 MeV within the LEAR facility of CERN, passed through approximately 3 mm of beryllium to slow them below 3 keV,

and caught in an ion trap for as long as 10 minutes. While this is a much shorter time than the 10 month confinement of a single electron in a Penning ion trap,[2] prospects are now excellent for holding antiprotons much longer under improved vacuum conditions.

An interesting possibility raised by the capture of antiprotons in an ion trap is that of producing antihydrogen in this environment. Separately trapped plasmas of electrons and protons are routinely studied at 4.2 K, with lower temperatures possible. Recently we have considered the antihydrogen production that would result from merging such plasmas.[3] In equilibrium at 4.2 K, merging 10^4 antiprotons with positrons at a density of $10^7/cm^3$ would yield an instantaneous antihydrogen production rate exceeding $10^6/sec$. This calculated rate is orders of magnitude higher than either the projected rate for merged beams of antiprotons and positrons in a storage ring[4,5] or for collisions between a positronium beam and trapped antiprotons.[6]

The Hydrogen Atom Editors: G.F. Bassani · M. Inguscio · T.W. Hänsch

With antihydrogen in thermal equilibrium below 4.2 K, intriguing experimental possibilities can be considered. It becomes energetically possible to confine the antihydrogen, because of its magnetic moment, in a minimum of a magnetic field as has been done with sodium atoms.[7] Since we began exploring this difficult scenario,[8] spin polarized hydrogen atoms at 0.04 K have also been confined this way.[9] A deeper well may be required than has been used so far to confine atoms (a 1.5 Tesla well is needed to trap antihydrogen at 1 K, for example). The coldest atoms in the thermal distribution can of course be caught in a shallower well. In fact, a trap environment is now considered to be most promising for more precise laser spectroscopy of hydrogen atoms.[10] Comparisons of the fine and hyperfine structure of hydrogen and antihydrogen would provide extremely precise tests of CPT. If the antihydrogen atoms are confined, even a weak, monochromatic Ly α source may be useful for further cooling. With low enough atom temperatures, the gravitational force on antihydrogen can be measured since this force shifts the location of the atoms within the trap.[11] Experimental probes of gravitation are scarce and there is current theoretical interest in possible scalar and vector contributions to gravity which would cancel for matter but not for antimatter.[12]

Only small numbers of positrons have been trapped and sutdied while a single component plasma of trapped electrons has been studied extensively.[13] The problem is that positrons are much more difficult to obtain and are necessarily produced outside of the trapping well. If they have sufficient energy to enter the trap, they have sufficient energy to get out as well, so they must lose energy while within the trap to be captured. A technique developed at the University of Washington is to damp positrons within a harmonic Penning trap by coupling the motions to a resonant tuned circuit which is cold.[14] The time available for damping is increased substantially by introducing the positrons off the center axis of the Penning trap, so that damping can occur over a complete orbit $E \times B$ drift motion. Up to 100 positrons were so trapped[15] directly into an ultra-high vacuum environment in which they could be stored indefinitely (months) and used for high precision experiments. Despite the small number, the density was of order $10^8/cm^3$.

In two experiments at Bell Laboratories,[16,17] positrons were damped within a trapping well via collisions with a background gas introduced into the trap for this purpose. In the first experment,[16] of order 100 positrons were trapped at one time and trapping times were very short, of order $30msec$. The eventual aim of the second of these experiments[17] is to produce trapped positron densities of $10^6/cm^3$, somewhat less than what is discussed here. In addition, the trap volume used is larger by more than 1,000 than what is discussed here. This requires accumulating 1,000 times as many positrons to get the same density. Also, there is no provision for cooling to lower temperatures or for getting rid of the background gas, as may be required for antihydrogen production, unless heroic differential pumping is employed and/or the positrons are reaccelerated through a window into a trap in high vacuum. This experiment is progressing nicely, but only has preliminary results at this time.

Our approach, which may lead to a significant increase in the density and number of trapped positrons, makes use of a harmonic Penning trap in a high magnetic field field (6 Tesla). We plan to increase the positron loading rate by 10^3 or more over previous positron trapping efforts at UW, by using modern positron moderator techniques[18,19,20]. (Sec. II). Most of this factor comes from the greatly improved damping which is possible with non-relativistic positrons. The use of a moderator and carefully designed shielding will allow a 100-fold increase in source intensity to a 100 mCi ^{22}Na source which is commercially available. Together, these measures should increase the number of positrons available for trapping by 10^5. We will initially use resistor damping of the positrons within a harmonic Penning trap.[21] (Sec. III). The use of a harmonic trap will allow us to use powerful, radiofrequency diagnostics. The initial studies will thus focus upon highest possible densities in relatively small volumes. Already, this should produce a density and number of trapped positrons such that a plasma description is clearly appropriate (Sec. IV). It is not immediately clear that plasma studies with positrons offer any advantages over such studies with electrons. A possible exception is that a plasma of positrons offers some interesting detection properties compared to electrons (Sec. V). The mentioned techniques for capturing positrons and our efforts to increase the number of trapped positrons all allow positrons to be loaded continuously, thus being matched to the continuous positron emission from a radiative source. With a pulsed positron source, positrons could be alternatively loaded into a trap which is pulsed open and closed, as was done to load pulses of antiprotons into a trap[1]. This is certainly possible[22]. A LINAC could be used[23] or positrons from a radioactive source could be bunched and stacked in an ion trap. We are investigating these possibilities. So far, it seems best to start with the simpler, continuous technique. The more complicated options can be pursued if necessary (Sec. VI.).

If sufficient positrons can be confined, studies of particle transport within the plasma, etc., similar to those conducted with electrons can be carried out. It may be possible to use the enhanced detection possibilities afforded since positron-electron annihilations can be detected. An ultra-cold source of positrons would also have a variety of other applications.[24] For example, it has been proposed to eject trapped positrons into a plasma as a diagnostic.[25] Also, positrons initially in thermal equilibrium at $4.2K$ within a trap would form a pulsed positron beam of high "brightness" when accelerated out of the trap.

II. Moderated Source

An unmoderated, 1 mCi source of ^{22}Na was used in the UW experiments. The positrons vary in total kinetic energy between 0 and 544 keV and travel along magnetic field lines into the trap. Positrons available for trapping are those whose kinetic energy parallel to the magnetic field is within a 1 meV window. Approximately 6×10^{-5} of the emitted positrons were estimated to be within this window.[26]

We will instead send the positrons into a moderator, for example, a single crystal of tungsten. Many of the positrons will thermalize before annihilating. Of order 10^{-3} will be desorbed from the moderator, ejected by the negative positron work function.[27] The moderator will be in a strong magnetic field which may enhance the efficiency over what has been observed. At 6 Tesla, the ejected positions will essentially be in a beam whose cross sectional area is the size of the illuminated area of the moderator. The moderated positrons are nearly monoenergetic with a very narrow energy width, depending upon the moderator material and its preparation and cleaniness. With a single crystal of tungsten cooled to 20 K, a width less than 65 meV has been observed.[28] Since we will work at 4.2 K, the width may be even narrower. The surface will be carefully cleaned by electron bombardment heating to help reduce surface oxidation that can increase the energy spread of the positrons to a few eV. We thus expect an efficiency comparable to that achieved in the UW experiment in a 1 meV trapping window. The difference is that the moderated

FIGURE 1

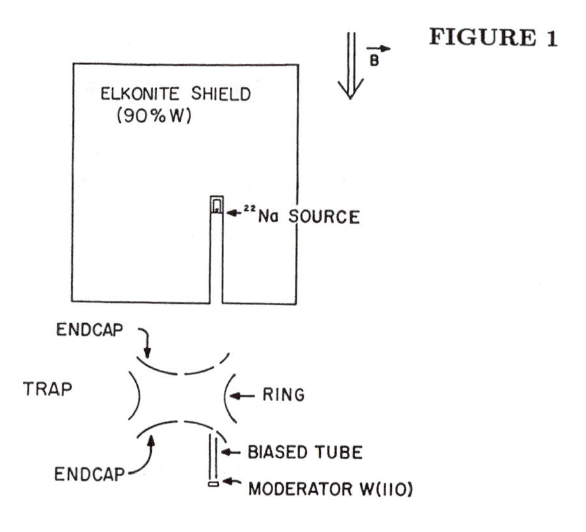

positrons will be nonrelativistic. We shall see that this has important consequences for damping and capturing the positrons.

The use of a moderator will also allow us to locate the positron source far enough from the trap to allow adequate shielding and minimze the effects of the decay gammas so that we will be able to use a 100 mCi source of ^{22}Na. This is a 10^2 increase compared to the UW experiment. A possible geometry of the source, traps and moderator is illustrated in Fig. 1.

III. Damping in a Harmonic Trap

We will first accumulate positrons in a harmonic Penning trap. A harmonic Penning trap is composed of electrodes which produce an electric quadrupole potential superimposed upon a uniform magnetic field.[21] The reason for this choice is that both single particle and center of mass motions are easily characterized and well understood in a harmonic Penning trap. In addition, well established radiofrequency techniques can be used as diagnostics, to nondestructively determine the number of particles for example, and for cooling. Once the difficulties of loading large numbers of positrons into a trap are understood and solved, it may then be possible to shift to the long, non-harmonic traps used for previous plasma experiments.[29]

The basic loading scheme was used in an earlier experiment, but made much more efficient by the low energy moderated positrons. Positrons are sent into the trap along a field line of a 6 Tesla magnetic field which makes focussing lenses, etc. unnecessary. The cyclotron radius of a 1 keV positron in a 6 Tesla field is only $15\mu m$. Positrons lose energy by inducing a current in a cold resistor to which they are coupled. The positrons must lose energy quickly or they will retain enough energy to escape along the path upon which they entered the trap. To make this time as long as possible, the positrons are introduced off the center axis of the trap. If a positron avoids leaving the trap after only one oscillation back and forth along a magnetic field line (by careful control of a bump in the trapping potential), further escape is thereby made impossible until the positrons complete an entire magnetron orbit. This orbit is caused by $E \times B$ drift within the trap. The time allowed for damping is thereby made of order 6,000 times longer for our parameters. Positrons will be cooled to the center axis via sideband cooling.

As mentioned, only positrons with kinetic energy along the magnetic field within a 1 meV window can possibly be trapped. In the UW experiment, most of the kinetic energy of such positrons was in the cyclotron motion. The high kinetic energies made a large shift in the "relativistic mass" of the positrons. The axial oscillation frequency, for the oscillation back and force along the magnetic field, depends upon the mass and is thus significantly shifted. The positrons within the 1 meV window thus oscillate with a large range of axial frequencies.

For the positrons to be trapped, the axial oscillation must be coupled to a cold resistor to lose energy. The cold resistor is actually part of a tuned circuit, required to tune out the effect of the unavoidable capacitance between trap electrodes. To get a high resistance, a high Q tuned circuit is required. The result in the UW experiment is that only 3×10^{-3} of the positrons in the 1 meV window have the right axial frequency to be damped via coupling to the resistor in the tuned circuit. With moderated positrons, however, the relativistic mass shift is completely negligible. This is an important advantage insofar as all of the positrons in the 1 meV window can be damped, giving an increase in efficiency of order 3×10^2. With the factor of 10^2 for increased source intensity, this suggests a positron loading rate higher by 3×10^4. We are optimistic that our use of lower temperatures and higher fields will allow us to actually realize a rate of loading positron which is thus larger than for the earlier experiment by 10^5.

Since previously up to 100 positrons were collected, an increase of 10^5 in loading rate suggests that at least 10^7 positrons could be collected in comparable times. Higher numbers may be possible in the future. Densities of $10^8/cm^3$ or more can be expected with reasonable trapping potentials. Long trapping times have been demonstrated so that it may be possible to collect positrons over longer times. In addition, since the point of these experiments is to accumulate large numbers of positrons rather than do precision measurements, further improvements seem likely with higher Q tuned circuits, higher trapping potentials, etc. It may even be possible to coherently bunch and inject the positrons as has been done with ions in a radiofrequency trap.[30]

The challenge of these experiments is to demonstrate that this loading scheme can be scaled up without introducing new problems. It is reassuring to note that the particles are loaded into the trap far off the center axis, where densities are low enough so that plasma effects should not be important during loading. They are then cooled to the center of the trap where a plasma is eventually formed. We are therefore optimistic that we can initially obtain more than 10^7 positrons at densities of order $10^9/cm^3$, with larger numbers available in the future.

IV. The Single Component Plasma of Positrons

Although our primary goal is to produce sufficient positrons for antihydrogen production, it is useful to examine the plasma that will be formed and to see if trapped positrons may also be useful for plasma studies. A single component plasma can be confined within a trap. As with two component plasmas, the Debye length

$$\lambda_D = \sqrt{\frac{kT}{4\pi n e^2}} \tag{1}$$

is a crucial parameter, where n is the number density of particles with charge e at temperature T. The minimal condition for collective plasma behavior is that λ_D be much less than a dimension of the trapped cloud of particles, D,

$$\lambda_D << D. \tag{2}$$

If we define the cloud dimension D such that the cloud occupies a volume D^3, then the total number of trapped particles N is related to the number density n by

$$nD^3 = N \tag{3}$$

The minimal condition for a trapped plasma may thus be written as

$$n\lambda_D^3 << N. \tag{4}$$

The total number of particles must be much greater than the number in a Debye volume, λ_D^3.

Properties of the plasma are often characterized by expansions in $(n\lambda_D^3)^{-1}$. Figure 2 shows $n\lambda_D^3$ plotted versus number densities for various plasma temperatures, T. The upper region in the figure, above the dashed line for $n\lambda_D^3 = 1$, is the classical plasma region, characterized by

$$n\lambda_D^3 >> 1 \tag{5}$$

This is the region where more than one particle is in a Debye volume λ_D^3, and expansions to lowest order in $(n\lambda_D^3)^{-1}$ are useful. A goal of one of the Bell experiments with positrons[17]

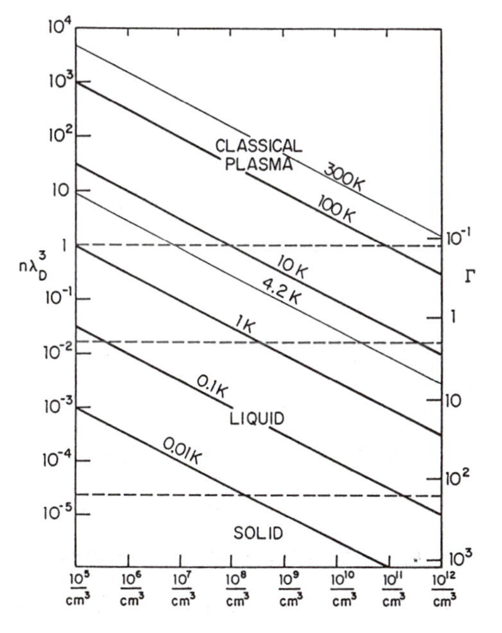

FIGURE 2

is to eventually get $n = 10^6/cm^3$ at $T \approx 300K$. This is is well within the the classical plasma region.

Below $n\lambda_D^3 = 1$ the plasma becomes increasingly correlated. In this region the plasma behavior is often equivalently characterized by the ratio Γ of the average Coulomb energy between particles, $e^2/n^{-1/3}$, to the thermal energy kT. Specifically, Γ is defined as

$$\Gamma = (\frac{4\pi}{3})^{1/3} \frac{e^2/n^{-1/3}}{kT} \tag{6}$$

where the constants are a convention. In terms of $n\lambda_D^3$,

$$\Gamma = [\frac{1}{4\pi\sqrt{3}}]^{2/3} [\frac{1}{n\lambda_D^3}]^{2/3}$$

$$\approx 0.128(n\lambda_D^3)^{-2/3} \tag{7}$$

Increasing Γ represents increasing correlation. The onset of liquid behavior[31] is expected at $\Gamma \sim 2$ while crystalization to form a solid[32] is expected at $\Gamma \sim 155$. In Fig. 2, the equivalent scale for Γ is indicated on the right. The liquid and solid regions are the the indicated regions between the dashed lines.

Experiments with trapped a "cryogenic plasma" of electrons at temperatures as low as 50K and densities as high as $10^{10}/cm^3$ have begun to probe plasma behavior below the classical plasma region in Fig. 2. Experiments with laser-cooled trapped ions have been underway to probe the liquid region.[33] Very recently crystalization was observed with only a few ions[34,35] and with more than 100 ions.[36] A very nice feature of the measurements is that one can learn the positions of shells of ions and even individual ions by imaging the fluorescence from the ions. A numerical simulation produced crystalization very similar to what was observed.[37]

In our initial experiments with trapped positrons, if we achieve densities of order $n = 10^8/cm^3$ with $N = 10^7$ positrons in the trap at T = 4.2 K, we will clearly have a plasma in the sense of Eqs. (2) and (4). The Debye length will be much smaller than the dimension of the positron cloud. Correlations will be important since $\Gamma \sim 0.3$. This may make it possible, for example, to study particle transport for a plasma with $n\lambda_D^3 \approx 1$. To produce a classical plasma for study, the positrons can easily be heated using radiofrequency drives. The range of accessible $n\lambda_0^3$ for positrons in the initial stage of these experiments is shown by the cross hatched region in Fig. 2.

Finally, it may be possible to lower the plasma temperature below 4.2 K. The temperature of 4.2 K is only a choice of convenience, being the boiling point of liquid ^4He at atmospheric pressure. By pumping on the ^4He, or pumping on a dewar with ^3He in it, or by using a dilution refrigerator, it should be possible to probe the liquid and even solid behavior of the highly correlated plasmas. The question here is whether one can develop tools for probing the liquid and solid regime as powerful as the optical techniques available for trapped ion experiments? This remains to be seen.

V. Detection of Positron Annihilations

One unique feature of a plasma of positrons (compared to an electron plasma) is that position annihilation can be detected in detectors placed outside the vacuum system of the trap. Each direct annihilation results in a pair of back-to-back gamma rays. With appropriately designed apparatus, it may be possible to identify the position or vertex of the annihilation and thus learn where and when losses are occurring. Detection of annihilation gammas can produce an extremely clean signal, with exceedingly low background.

To illustrate the detection, we consider positrons annihilating at a point. We assume that two 7.5 cm diameter sodium iodide detectors can be located 13 cm away to either side of the trap, outside a copper vacuum system 0.5 cm thick. The photopeak efficiency for 511 keV annihilation, including geometry and scattering, is 2×10^{-2} for a single detector and 5×10^{-3} for coincident pair annihilation. More pairs or larger detectors could, of course, be used as space permits. To determine the verticle position of the initial annihilation to within a centimeter, it is necessary to mask off part of each detector. With a 0.5 cm slit collimator, the efficiency would be 4×10^{-4} and 10^{-4} for single and coincident detector efficiencies. The spacial resolution would be 0.4 cm full width at one-tenth maximum.

We will not immediately rely on annihilation detection, since initial experiments will be carried out inside an existing superconducting magnet. This means that the detectors must be located 45 cm from the annihilation region, thus reducing the solid angle by a factor of 10. In addition, the magnet windings and dewar effectively present 5 cm of copper and 5 cm of aluminum through which the gamma rays must pass, thus reducing by a factor of 200 the probability that an annihilation gamma will be detected. For the two detectors in the illustration, this reduces the detection efficiency to 2×10^{-5} for detection of one of the gammas and to 7×10^{-7} for detection of coincident gammas.

Radiofrequency detection is thus a much more effective diagnostic for the initial experiments in existing magnets. In the future, however, we are interested in a magnet which is specially designed to get the detectors very close to the trap. Then detection of annihilations could be very useful. Meanwhile, we estimate that the detection of annihilation gammas will be useful initially for steering the positrons from the ^{22}Na source into the moderator and into the trap region. We will assemble the detectors needed for this purpose. The detection system and experience we acquire will make us ready for a new magnet system at a later date, if this proves to be advantageous.

VI. Pulsed Positron Sources

Discussion so far has been upon continuous loading of positrons into a trap. This is because positrons are ejected continuously from radioactive sources. Such continuous loading of positrons is, in fact, a way of bunching the positrons since all of the positrons in the trap can be ejected in a very small pulse.

On the other hand, if a pulsed source of positrons was available, pulses of positrons could be loaded into a trap much differently, the way we loaded pulses of antiprotons into a trap, for example.[1] The positrons could enter the trap electrodes through one electrode which is grounded to allow entry. While the positrons are inside, the potential on this electrode would be quickly changed to prevent their escape. This is manageable for moderated $\approx 1eV$ positron beams.

An appropriate pulsed source exists. An electron LINAC has produced[38] 4×10^6 positrons in a $3\mu sec$ pulse with a repetition rate of 300 Hz and has produced 4×10^5 positrons in a 1 ns pulse with a 1.4 kHz repetition rate. An order of magnitude increase in intensity may be possible.[39] Bell Labs presently has a microtron based beam producing moderated positrons at a rate of 5×10^4 e^+ per pulse,[40] with large intensity increases expected.

The obvious drawback to this approach is that an electron LINAC (or comparable accelerator) is a fairly major facility which does not easily fit in our laboratory. We may attempt some experiments, nonetheless, at an appropriate facility, but initially it seems appropriate to get more experience with higher density positrons plasmas using more modest and manageable sources. Moreover, even though the LEAR facility at CERN (where we trap antiprotons) does have a LINAC, it is not very easy to get positrons and antiprotons from two very substantial and separate facilities to the same point.

Another possibility is to efficiently bunch the positrons from a radioactive source in a first apparatus, then transport the positrons into a trap apparatus where they are loaded. A bunched positron source has been made[24] but the efficiency of loading a positron from the source into a bunch was only 4×10^{-2}. Up to 100 positrons were in each 7 ns bunch, with a repetition rate of 1 KHz. We believe that intense bunched positron sources could be built. For initial experiments, however, it seems desireable to avoid separating the experiment into two regions, for separately bunching and loading the positron, making it necessary to transport the intense bunch of positrons into a high field region between. Our initial experiments will take place within a 6 Tesla field which effectively guides the positrons without need for lenses, etc. normally required.

References

[1] G. Gabrielse, X. Fei, K. Helmerson, S. L. Rolston, R. Tjoelker, T. A. Trainor, H. Kalinowsky, J. Haas and W. Kells, Phys. Rev. Lett. **57**, 2504 (1986).

[2] G. Gabrielse, H. Dehmelt and W. Kells, Phys. Rev. Lett. **54**, 537 (1985).

[3] G. Gabrielse, L. Haarsma, S. L. Rolston and W. Kells, in *Laser Spectroscopy VIII*, p. 26, edited by W. Persson and S. Svanberg (Springer-Verlag, New York, 1987). A preliminary version of this work was presented by G. Gabrielse, L. Haarsma, S. L. Rolston and W. Kells, in *Proc. Symp. on the Production and Investigation of Atomic Antimatter*, edited by H. Poth and A. Wolf (Scientific, Basel, 1988).

[4] H. Herr, D. Mohl and A. Winnacker, *Physics at LEAR with Low-Energy Cooled Antiprotons,* Erice (1982) 659.

[5] R. Neumann, H. Poth and A. Winnacker, A. Wolf, Z. Phys. A **313**, 253 (1983).

[6] J. W. Humberston, M. Charlton, F. M. Jacobsen and B. I. Deutch, J. Phys. B **20**, L25 (1987).

[7] A. L. Migdall, J. V. Prodan, W. D. Phillips, T. H. Bergeman and H. J. Metcalf, Phys. Rev. Lett. **54**, 2596 (1985).

[8] G. Gabrielse, *Penning Traps, Masses and Antiprotons*, Invited Lecture at the International School of Physics with Low Energy Antiprotons: Fundamental Symmetries, Erice, Italy (1986).

[9] H. F. Hess, G. P. Kochanski, J. M. Doyle, N. Masuhara, D. Kleppner and T. J. Greytak, Phys. Rev. Lett. **59**, 672 (1987).

[10] T. Hansch, R. G. Beausoleil, B. Couillaud, C. Foot, E. A. Hildum and D. H. McIntyre, in *Laser Spectroscopy VIII*, p. 2, edited by W. Persson and S. Svanberg (Springer-Verlag, New York, 1987).

[11] G. Gabrielse, in *Proc. Symp. on the Production and Investigation of Atomic Antimatter*, edited by H. Poth and A. Wolf (Scientific, Basel, 1988).

[12] T. Goldman, M. N. Nieto and R. J. Hughes, Phys. Lett. B **171**, 217 (1986).

[13] J.H. Malmberg, T.M. O'Neil, Proceedings of 1984 Sendai Symposium on Plasma Nonlinear Phenomena, *32* (1984)

[14] P.B. Schwinberg, R.S. Van Dyck, Jr. and H.G. Dehmelt, Phys. Lett. **81A**, 119 (1981).

[15] R.S. Van Dyck, Jr., private communication.

[16] B.L. Brown and M. Leventhal, Phys. Rev. Lett. **57**, 1651 (1986).

[17] F.J. Wysocki, M. Leventhal A. Pasner, C.M. Surko, Hyperfine Int., in press (1988).

[18] A. Vehanen, K.G. Lynn, P.J. Shultz and M. Eldrup, Appl. Phys. **A32**, 2572 (1983).

[19] A.P. Mills, Jr., Appl. Phys. Lett. **37**, 667 (1980).

[20] E.M. Gullikson, A.P. Mills, Jr., W.S. Crane and B.L. Brown, Rapid Comm., Phys. Rev. **B32**, 5484 (1980).

[21] See the detailed review by L.S. Brown and G. Gabrielse, Rev. Mod. Phys. **58**, 233 (1986) and references therein.

[22] W. Kells, Proc. of 1987 Workshop on Intense e^+ Beams, Idaho Falls, June 1987.

[23] An introduction to LINAC and high intensity radioactive positron beams is given by R. Howell, et al. and K. Lynn, et al. respectively in "Positron Scattering in Gases", ed. John W. Humberston and M.R.C. McDowell (Plenum, New York), 1983, p. 155.

[24] A.P. Mills, Jr., in "Positron Scattering in Gases," ed. John W. Humberston and M.R.C. McDowell, (Plenum, New York), 1983 p. 126.

[25] C. Surko, M. Leventhal, W.S. Crane, A. Passner, F. Wysocki, T.J. Murphy, J. Strachan and W.L. Rowan, Rev. Sci. Instrum. **57**, 1892 (1986).

[26] P.B. Schwindberg, thesis (unpublished).

[27] A.P. Mills, Jr., P.M. Platzman and B.L. Brown, Phys. Rev. Lett. **41**, 1076 (1978).

[28] B.L. Brown, W.S. Crane and A.P. Mills, Jr., Appl. Phys. Lett. **48**, 739 (1968).

[29] C.F. Driscoll, K.S. Fine and J.H. Malmberg, Physics Fluids, **29**, 2015 (1986).

[30] R. B. Moore and S. Gulick, Phys. Scripto. **T22**, 28 (1988).

[31] J.H. Malmberg and T.M. O'Neil, Phys. Rev. Lett., **39**,

[32] J.H. Malmberg, T.M. O'Neil, A.W. Hyatt and C.F. Driscoll, in Proc. of the 1984 Sendai Sympsoium on Plasma Nonlinear Phenomena, **32**, (1984).

[33] J.J. Bollinger and D.J. Wineland, Phys. Rev. Lett. **53**, 348 (1984).

[34] F. Diedrich, E. Peik, J.M. Chen, W. Quint and H. Wather, Phys. Rev. Lett. **59**, 2931 (1988).

[35] D. J. Wineland, J. C. Bergquist, Wayne M Itano, J. J. Bollinger, and C. H. Manney, Phys. Rev. Lett. **59**, 2935 (1988).

[36] S.L. Gilbert, J.J. Bollinger and D.J. Wineland, Phys. Rev. Lett. **60**, 2022 (1988).

[37] D.H.E. Dubin and T.M. O'Neal, Phys. Rev. Lett. **60**, 522 (1988).

[38] R. Howell, M.J. Fluss, I.J. Rosenberg and P. Meyer, Nucl. Inst. and Meth. **B10/11**, 373 (1985).

[39] R. Howell, private communication.

[40] A.P. Mills, Jr., private communication.

Antihydrogen by Positronium-Antiproton Interaction

B.I. Deutch

Institute of Physics, University of Aarhus,
DK-8000 Aarhus C, Denmark

A method for producing antihydrogen, $\bar{H}$, by positronium-antiproton collisions in an antiproton ion-trap (see Fig.1) has been published by us earlier in Ref.1. The reaction

$$\bar{p} + Ps \rightarrow \bar{H} + e^-$$

involves the transfer of the positron, e^+, from positronium. This reaction has a large cross section [2-4] such that useful fluxes of $\bar{H}$ can be achieved. Included in the above article are the calculated capture cross sections, short descriptions of the slow positron beam, positronium-formation, and the antiproton ion-trap to be used for antihydrogen production. With presently available technology, collimated monoenergetic $\bar{H}$ beams with an energy of a few KeV and intensity of the order of one per second can be produced by this method. Possible enhancements of this rate are also discussed.

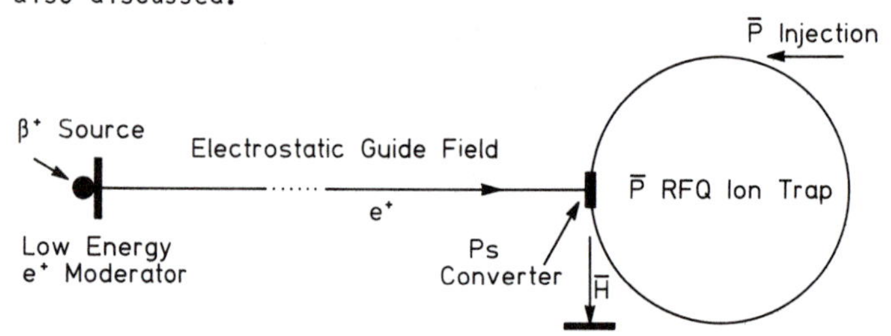

Fig.1. Schematic arrangement for production of antihydrogen. The β^+ particles from a radioactive source (^{22}Na) are moderated; the slow e^+ are electrostatically focused onto the Ps converter at the walls of a $\bar{p}$ ion trap (here shown as an RFQ race-track design), from which a collimated $\bar{H}$ beam emerges.

References

1. B.I. Deutch, F.M. Jacobsen, L.H. Andersen, P. Hvelplund, H. Knudsen, M.H. Holzscheiter, M. Charlton, and G. Laricchia, Phys.Scripta T 22 (1988) 248
2. J.W. Humberston, M. Charlton, F.M. Jacobsen, and B.I. Deutch, J.Phys.B: At.Mol.Phys.20 (1987) L25
3. A.E.Ermolaev, B.H.Bransden, and C.R. Mandel, Phys.Lett.A 25 (1987) 44
4. A.E.Ermolaev, Proc. of the Symp. on the Production and Investigation of Atomic Antimatter, Karlsruhe, 1987 (to be published in Hyp.Int.)

The Hydrogen Atom Editors: G.F. Bassani · M. Inguscio · T.W. Hänsch

Antiprotonic Hydrogen

E. Klempt

Institut für Physik, Universität Mainz,
D-6500 Mainz, Fed. Rep. of Germany

Antiprotonic hydrogen is a positronium-like system, but its reduced mass is larger by a factor m_p/m_e and nuclear interactions play a significant role. Recent experiments at CERN have provided data which allow to test our understanding of the nuclear forces and of the atomic effects which govern the cascade of antiproton hydrogen from capture to annihilation.

1. Introduction

The discovery of antiprotons in 1955 by CHAMBERLAIN, SEGRE, WIEGAND, and YPSILANTIS /1/ was a great triumph for field theory and was honoured by the Nobel prize. Today antiproton beams of several 10^{11} antiprotons can be stored in the Antiproton Accumulator at CERN. The antiproton beam can be extracted and further accelerated. In collisions with protons at a center of mass energy of 540 GeV/c² particle jets are observed revealing the fundamental antiquark-quark scattering in Rutherford-like experiments; energetic electrons or muons manifest the existence of the gauge particles of weak interactions. On the other hand, the antiprotons can also be decelerated to 600 MeV/c and transferred to the Low Energy Antiproton Ring (LEAR). This ring, in operation since the end of 1983, allows to store 10^{10} antiprotons, to accelerate or decelerate antiprotons to all momenta between 70 MeV/c and 2000 MeV/c, to cool antiprotons in phase space, and to extract the particles into external beam lines. This operation results in intense ($>10^6$ $\bar{p}$/s) high-quality antiproton beams permitting experiments which were unfeasible before LEAR was built.

In the initial phase of the LEAR program there were three experiments (PS171, PS174 and PS175) searching for the characteristic lines of antiprotonic hydrogen atoms (protonium) /2-4/. Antiprotonic hydrogen atoms are formed by stopping antiprotons in H_2. The intensities of

The Hydrogen Atom Editors: G.F. Bassani · M. Inguscio · T.W. Hänsch

the lines were expected to depend strongly on target density. Since the optimum choice of the target density was not known, different densities were used in these experiments. The energies of the lines are dominantly given by the Coulomb interaction between proton and antiproton; the strong interaction leads to an additional energy shift and a line broadening. The measurement of these strong interaction effects was the primary motivation of these experiments. The strong density dependence of the line intensities indicates the presence of important atomic effects during the cascade which deserve interest on their own right.

Table I shows the basic properties of protonium, Fig. 1 the energy levels. Due to the change in reduced mass, the binding energy and

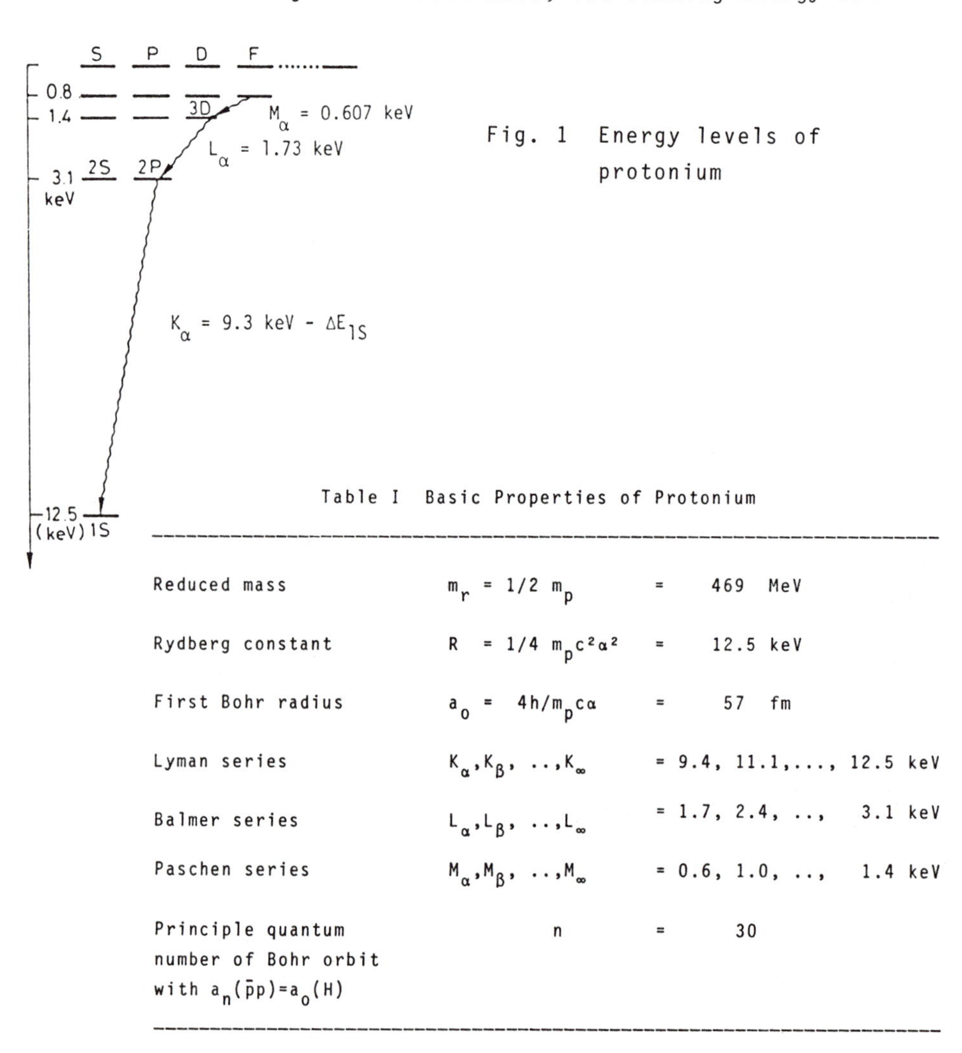

Fig. 1 Energy levels of protonium

Table I Basic Properties of Protonium

Reduced mass	$m_r = 1/2\ m_p$	=	469 MeV
Rydberg constant	$R = 1/4\ m_p c^2 \alpha^2$	=	12.5 keV
First Bohr radius	$a_o = 4h/m_p c\alpha$	=	57 fm
Lyman series	$K_\alpha, K_\beta, .., K_\infty$	=	9.4, 11.1,..., 12.5 keV
Balmer series	$L_\alpha, L_\beta, .., L_\infty$	=	1.7, 2.4, .., 3.1 keV
Paschen series	$M_\alpha, M_\beta, .., M_\infty$	=	0.6, 1.0, .., 1.4 keV
Principle quantum number of Bohr orbit with $a_n(\bar{p}p)=a_o(H)$	n	=	30

transition energies are larger by nearly a factor of 1000 compared to ordinary hydrogen, the Bohr radii correspondingly smaller. The small size of the system allows the strong interaction to play a major role: the S and P wave atomic levels are shifted, and annihilation leads to a rather large width of these states. In states with higher angular momenta the strong interaction is too weak to lead to any observable consequences.

2. Identification of the Lyman Series

Fig. 2 shows the X-ray spectrum observed by PS171 in coincidence with antiprotons stopping in a H_2 gas target at NPT. The large peak at low energies is due to the unresolved Balmer series and some contamination from Argon fluorescence at 3 KeV. At higher energies a contribution from the Lyman series may be seen above a smooth background from Bremsstrahlung. The solid line represents the calculated Bremsstrahlungs-spectrum (no fit). The data show the difficulties of protonium spectroscopy even using the excellent antiproton beams at LEAR: the yield of the Balmer series is 10% per antiproton stop, but a Balmer X-ray feeding one of the 2P levels is mostly not followed by radiative

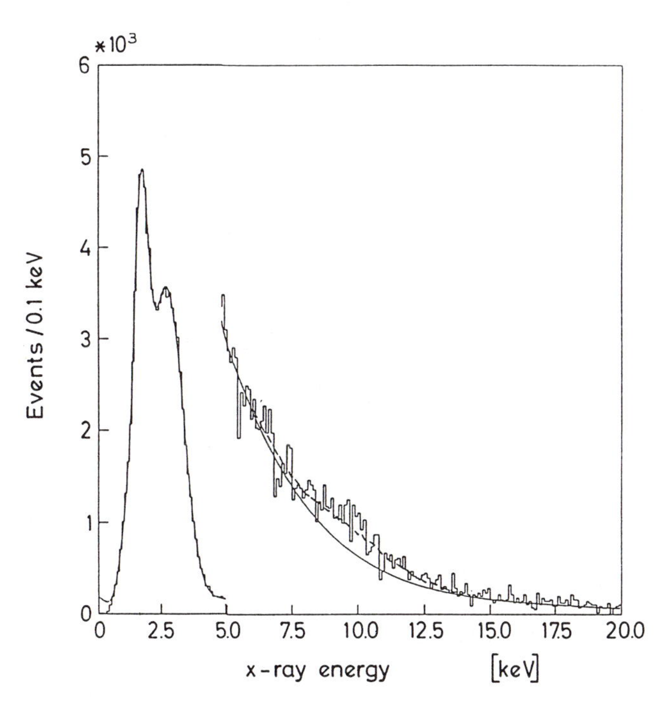

Fig. 2 X-ray energy spectrum of protonium in PS171. The solid line corresponds to the fit of the Balmer series and the calculated Bremsstrahlung. The dashed line represents possible Lyman contributions

emission of a Lyman-α line but by annihilation. The fraction of protonium atoms emitting a Lyman-α line is 1:1000; hence the experiments are forced to struggle against a high background produced in secondary reactions. In $\bar{p}p$ annihilation charged and neutral pions are produced, high energy γ-rays simulate low-energy X-rays by Compton scattering, negative pions lead to spectral lines of pionic atoms and to a neutron background by nuclear disintegration. In addition Brems-strahlungs-quanta are emitted in the sudden acceleration of charged particles in the annihilation process.

The distinctive feature of PS171 is its capability to practically eliminate all these sources of background. This is achieved by a gaseous detector of very low mass. Compton scattering is correspondingly reduced, and solid material in which pions may stop is well separated from the target. The main source of background is the Bremsstrahlung. Energy and angular distributions of Bremsstrahlung can be calculated absolutely /5/, the agreement with the data is very good demonstrating the low background level which was achieved /6/. Bremsstrahlung can be avoided by triggering events in which protonium annihilate into neutral particles only. The residual background can then be eliminated by requiring that two X-rays should be observed in coincidence /7/.

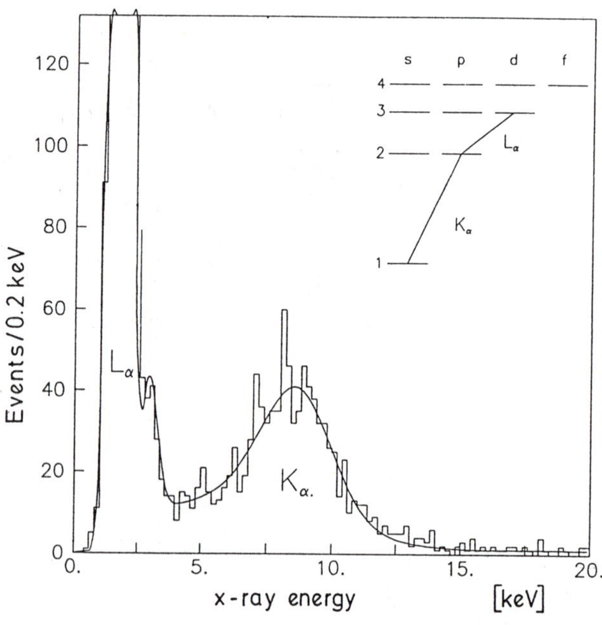

Fig. 3 X-ray spectrum of protonium. The energy of the more energetic X-ray of two coincident X-rays is shown. The peak at 8.5 keV is due to Lyman α lines (PS171). The line shape theory of ERICSON and HAMBRO /8/ is used

Fig. 3 shows the energy spectrum of X-rays for which a less energetic X-ray is observed in coincidence. The $\bar{p}p$ atom is required to annihilate into neutral particles only. This annihilation mode occurs with a rate of 4 %. The K_α peak is clearly seen. A number of checks confirm this interpretation: the distribution of X-ray energies observed in coincidence with the K_α peak shows the full Balmer series, the distribution of X-ray conversion points demonstrates that the X-rays originate out of the target, and gas analysis excluded the presence of contaminating lines. Hence the interpretation of the observed peak as K_α line with at most very few background events is unambiguous.

PS174 uses a conventional gas target of variable temperature in order to change the H_2 density. Extremely low pressures were achieved in experiment PS175. An inverse cyclotron was used to decelerate antiprotons to very low momenta, and antiprotons could be stopped with high efficiency at H_2 pressures as low as 30 mbar. Si(Li) solid state detectors (PS174, PS175) and gas scintillation proportional chambers (PS174) record the X-ray energy spectrum associated with antiprotons stopping in the target. Fig. 4 shows the data of PS174 recorded with two gas scintillation proportional chambers at 0.25 and 0.92 times the gas density at normal temperature and pressure /9/.

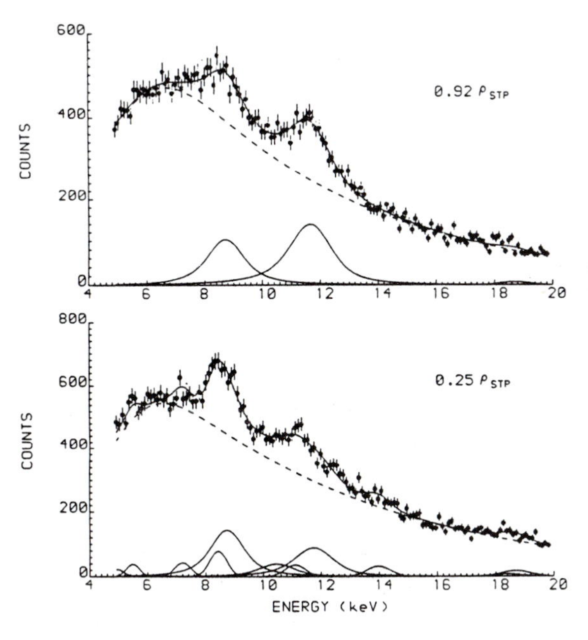

Fig. 4 X-ray spectrum of protonium (PS174). The solid line corresponds to the best fit taking into account the shape of the background from D_2 runs (dashed) and two Voigtian functions

H_2 and D_2 were used as target gases. Due to the higher absorption probability the yield of the Lyman series of $\bar{p}D_2$ is negligibly small, but the data can be used to identify the shape and the size of the background for the data taken with H_2. The data demonstrate clearly the existence of the K_α line.

The data sets gathered with Si(Li) detectors /10,11/ have a worse signal-to-background ratio but the results are consistent with the data shown above (see Fig. 6). Hence one can conclude that the same line is seen in different data sets.

3. Energy Shift and Width of the Ground State of Protonium

The line shape of the Lyman α radiation as observed in the detector is influenced by several effects:

1. The strong interaction leads to a finite width of the 1S state which is described by a Lorentzian line shape. The hyperfine splitting is estimated to be 200 eV. Therefore two lines with unknown fractional intensities are expected. Since the line width is much larger this splitting is not taken into account.

2. The electric dipole moment is not constant over the line. This effect requires a special lineshape theory which was developed by ERICSON and HAMBRO /8/. Not taking this effect into account leads to an apparent shift of the peak to higher energy.

3. The detection efficiency over the line spectrum is not constant as a function of energy. For gaseous targets it falls with increasing energy resulting in an apparent shift of the observed peak to lower energies.

4. The finite detector resolution leads to an additional broadening of the line.

Neglecting point 2 and 3 leads to the well known Voigt function which is used by PS174 and PS175. However, PS171 finds a large dependence of the fit results on line shape theory and on the energy dependence of the detection efficiency; therefore the data of PS171 are fitted with a convolution of all four effects. The final results

Table II ΔE_{1S}, Γ_{1S}, and Γ_{2P} of protonium

Experiment	Energy shift ΔE_{1S}(keV)	Width Γ_{1S}(keV)	Width Γ_{2P}(MeV)
PS171	-0.70(15)	1.60(40)	
PS174 a	-0.73(5)	1.13(9)	
PS174 b	-0.75(6)	0.90(18)	45(10)
PS175	-0.66(13)	1.13(23)	40(11)
Theory			
Richard and Sainio	-0.73	0.94	26
Moussallam	-0.75	1.00	27

of all three experiments are presented in Table II and compared to two recent theoretical results /12,13/. The agreement between experiment and theory and in between the different theoretical models is rather good.

Very likely the essential ingredients for an understanding of the strong interaction effects are the Coulomb potential, π exchange and annihilation. The annihilation suppresses the central part of the protonium wave function with a large contribution to the binding energy. This leads to less binding even though the nuclear interaction is basically attractive. The theoretical models use the same π exchange potential amd phenomenological annihilation potentials fitting low-energy cross-sections; since short range interactions (in which different One-Boson-Exchange models differ) are suppressed by annihilation, all models give very similar results.

4. The 2P Strong Interaction Width

The data of Fig. 2 demonstrate that the 2P level has a strong interaction width Γ_{2P} exceeding the radiative width Γ_{rad} by about two orders of magnitude. These data can, however, not be used to determine Γ_{2P} since the fraction of neutral annihilation of the hyperfine levels of the 2P and 1S levels is not known. An average Γ_{2P} can be obtained by not triggering on specific annihilation modes. Fig. 5 shows the Balmer series, Fig. 6 the Lyman series of PS174 taken with a solid state Si(Li) detector at two different H_2 gas densities. The

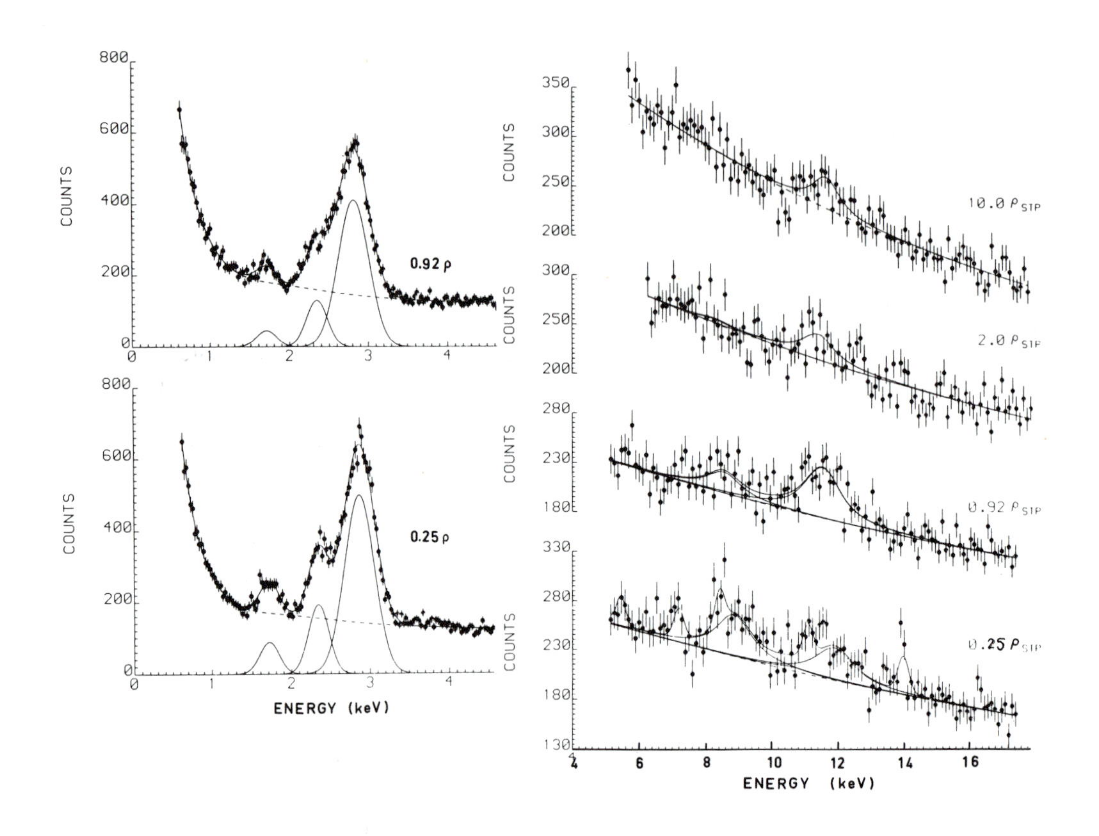

Fig. 5 and 6 X-rays from protonium (PS174), Si(Li) detector. Balmer series (left) and evidence for the Lyman series are shown as a funtion of H_2 gas density

detection efficiency at these low energies is a steeply rising function; it was calculated by Monte Carlo techniques. From this the strong interaction width of the 2P levels weighted with the (unknown) fractional populations of the 2P levels were determined. The result is also shown in Table II. The large strong interaction width of the 2P levels is provoked by the tunnel effect: the protonium wave function may tunnel through the centrifugal barrier into the region of strong interaction in which annihilation is extremely strong. This simple picture is capable of a quantitative description of Γ_{2P}.

5. Cascade

Fig. 6 shows that the X-ray intensities depend strongly on the density of the H_2 target. When antiprotons are stopped in H_2 they are captured into high Rydberg states of protonium: the H_2 molecule dissociates and the electron of one H atom is ejected by Auger emission. The subsequent cascade results from a competition of density dependent and density independent effects: in collisions $\bar{p}p$ atoms experience high electric field strengths of 10^{12} V/m leading to a reshuffling of states with different angular momenta; the Coulomb interaction between $\bar{p}$ and atomic electrons may lead to external Auger effect resulting in a de-excitation of the $\bar{p}p$ atom; radiative transition may occur, and annihilation from protonium states with angular momenta 0 or 1 may stop the atomic cascade. Recently, we have calculated these processes /14/. Results of the calculations are shown in Fig. 7. The

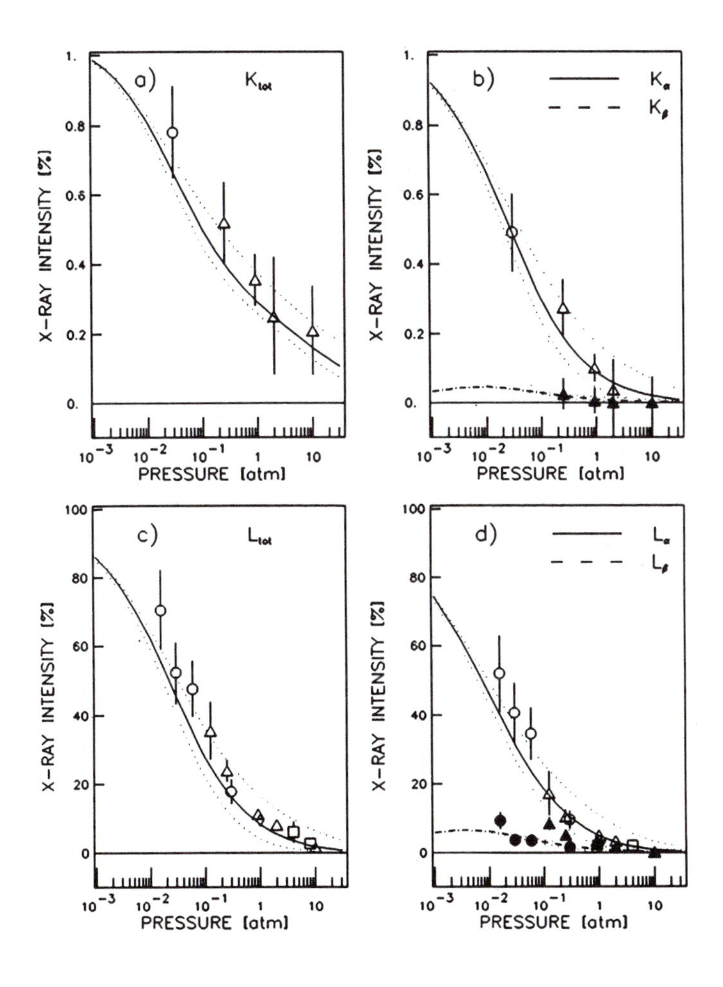

Fig. 7 Cascade of protonium in H_2 gas. The data are from PS171, 174 and PS175. The solid line represents an absolute calculation, the dotted lines two extreme variations of the calculated Auger effect

good agreement between the experimental result and the calculation - which uses no fit parameter - demonstrates the good understanding of the atomic physics in antiprotonic hydrogen atoms.

References

/1/ O. Chamberlain, E. Segrè, C. Wiegand, Th. Ypsilantis: Phys.Rev. 100, 947 (1955)

/2/ R. Armenteros et al.: A study of $\bar{p}p$ interactions at rest in a H_2 gas target at LEAR. CERN, PS171 (1980). In Experiments at CERN, Geneva (1987)

/3/ E.W.A. Lingman et al.: Precision Survey of X-rays from $\bar{p}p$ ($\bar{p}d$) Atoms. CERN PS174 (1980) ibid.

/4/ R. Bacher et al.: Measurement of the Antiprotonic Lyman- and Balmer X-rays of $\bar{p}H$ and $\bar{p}D$ Atoms. CERN, PS175 (1980)

/5/ R. Rückl and C. Zupancic: Phys.Lett. B150, 225 (1985)

/6/ U. Schäfer et al.: "X-rays from proton antiproton annihilation into charged final states", submitted to Phys.Lett. B

/7/ M. Ziegler et al.: Phys.Lett. B206, 151 (1988)

/8/ T.E.O. Ericson and L. Hambro: Annals of Physics 107, 44 (1977)

/9/ C.W.E. van Eijk et al.: Nucl.Phys.A (in print)

/10/ C.A. Baker et al.: Nucl.Phys. A483, 429 (1988)

/11/ L. Simons: "Results from antiprotonic atoms at LEAR", in Physics at LEAR with low energy antiprotons, eds. C. Amsler et al., (Harwood Academic Publishers, Chur 1988), p. 703

/12/ J.M. Richard and M.E. Sainio: Phys.Lett. B110, 349 (1982)

/13/ B. Moussallam: Z.Physik A325, 1 (1986)

/14/ G. Reifenröther and E. Klempt: Antiprotonic Hydrogen: From Capture to Annihilation, MZ-ETAP/88-11

The Spectroscopy of Hydrogen-Like Highly Ionised Atoms

J.D. Silver

University of Oxford, New College, and The Clarendon Laboratory,
Parks Road, OX1 3PU, U.K.

When the organisers of this meeting on the hydrogen atom very kindly asked me to give a lecture, they asked me to give something of an *overview* of the spectroscopy of hydrogen-like highly ionised atoms. This is now a fairly active field of study, so I shall concentrate on some recent work which has been carried out by my group in collaboration with Daniel Dietrich of Lawrence Livermore National Laboratory, and I shall only give brief references to the work of other groups. I hope they will forgive me, and not think that by magnifying our work, I am seeking to minify theirs. I shall also speculate a little on how the field might develop in the reasonably near future.

Our primary aim in studying the hydrogenic ions is to make good measurements of the Lamb shift. We are interested in the Lamb shift because it is a purely quantum-electrodynamic (QED) quantity. One of the inventors of the modern theory of QED, Richard Feynman, recently wrote " ... The theory of quantum electrodynamics has now lasted for more than fifty years, and has been tested more and more accurately over a wider and wider range of conditions. At the present time, I can proudly say that there is *no significant difference* between experiment and theory! ... We physicists are always checking to see if there is something the matter with the theory. That's the game, because if there is something the matter, it's interesting! But so far, we have found nothing wrong with the theory of QED ... " [1]

There is nonetheless something a a little bit strange about the apparent success of the QED theory. This relates to the fact that certain of the procedures used to calculate experimentally measurable quantities such as the Lamb shift do not appear to be mathematically sound. The problem has been stated by Dirac [2] , who suggested that the remarkable agreement between the results of renormalisation theory in QED, and experiment, should be looked on as a "fluke", in much the same way that the agreement between the Bohr theory and experiment for the spectrum of hydrogen turned out to be a fluke.

The Hydrogen Atom Editors: G.F. Bassani · M. Inguscio · T.W. Hänsch

The "holy grail" of those that test QED is that they will make a measurement of a QED effect which will not agree with the predictions of QED theory, thereby indicating the need for a different theory. So far,however, most measurements seem to agree well with the theory [3].

Our reason for studying hydrogenic *ions*, rather than the neutral hydrogen atom, is that the single electron is subject to a coulomb field whose strength is determined by Ze for an ion of atomic number Z. The Lamb shift scales with Z approximately as Z^4, becoming larger relative to the electron binding energy as Z^2 for increasing Z. So for an ion such as U^{91+}, the Lamb shift is approximately 10^4 times as large as in neutral hydrogen, for a bound electron of given principal quantum number, n . In addition, it is conceivable that the sought for "breakdown in QED" might be detectable for "strongly bound" electrons , ie when Z is large, and yet not show up in the very precise measurements which exist for free electrons (for example g-2) or low Z atoms . Unfortunately, the study of highly charged ions is experimentally difficult. The most widely used technique is the fast beam, or beam-foil method [4]. In this method, ions are accelerated to high energy, and electrons are removed by fast collisions with solid or gaseous targets. The method has the advantage that relatively large numbers of ions may be produced, and essentially *any* charge state can be reached. However, the disadvantage is that the charge state reached is determined by the beam velocity, with high charge states being necessarily associated with high velocities. Spectroscopic measurements on fast beams of highly charged ions are therefore generally prone to errors arising from the Doppler effect, although methods have been devised to overcome these problems, and I shall describe some of these.

Highly charged ions may also be produced in plasmas of various kinds. In general, these plasmas must be heated to generate hydrogenic ions of high Z, so that, again, Doppler effects can become problematic. Another source of highly ionised atoms is the recoil-ion source [5]. This is a most interesting source spectroscopically, since the highly charged ions are produced with *directed*,low velocities. The low recoil ion velocities have already been exploited to make precise measurements of Lyman α wavelengths in recoil ions of few-electron argon [6] although these measurements demonstrate one of the apparent limitations of the technique, namely that the transitions of interest are very often surrounded by "satellites" arising from the lower charge state ions which are also present in a recoil-ion source. The *directed* nature of the recoil-ion source means that almost Doppler free spectra could be obtained by *axial* observation, as suggested by Martin Laming and myself[7]. Spectral resolutions of order 10,000 might be expected for resonance transitions in hydrogenic or helium-like argon, so that there is some hope that the satellite problem might be overcome with this new method.

Finally, there have been recent significant advances in techniques for the production of highly charged ions in what one might term "novel ion sources". There are three important types of device which have received attention, the Electron Cyclotron Resonance (or ECR) source, invented by R.Geller, the Electron Beam Ion Source (or EBIS), invented by E.D. Donets[8], and the Electron Beam Ion Trap (or EBIT) which is a derivative of the EBIS due to M.A. Levine [9]. This latter device is currently of serious interest, since useful numbers of very highly charged ions, such as Au^{69+}, have already been trapped, and it is believed [10] that an improved device (the SuperEBIT) can be built within the next year or two which will be able to generate and trap the most highly charged species, such as U^{92+}.The SuperEBIT is also planned to trap a larger number of ions than the original EBIT device, so that it will more nearly approach the spectroscopic brightness of an accelerator source.

The 1s Lamb Shift

There is a little unclarity in the literature concerning what is meant by "the Lamb shift". Some would rather narrowly understand the Lamb shift to be the 2s $^2S_{1/2}$ - 2p $^2P_{1/2}$ splitting in atomic hydrogen, and the hydrogenic ions. There is, however, a Lamb shift of the 1s $^2S_{1/2}$ state, which is about eight times larger than the 2s $^2S_{1/2}$ - 2p $^2P_{1/2}$ splitting, and when using the term Lamb shift, I have in mind the definition of Johnson & Soff [11]. In particular, the energy shift arising from the finite size of the nucleus is, by convention, included in the Lamb shift, although it is not a QED effect. A discussion of this point is given by R.G. Beausoleil [12]. The 1s Lamb shift is of some historical interest, since it appears to have been the *first* Lamb shift to be observed, some fifteen or so years prior to the Lamb and Retherford experiment[13]. The method of measurement is also interesting in that the Lamb shift in question (in Li^{++}) was made via a careful absolute Lyman α wavelength measurement; a comparison of the measured wavelength with the theoretical (ie Dirac) value gave a difference, and that difference was the 1s Lamb shift. This method, which I will call the method of absolute spectroscopy, is still in use by some researchers. It has the obvious drawback that since the 1s Lamb shift is only a small fraction of the 1s - 2p energy separation, wavelength measurements must be extremely accurate in order to yield an interesting measurement of the 1s shift. It will probably be helpful to qualify what is meant by the terms "extremely accurate", and "interesting" in this context. In the ideal experimenter's situation, there would be some predicted level at which QED would break down, so that an accurate measurement of the Lamb shift at this level could be used to establish whether or not the predicted breakdown had occurred. At present, there is no such predicted breakdown, so that we (that is the Oxford group and our collaborators) aim to make our experiments as accurate as the claimed uncertainty in the most reliable calculations of the Lamb shift [14]. It turns out that for ions with nuclear charge Z in the range 10 to 92, the theoretical uncertainty in the order α self energy correction is about one part in 1000 for both the 1s and 2s Lamb shifts, though the uncertainty in the nuclear charge radius must of course be taken into account in estimating the total theoretical uncertainty. To take a specific example, for the hydrogenic ion Fe^{25+}, the 1s Lamb shift is $\sim 6\times10^{-4}$ of the 1s-2p separation. So a 1ppm measurement of the Lyman α wavelengths will yield the 1s shift to ~+/- 0.17 %, just about at the level I call "interesting".

An alternative approach to the problem of 1s Lamb shift measurement is to utilise the simple integral relations between the wavelengths of certain transitions in the hydrogen soectrum. In particular, in the Schrodinger approximation, the Balmer β wavelength is four times the Lyman α wavelength; in a hydrogenic ion of high enough Z, both Lyman α and Balmer β lie in the x ray region of the spectrum, and will "overlap" if studied in first and fourth order of Bragg diffraction respectively. This gives what is effectively a differential method for the measurement of the 1s ground state Lamb shift, as I realised in 1981 after reading about Ted Hänsch's related work on the two photon 1s - 2s transition in atomic hydrogen. The Lyman α / Balmer β technique [15,16,17] and its derivatives might be termed "differential spectroscopy" since the 1s Lamb shift may be extracted as a difference rather than a small change in a large quantity. The method is rather useful when applied to a fast beam source, since it effectively eliminates Doppler shifts, which are often the dominant source of error in such experiments. The need for accurately known external calibrations is also obviated.

In recent years, several experimenters have tried to make interesting measurements of the 1s Lamb shifts in hydrogenic argon, chlorine, iron, germanium and krypton.There have been three experiments on hydrogenic chlorine, two with fast beam sources [18,19] and a third with a tokamak source [20]. There have been two experiments with hydrogenic argon, one with a recoil-ion source [6] and the other with a tokamak source [21]. There have been two fast beam experiments with hydrogenic iron [22,16,17] , a fast beam experiment with hydrogenic germanium [23, 24], and a fast beam experiment with hydrogenic krypton [25] . Of these experiments, only two (those carried out by the Oxford group in collaboration with the Livermore group [16,17,23,24]) used the method of differential spectroscopy.

The number of experiments is still sufficiently small that it is possible to comment on each of them; the earliest fast beam experiment with hydrogenic chlorine [18] was a very straightforward measurement for which the conceptual set-up is shown in figure 1 of Silver [17a] .

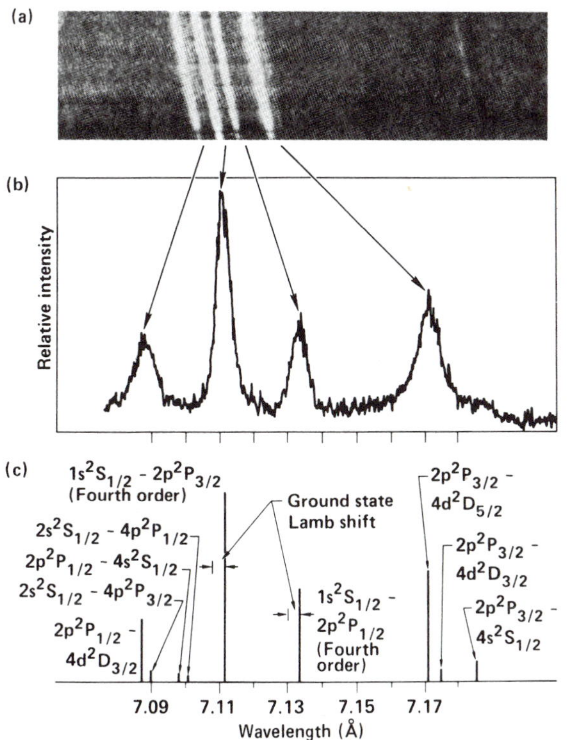

Figure 1: a) A print of a typical film in the Lyman α / Balmer β region for hydrogenic iron [ref 17]. b) A microdensitometer scan of the Lyman α /Balmer β region. c) The predicted spectrum using statistical relative intensities

Experiments of this type suffer from errors arising from the Doppler effect inherent with fast beam sources, and they can also suffer from contamination of the spectrum with satellites to Lyman α from ions of lower charge, and these effects seem to have dominated the final errors in the measurement; the Lyman α wavelengths were obtained with an accuracy of~ 34 ppm , giving the 1s Lamb shift to ~12% .The second measurement of the Lyman α wavelength in hydrogenic chlorine used the accel-decel method originated by Paul Mokler [26]. Here the ions to be studied are accelerated to high energy, stripped, and then decelerated. In the experiment with hydrogenic chlorine [19] decelerated bare chlorine nuclei captured a single electron in a gas target, and it was suggested that the method in principle overcomes both the satellite problem, and the Doppler problem. Unfortunately, the decelerated ions still have appreciable velocity so that residual Doppler shifts are not negligible. Measurements must therefore be made over a range of beam energy,but the capture probability for higher beam energies is small, so that spectra are weak. This in turn means that the Doppler correction cannot be extracted as accurately as hoped . Nonetheless, some of the spectra obtained with this technique are of very high quality and apparently free from serious satellite contamination, the Lyman α wavelengths were obtained with an accuracy of ~41 ppm, giving the 1s Lamb shift to ~15%.The third measurement of the Lyman α wavelengths in hydrogenic chlorine[20] was carried out using a tokamak as a light source. The use of a tokamak as a spectral "lamp" for precision spectroscopy of highly ionised atoms appears to have originated with Peacock and Silver [27] . We were interested in making accurate measurements of the wavelengths of the 1s2s ^3S - 1s2p ^3P transitions in moderate Z helium-like ions, since such measurements act as a sensitive test of relativistic and quantum electrodynamic effects in the two-electron system. Following this early work with helium-like ions , a tokamak was used to study the Lyman α wavelengths in hydrogenic chlorine . High quality spectra were obtained, but there were uncertainties in the extracted wavelengths arising from instrumental effects, and from the presence of overlapping satellites in lower charge.The Lyman α wavelengths were measured with an accuracy of ~34 ppm, giving the 1s Lamb shift to ~12% . A potential source of error in such measurements arises from possible net motions of the plasma, though Kallne et al pointed out [20] that a dual spectrometer system could possibly overcome this effect.

Of the experiments in hydrogenic argon, the first used the relatively new recoil-ion technique [5,6]. As explained above, the recoil ion technique suffers from the satellite problem. For the experiments in argon[6] , high quality spectra were obtained, and a rather good precision of 5ppm was claimed for the Lyman α wavelengths, which corresponds to 1.5% accuracy for the 1s Lamb shift. A visual inspection of the spectra, however, shows serious satellite contamination, and proper interpretation of the spectra was clearly rendered difficult by this effect. The second experiment in hydrogenic argon [21] was again carried out with a tokamak source. Improved accuracy was claimed (11ppm for Lyman α_2 giving 3% for the 1s shift) compared with the earlier experiment with hydrogenic chlorine, and it was suggested that the influence of satellite lines in the tokamak source was only of secondary importance, although the spectra do show evidence of some satellite contamination.

The first experiment in hydrogenic iron to be published [20] used the straightforward "absolute spectroscopy" technique [17a] as for hydrogenic chlorine. The authors claimed that their precision was for the first time sufficient to observe the 1s Lamb shift for any hydrogenlike atom heavier than neutral hydrogen, apparently unaware of the measurement in hydrogenic lithium made by Edlen in 1933 [13] . Rather weak spectra with no apparent satellite contamination were obtained, the dominant source of error appears to have been the Doppler effect , and an accuracy of about 100ppm was claimed for the Lyman α wavelengths, giving ~ 17% accuracy for the 1s Lamb shift. A small disagreement was found between the measurements and the theoretical (Dirac) value of the 2p fine-structure splitting, but this does not seem to have been substantiated by more recent work. The second experiment in hydrogenic iron was carried out by the Oxford group in collaboration with Daniel Dietrich and his group from Livermore, and it was the first succesful demonstration of the "differential spectroscopy" technique using the overlap of Lyman α with Balmer β [16,17].Figure 1 shows one of the first spectra observed using the new technique. This experiment was carried out using the SuperHILAC at Lawrence Berkeley Laboratory in 1986. A beam of bare iron nuclei (Fe^{26+}) was passed through a thin carbon foil, and the n=2 and n=4 levels were populated by capture.The spectra obtained were interesting in that they demonstrated that the technique was actually experimentally feasible. The Lyman α wavelengths were measured to ~107 ppm, leading to ~17% error in the 1s Lamb shift. However, a hundredfold improvement in the accuracy is required in order to get to the level defined above as "interesting" in the sense that the error is the same as the theoretical uncertainty in the Lamb shift.

The only 1s Lamb shift measurement in hydrogenic germanium [23,24] also used the Lyman α / Balmer β technique, and was a development which grew out of the first experiment with iron. It was carried out at GSI Darmstadt in September 1986. Figure 2 shows a typical Lyman α / Balmer β spectrum of Ge^{31+} , obtained with a 1.108 GeV beam of hydrogenic germanium ions. A comparison of the data between figure 1 and figure 2 shows that the resolution and statistical quality is better for germanium than for iron in our experiments; one reason for this is that the germanium transitions are at shorter wavelength, so that a silicon III diffraction crystal could be used. Another is that the ion beam quality was better with the germanium beams than with the iron beams. Unfortunately, at the time of writing, it is not possible to give final values for the Lyman α wavelengths in Ge^{31+} from our data. Although the purely statistical errors in the best data are at the level of only a few ppm, there are systematic errors which are still under investigation. One source of error which is under examination is an *instrumental* shift between first and fourth orders. One contribution to this shift is the so called refractive index correction [16,23], but for the small radius Rowland circle crystals used in the iron and germanium experiments, we believe that there are other factors to be considered if we are to reach a full understanding of this systematic effect.

Another source of error arises from the uncertainty in the relative intensities of fine structure components of the Balmer β transitions in our spectra. During an analysis of the data from the experiment with hydrogenic Germanium,Martin Laming realised that the assumption of statistical population of the n=4 states which had been made previously is probably not justified [23,28]. One hint that this may be the case comes from the fact that in the Lyman β spectrum, the relative intensities of the two components are not in the statistical ratio of 2:1. We now believe that account must be taken of the presence of the large electric field which the bare nucleus experiences in the carbon target *during* the single electron capture process which populates the n=2 and n=4 states.The eigenstates of the one electron ion formed during capture are those appropriate to a stark mixed system, and in particular the electric field can be strong enough to completely mix the 4p $^2P_{3/2}$ and 4d $^2D_{3/2}$ states. The relative intensities in the Balmer β spectrum are affected by this , and since not

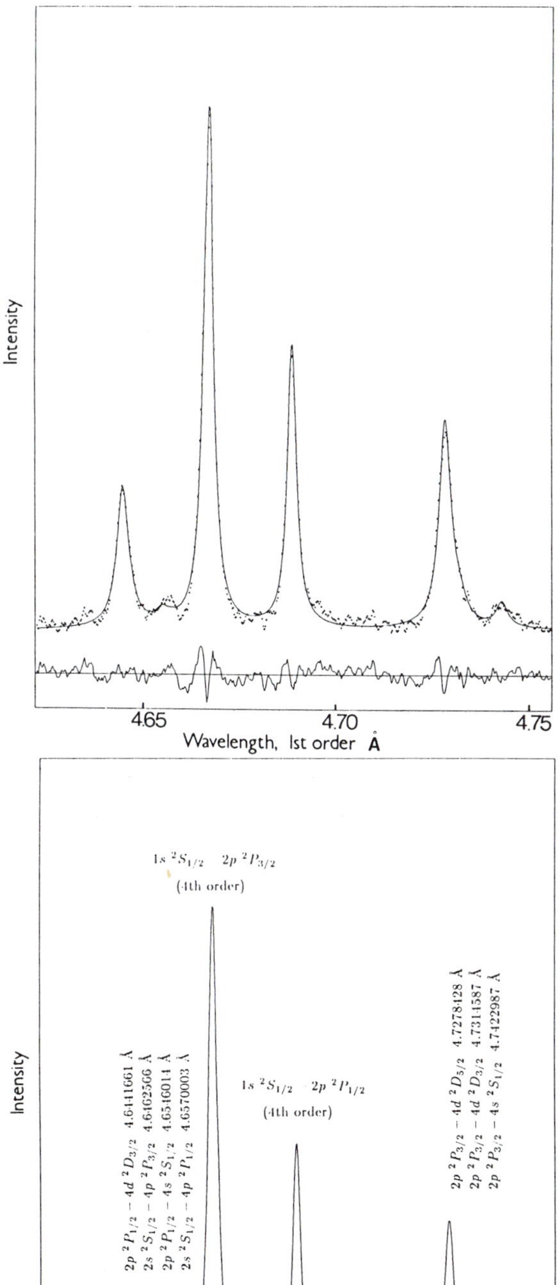

Figure 2: a) Typical spectrum of the Lyman α / Balmer β region for hydrogenic germanium [ref 24] b) Fitted line profiles for the individual transitions of the Lyman α / Balmer β region

all the components are resolved in our spectra, this phenomenon can lead to shifts in the spectrum if statistical rather than stark mixed relative intensities are used in the profile analysis. We currently think that the instrumental first order-fourth order shift, and the non-statistical relative intensity effect taken together can account for the apparent discrepancies noted in table 2 of Laming et al [24] . The germanium data is undergoing further analysis, and we expect final errors for the Lyman α wavelengths in Ge^{31+} from our experiment will probably be about 30ppm, (corresponding to an accuracy of ~3% for the 1s Lamb shift) when these further systematic errors are properly taken into account. We have also more recently carried out a further Lyman α /Balmer β experiment on hydrogenic iron with improved diffraction crystals, and the data from this experiment are currently under analysis by Chris Chantler. It is hoped that this more recent experiment will yield better resolved spectra of Fe^{25+}, giving an improved value for the 1s Lamb shift.

The highest Z system for which the Lyman α wavelength has been measured is currently hydrogenic krypton [25]. Again the measurement was carried out by the absolute spectroscopy method. Discussion of errors is unfortunately a little brief, and no spectra are shown, but it is claimed that 26 ppm error in the Lyman α wavelengths arises from Doppler effects, with a similar error being attributed to the spectrometer itself, including angular errors, statistics and linearity effects. These effects are then combined to give a claimed 36ppm accuracy for the Lyman α_1 line, leading to 4% claimed accuracy for the 1s Lamb shift.

Comparison between experiment & theory, and future work

Figure 3 shows a comparison between experiment and theory for the 1s Lamb shift for the hydrogenic ions of chlorine, argon, iron and krypton. No results are given for germanium since our data are still under analysis. It will be seen that none of the existing measurements can be said to test the theory, in the sense that in no case does the accuracy of the experiment approach the claimed theoretical uncertainty. If the aim of this sort of experiment is to provide a test of QED,then improved measurements are clearly required; in general terms, experimental accuracies for the Lyman α wavelengths of ~1ppm are needed. This might prove possible with the "absolute wavelength measurement" approach using the accel-decel or possibly even the fast-beam method, though extreme care will be needed to make adequate allowance for Doppler shifts and satellite effects. It is also possible that measurements of

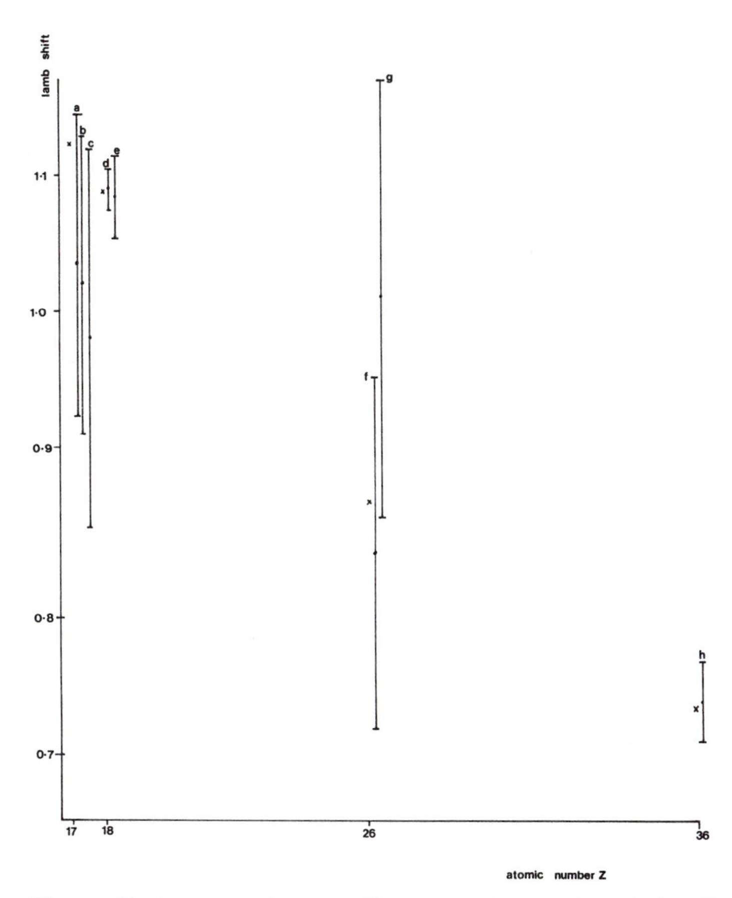

Figure 3: A comparison of theory and experiment for the Lamb shift contribution to the 1s - 2p transitions in hydrogenic ions with Z =17,18,26 and 36. The small crosses give the theoretical values [14], with the theoretical uncertainty too small to show on the scale used. The points with bars are measured values, references are a: 18],b:[20], c:[19], d: [6], e:[21], f: [22], g:[17], h:[25]. For convenience of display, the Lamb shifts are given in eV scaled by $10x\,(Z/10)^4$.

sufficient accuracy might be made with a tokamak source, though again great care will be needed with the same error sources. The "differential" method using Lyman α / Balmer β comparison has already demonstrated *statistical* precisions at the required few ppm level. The purely instrumental problems involved in a comparison of first and fourth order wavelengths at the ppm level appear tractable, and the problem associated with uncertainties in the relative intensities of different fine-structure components can be tackled if we can achieve improved resolution. With this in mind, we have recently investigated experimentally a rather old suggestion to improve spectral resolution using *axial* viewing of a fast beam source. Axial viewing of a fast beam source under appropriate conditions gives a significant reduction in Doppler broadening , we recently used this method to obtain the best resolved spectra in hydrogenic neon ions and helium-like silicon ions[29], in a collaboration involving Oxford, Leicester and Culham Laboratory, and we believe that this additional "new" technique, when combined with the Lyman α /Balmer β comparison method, offers the promise of eventual ppm accuracies for the Lyman α wavelengths, and, correspondingly, ~0.1% accuracies for the 1s Lamb shifts. We plan experiments with axial viewing of fast beams of Ge^{31+} at GSI Darmstadt within the coming year, to see whether this optimism is justified, and we also anticipate an extension of this series of experiments to the hydrogenic ions of very high Z atoms (ie Pb^{81+}, U^{91+}) when the new SIS high energy heavy ion accelerator is operating at Darmstadt. This new accelerator will give beams of such exotic ions at about the same intensity as is currently available for the systems we have already studied, such as Fe^{25+} and Ge^{31+}.

Another very interesting new development concerns the use of the newly developed EBIT [9] device for 1s Lamb shift measurement in hydrogenic ions. The EBIT has already been demonstrated to produce useful quantities of ions of extremely high charge, such as Au^{69+}. The source appears quite well suited to precision spectroscopy; the highly charged ions form a "filament" ~ 50μm in diameter and ~ 1cm long, and the source can be of comparable brightness to a fast beam . Relatively clean spectra of the one electron ion can be expected, free from satellite contamination, and one of the main problems associated with fast beams, namely the Doppler shift, will be much less serious in the EBIT, since ion velocities, though not directed, are many orders of magnitude lower than a fast beam for a given charge. The electric and magnetic fields present in the EBIT , though appreciable, will also only give negligible shifts in the low n levels of highly charged hydrogenic ions. Accurate measurement of Lyman alpha wavelengths in ions such as Fe^{25+}, Ge^{31+} , etc may therefore be expected using the EBIT as a source, and one can also envisage the application of an improved EBIT, (or "SuperEBIT") operating with higher electron currents and voltages than the original EBIT device to the problem of 1s Lamb shift measurement in hydrogenic ions of very high charge such as Pb^{81+}, or U^{91+}

The 2s Lamb shift

When considering the Lamb shift in hydrogenic ions as a test of QED, it is not sensible to look at a comparison of theory and experiment for the 1s Lamb shift alone; indeed, the situation for the 2s Lamb shift is probably the more interesting at the time of writing, since the measurements, at least in the range of Z between 15 and 17, are somewhat more accurate than the measurements of the 1s Lamb shift, and in addition, there is a hint of a discrepancy between experiment and theory, as shown in figure 4. The most recent measurement is that of Muller et al in hydrogenic phosphorus, [30,31] and this measurement lies 1.6 standard deviations below the theoretical result for the 2s Lamb shift. This article is not the

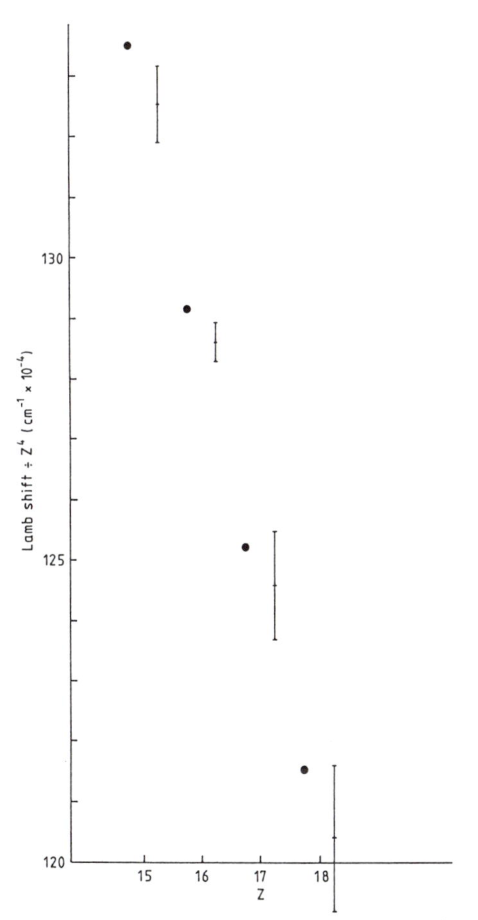

Figure 4: A comparison of theory [ref 14](solid dots with the theoretical uncertainty given by the size of the dot) with experiment [see refs 30,31] (bars give experimental errors) for the 2s Lamb shift in hydrogenic ions with Z = 15,16,17,18.

place for an exhaustive review of measurements of the n=2 Lamb shift, but it is clear that, as in the case of the 1s shift, further experiments are really required. The two most accurate 2s Lamb shift experiments shown in figure 4, for phosphorus [31] and sulphur [32] were carried out using fast beam resonance spectroscopy, and may be viewed in a sense as an extension of the 2s Lamb shift measurements started by Pipkin & collaborators on fast beams of neutral hydrogen [33].

In our group in Oxford we have been interested in the application of accurate laser spectroscopy to highly charged ions for an extended period [34,35,36], and indeed we have made several attempts over the years to secure support from SERC to make a "definitive" laser resonance measurement of the 2s Lamb shift for hydrogenic ions with Z in the range 10 to 20. These attempts have not been very successful, and my own thinking on the problem of how to measure the 2s Lamb shift in medium Z hydrogenic ions to 0.1% accuracy has evolved somewhat, so that we currently have three experiments under consideration which may prove promising.

The first of these is based on the fact that the 2s $^2S_{1/2}$ - 2p $^2P_{3/2}$ transition (from which transition the 2s Lamb shift is extracted using a theoretical value for the 2p finestructure splitting) for hydrogenic ions in the range of Z of interest has a very large natural width on account of the short 2p state lifetime. One is therefore led to ask whether an expensive and specialised laser of the sort used by us[34,35] and others [37,30,32] in earlier experiments is really necessary in order to carry out resonance spectroscopy, or whether indeed a bright lamp might perhaps do instead ! As a further step, one can also ask whether there might perhaps be some sort of coupling optics which could be used to match the spectrally dispersed radiation from our bright lamp to a fast beam in such a way that the beam "sees" a wide range of the spectrum as if it were monochromatic !! It turns out that this apparently rather strange situation seems quite easy to arrange in the laboratory, suitable coupling optics having been investigated some years ago in Oxford [38], and more recently in Canterbury [39]. At the time of writing, we have just set up such a system, which is described more fully elsewhere [40,41], at the SuperHILAC in Berkeley for an experiment whose aim is to measure the 2s Lamb shift in hydrogenic silicon and aluminium via a resonance measurement of the wavelength of the 2s $^2S_{1/2}$ - 2p $^2P_{3/2}$ transition of a fast beam of hydrogenic silicon or aluminium ions. We hope to try this experiment in the next few months. A feasibility study of this new

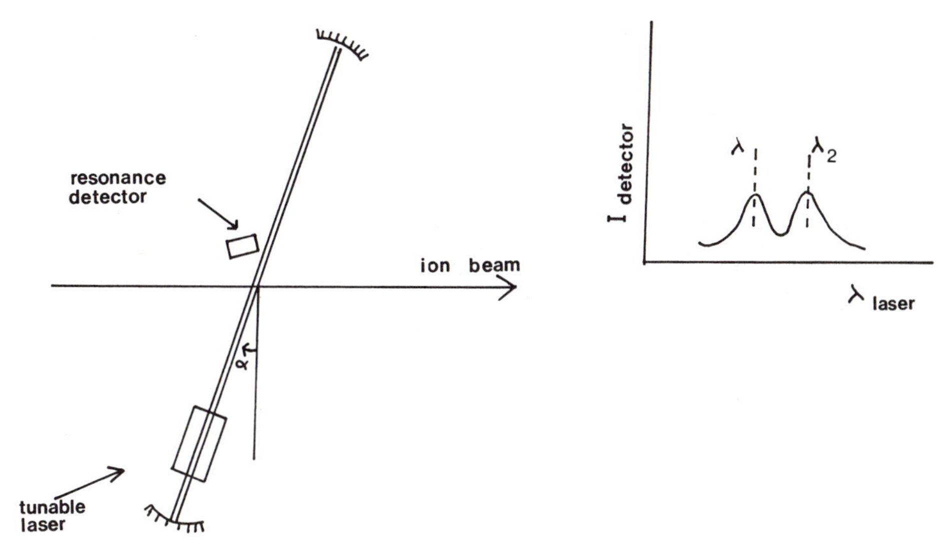

Figure 5: Schematic diagram for a proposed intracavity laser resonance measurement of the 2s $^2S_{1/2}$ - 2p $^2P_{3/2}$ interval in a hydrogenic ion such as Si^{13+}. In the resonance spectrum, $\lambda_1 = \gamma\lambda_0(1+\beta\sin\alpha)$, and $\lambda_2 = \gamma\lambda_0(1-\beta\sin\alpha)$, where λ_0 is the wavelength of the resonance in the ion's rest frame, the ion velocity $v = \beta c$ with c the speed of light, and $\gamma = (1-\beta^2)^{-1/2}$. So $\lambda_1 + \lambda_2 = 2\gamma\lambda_0$, and $\lambda_1 - \lambda_2 = 2\gamma\lambda_0\beta\sin\alpha$. If α can be varied precisely, β and λ_0 may be found.

technique [41] (which one might perhaps call laserless resonance) does suggest, however, that we will still encounter problems associated with the large Doppler effect present with fast beam sources. A way to overcome this is to return to the use of laser radiation, but to use intracavity irradiation of the fast beam. The situation will then be as depicted in figure 5, with two counterpropagating waves interacting with the ion beam. If the laser beam direction is slightly offset (by an angle α) from the normal to the ion beam, and the laser is scanned so that the resonance driven by the forward and return beams may each be observed, as depicted schematically in the figure, then a first-order Doppler free result may be obtained for the resonance of interest. If in addition we set the system up so that the angle α may be precisely varied, then a completely Doppler free result may be obtained from a series of spectra taken as a function of α. This suggested method is conceptually similar but not quite the same as a proposal for accurate Lamb shift measurements made for the new SIS accelerator [42] as well as earlier accurate laser resonance experiments in moleclues, and ions carried out by Dick Holt and Dave Rosner and colleagues [43,44,45].

The third technique which we have under consideration for a resonance measurement of the 2s Lamb shift involves the use of an EBIT. The experiment would be not unlike that proposed for application to a recoil-ion source by McClelland et al[36] , and an analysis similar to that presented by those authors suggests that the 2s $^2S_{1/2}$ - 2p $^2P_{3/2}$ resonance transition should be observable in hydrogenic ions with Z~14 - 17 using intracavity laser radiation of these ions in an EBIT, and that the perturbations arising from the fields in the EBIT would not prevent ~0.1% measurements of the 2s Lamb shift for these ions.

References

1. R.P. Feynman: in QED, The Strange Theory of Light and Matter, Princeton University Press (1985)
2. P.A.M. Dirac: Scientific American 208 , 45 (1963)
3. Attention is drawn, however, to A. Rich's contribution to this symposium regarding a discrepancy between QED theory and experiment for the positronium decay rate
4. L.Kay: Nucl. Instr. & Meths. B9 544 (1985), and S. Bashkin, Nucl.Instr. & Meths. B9 546 (1985)
5. I.A. Sellin et al, Phys Lett A61 107 (1977)
6. H.F. Beyer et al, J.Phys B18 207 (1985)
7. J.M. Laming and J.D. Silver, Phys Lett A123 395 (1987)
8. E.D. Donets and V.P. Ovsyannikov, JETP 53 466 (1981)
9. M.A. Levine et al, Physica Scripta, T22 157 (1988)
10. M.A. Levine, private communication
11. W.R. Johnson and G.Soff, Atomic Data and Nuclear Data Tables 33 406 (1985)
12. R.G. Beausoleil, Thesis, Stanford University (1986)
13. B. Edlen, Nova Acta Reggiae Societatis Scientarium Uppsaliensis Ser IV 9 28 (1934)
14. P.J. Mohr, Atomic Data and Nuclear Data Tables 29 435 (1983), see also reference [11]
15. A.F. McClelland et al,Nucl. Instr.& Meths. B9 706 (1985)
16. A.F. McClelland, Thesis, University of Oxford (1988)

17. J.D. Silver et al, Phys. Rev. A36 1515 (1987), 17a. J.D. Silver, in Physics of Strong Fields, 655 Ed W. Greiner Plenum Press (1987)
18. P. Richard et al, Phys. Rev. A29 2939 (1984)
19. R.D. Deslattes et al, Phys.Rev. A32 1911 (1985)
20. E. Kallne et al, J. Phys. B17 L115 (1984)
21. E.S. Marmar et al, Phys. Rev. A33 774 (1986)
22. J.P. Briand et al, Phys. Rev. Letts. 50 832 (1983)
23. J.M. Laming, Thesis, University of Oxford (1988)
24. J.M. Laming et al., Nucl. Instr. & Meths. B31 21 (1988)
25. M. Tavernier et al, J. Phys. B18 L327 (1985)
26. P.H. Mokler et al, Nucl. Instr. & Meths. B10 58 (1985)
27. M.F. Stamp et al, J. Phys. B14 3551 (1981)
28. J. M. Laming, to be published
29. J. M. Laming et al, Phys. Lett. A126 253 (1988)
30. D. Muller et al, Europhys. Lett. 5 (6) 503 (1988)
31. D. Muller, Thesis,University of Cologne, (1988)
32. A.P. Georgiadis et al, Phys. Lett. A115 109 (1986)
33. S.R. Lundeen and F.M. Pipkin, Metrologia 22 9 (1986) and earlier references therein.
34. J.D. Silver et al, Appl. Phys. Lett. 31 278 (1977)
35. E.G. Myers et al., Phys. Rev. Lett. 47 87 (1981)
36. A.F. McClelland et al, Nucl. Instr. and Meths. B9 710 (1985)
37. H.W. Kugel and D.E. Murnick, Repts. on Progr. in Phys. 40 299 (1977)
38. N.A. Jelley et al, J. Phys. B10 2339 (1977)
39. L.Kay and M.A. Wood, J. Phys. E19 830 (1986)
40. J.D. Silver, Physica Scripta 37 720 (1988)
41. S. Lea, Report, University of Oxford (1988)
42. P. von Brentano et al, Proceedings of the Workshop on Experiments and Experimental Facilities at SIS/ESR, GSI Report 87-7 (1987)
43. S.D. Rosner et al, Phys. Rev. Lett. 35 785 (1975)
44. S.D. Rosner et al, Phys. Rev. Lett. 40 851 (1978)
45. R.A. Holt et al, Phys Rev A22 1563 (1980)

Part III

Quantum Electrodynamics and Beyond

The Bound State Problem in QED*

A. Hill, F. Ortolani, and E. Remiddi

Dipartimento di Fisica, Università di Bologna and
INFN, Sezione di Bologna, Bologna, Italy

I. Introduction.

In ordinary quantum mechanics there is a Hamiltonian H and one tries to solve the corresponding Schrödinger eigenvalue equation

$$H\,|\mathrm{n}\rangle = E_n\,|\mathrm{n}\rangle \ , \tag{1}$$

where the notation for the case of discrete eigenvalues is used. From the full knowledge of eigenvalues E_n and eigenstates $|\mathrm{n}\rangle$ one can then construct the Green function

$$G = i\sum_n \frac{|\mathrm{n}\rangle\langle\mathrm{n}|}{E-E_n+i\varepsilon} \ . \tag{2}$$

In most cases H is too complicated and eq. (2) cannot be solved in closed form. If one writes

$$H_1 = H + V, \tag{3}$$

where H is a simpler Hamiltonian, whose solutions are supposed to be known, and V a perturbation which is small compared to H, standard perturbation theory can be used to get the eigenvalues $E_n^{(1)}$ of H_1 in the customary form

$$E_n^{(1)} = E_n + \langle V\rangle_n + \cdots. \tag{4}$$

Similarly, the Green function for the Hamiltonian (3) becomes

$$G^{(1)} = G + GVG + \cdots. \tag{5}$$

In QED the situation is quite different. The starting point for the Coulomb bound state problem is the (renormalized) perturbative expansion in the fine structure constant $\alpha \approx 1/137$ of the Green function for the relevant two body scattering problem (μ^+e^- scattering for muonium, the μ^+e^- bound state, etc.),

$$G = \sum_{n=0}^{\infty}\left(\frac{\alpha}{\pi}\right)^n G^{(n)}, \tag{6}$$

where the $G^{(n)}$ are sums of the renormalized off mass–shell Feynman graphs for that scattering process. The kernel for the expansion (6) is then defined as

* Partly supported by Ministero Pubblica Istruzione.

The Hydrogen Atom Editors: G.F. Bassani · M. Inguscio · T.W. Hänsch

$$K = G_0^{-1} - G^{-1} , \tag{7}$$

where G_0 is a zeroth order Green function for the problem. If G_0 is the product of the free propagators of the two scattering particles, then, since G and G^{-1} are (formal) power series in $\left(\frac{\alpha}{\pi}\right)$, K itself can also be obtained as a (formal) power series in $\left(\frac{\alpha}{\pi}\right)$:

$$K = \sum_{n=0}^{\infty} \left(\frac{\alpha}{\pi}\right)^n K^{(n)}. \tag{8}$$

Combinatorics shows that K is the sum of all two-fermion irreducible graphs.

From (7) one obtains the Bethe-Salpeter equation

$$G = G_0 + G_0 K G ; \tag{9}$$

the corresponding equations for the wave function and its conjugate are

$$\begin{cases} \psi_n = G_0 K \psi_n , \\ \bar{\psi}_n = \bar{\psi}_n K G_0 . \end{cases} \tag{10}$$

Note that according to eq. (7), eq. (9) is true almost by definition; its justification derives from the fact that by trivial perturbative iteration it reproduces the (formal) expansion eq. (6). The kernel, as defined through eq. (7), plays the role of the interaction in that approach. With respect to ordinary quantum mechanics the situation is therefore almost reversed. The Green function eq. (6) is known and the kernel is obtained from it; the problem is that the (formal) expansion eq. (6) does not exhibit a bound state at any order in α; eqs. (7),(8),(9) are then used as a guideline for a suitable resummation of (6). The kernel is an infinite (formal) series in itself that depends explicitly on the energy. An exact solution of eq. (9) in closed form is obviously out of reach; it is almost mandatory to pick up a leading Coulomb kernel K_c and a perturbation kernel δK defined by

$$K = K_c + \delta K , \tag{11}$$

where δK is again a power series in $\left(\frac{\alpha}{\pi}\right)$. Then one can consider the lowest order equation

$$G_c = G_0 + G_0 K_c G_c ; \tag{12}$$

the solution of (12) gives the unperturbed bound state energy levels as the position E_n^c of its singularity in the complex energy plane and the corresponding wavefunctions are obtained from the residues. Standard perturbation theory then gives the corrections to the energy levels

$$E_n = E_n^c + \langle \delta K \rangle_n + \langle \delta K \hat{G}_c \delta K \rangle_n + \langle \delta K \rangle_n \langle \delta K' \rangle_n + \cdots , \tag{13}$$

where $\hat{G}_c$ is equal to G_c without the nth pole, $\delta K'$ is the derivative of δK with respect to the energy and $\langle\rangle_n$ is the average over the nth wave function. The expansion (13) can then be rearranged as a series in $\left(\frac{\alpha}{\pi}\right)$.

It took some time to invent a suitable way of writing a Coulomb kernel K_c for which a solution in closed form can be derived[1]. Most of the results which were found so far in bound state calculations, have indeed been obtained with very little or no direct use of the formalism sketched above. In hydrogen and $\mu^+ e^-$, e. g., it is natural to expand in the ratio of the masses of the electron and the positive particle and therefore to use a simplified formalism from the beginning. In positronium, however, the masses of the constituents are equal and simple approximations are not adequate (see, however, the recent approach by Caswell and Lepage[2] who propose an interesting way to circumvent the Bethe–Salpeter covariant formalism in favor of a purely nonrelativistic treatment).

II. Brief Review of Results.

Not very much has changed since the 1984 Washington conference on Atomic Physics, to which we refer for further details[3].

i. Hydrogen Lamb Shift. Historically this is the first bound state problem which was attacked while the foundations of QED were at the same time being laid. In most of the classical results the proton can be regarded as an external field and one deals with mean values of relativistic electron self mass radiative corrections averaged over the nonrelativistic Schrödinger wave function. Unfortunately the accuracy of Lamb shift results is, by present standards, very limited for both experimental and theoretical reasons: the natural line width of the $2P_{1/2}$ state (about 100MHz) makes it almost impossible to measure the 1.058 GHz of the Lamb shift to better than 10 ppm or 10 kHz and the contribution of proton structure (finite size) effects has a 7≈10 kHz error which theory cannot compute.

ii. H hyperfine splitting. A similar situation occurs in the H hfs. Here the experimental results are extremely good (with a relative error of 10^{-13}) but again proton finite size effects spoil the theoretical prediction at the 1 ppm level.

iii. Muonium hyperfine splitting. No theoretical limitations are present to μ^+e^- which can, in principle, be evaluated to any desired accuracy. Indeed the precision of the theoretical prediction is very good, 0.5 ppm. The bound state part of the problem is greatly simplified by an expansion in $\frac{m_e}{m_\mu}$ (as well as in α). $O(\alpha^2)$ and $O(\alpha\frac{m_e}{m_\mu})$ corrections are known and only some $O(\alpha^3)$ corrections remain to be calculated. Unfortunately, because of this dependence on $\frac{m_e}{m_\mu}$ besides the Rydberg constant and α, the muonium hfs is used to obtain a more accurate value for this ratio rather than being a stringent test of QED or an independent, precise determination of α. An update on these results can be found in ref[4].

III. Positronium.

As for muonium, no a priori limitations prevent positronium from being used as an accurate test of bound state QED; in practice, however, there are serious difficulties to surmount before a competitive level of precision can be reached. The system is indeed so light that Doppler broadening and similar effects are experimentally particularly difficult to control, while on the theoretical side calculations are an order of magnitude harder than for the previously discussed systems. The most accurate prediction is in fact currently the ground state hfs:

$$(15) \qquad \Delta\nu(\text{hfs,th}) = \text{Ry}\,\alpha^2\left[\frac{7}{6} - \left(\frac{\alpha}{\pi}\right)\left(\frac{16}{9} + \ln 2\right) - \frac{5}{12}\alpha^2\ln\alpha\right],$$

where the $O(\alpha^2)$ corrections are still to be calculated.

Another important problem which has always been around has recently acquired new importance: the decay rates. New experimental results from 1987[5]:

$$(16) \qquad \Gamma(3\gamma,\text{exp}) = 7.0516(13)\times 10^6 s^{-1}$$

showed a 10σ(exp) disagreement with the theoretical prediction, lowest order plus one–loop $(\frac{\alpha}{\pi})$ corrections[6]:

$$\text{(17)} \qquad \Gamma(3\gamma, \mathrm{th}) = \frac{2}{9\pi}(\pi^2 - 9)m\alpha^6 \times \left\{1 - 10.282(3) \times \left(\frac{\alpha}{\pi}\right)\right\} \approx 7.0389 \times 10^6 s^{-1} .$$

In order to recover agreement between theory and experiment, one needs either a value of about $-9\left(\frac{\alpha}{\pi}\right)$ for the first order correction, i. e. a difference of $+1 \times \left(\frac{\alpha}{\pi}\right)$ with respect to the reported value $-10.3 \times \left(\frac{\alpha}{\pi}\right)$, or an extremely large $O\left(\frac{\alpha}{\pi}\right)^2$ term.

In an attempt to understand whether the discrepancy might be due to a theoretical error, we have carried out an ab initio recalculation of the one–loop correction to the decay rate using the full Bethe–Salpeter formalism of ref.[1] and exercising particular care in studying the approximations that are anyhow needed in the calculation. To focus attention on binding problems, we considered the simpler case of para–positronium (which is less well known experimentally). It is obvious that most of the following discussion applies to ortho–positronium as well. We have used Feynman gauge throughout.

Quite in general, decay rates are the imaginary parts of radiative energy shifts; the lowest order kernel $\delta K^{(2)}$ contributing to para–Ps decay is depicted in Fig.1:

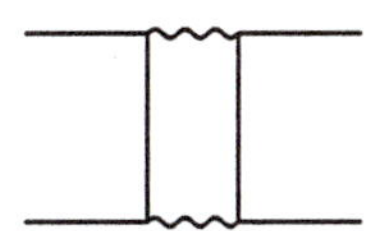

Fig. 1: The lowest order kernel $\delta K^{(2)}$ for parapositronium decay.

Following reference[1] we use the wave function

$$\text{(18)} \qquad \psi(\mathrm{W_0}, \mathrm{p_0}, \mathbf{p}) = \frac{i}{\mathrm{E_p}}\left[\mathrm{E_p} + (m - \vec{\mathbf{p}}\cdot\vec{\gamma})\gamma_0\right]\gamma_5 \frac{(\mathrm{E_p} - \mathrm{W_0})\sqrt{\mathrm{E_p} + \mathrm{W_0}}}{\mathrm{p_0^2} - (\mathrm{E_p} - \mathrm{W_0})^2 + i\varepsilon} \times \phi(\mathbf{p}),$$

where

$$\text{(19)} \qquad \phi(\mathbf{p}) = \frac{8\pi\gamma}{(\mathbf{p}^2 + \gamma^2)^2}\sqrt{\frac{\gamma^3}{\pi}},$$

with the notation

$$\text{(20)} \qquad \gamma = \frac{m\alpha}{2}, \qquad \mathrm{W_0} = \sqrt{m^2 - \gamma^2},$$

is the nonrelativistic Schrödinger wave function. As the very first step, we integrate on the relative energy p_0, then on $\mathbf{p}$. The contribution corresponding to Fig.1 is found to be

$$\text{(21)} \qquad \Gamma^{(2)} = \langle \delta K^{(2)} \rangle = \Gamma_0 \cdot \left\{1 + 2a_0\left(\frac{\alpha}{\pi}\right)\right\},$$

where

$$\text{(22)} \qquad \Gamma_0 = \frac{1}{2}m\alpha^5$$

is the lowest order decay rate, well known in the literature. For the coefficient of the $O\left(\frac{\alpha}{\pi}\right)$ correction in (21), we find $a_0 = -3.1137...$; its explicit value is, however, somewhat irrelevant, as it is found to be reabsorbed by the one-loop contributions to be discussed below.

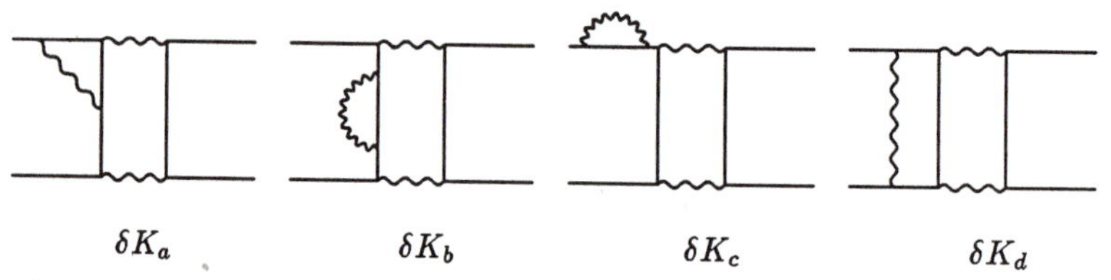

Fig. 2: Next to leading order kernels for parapositronium decay.

The kernels contributing to next to leading order in $\left(\frac{\alpha}{\pi}\right)$ are usually depicted as in Fig. 2; they contribute to the actual decay rate with multiplicities 2,4,2,2 respectively. Kernels $\delta K_{a,b,c}$ are UV divergent; to evaluate them we use a Pauli–Villars regulator with square mass A and then perform on–shell mass renormalization; we do not carry out on–shell wave-function and vertex renormalization because on the one hand that would introduce spurious IR divergences while, on the other hand, those remaining UV divergences cancel out in the sum anyhow.

We find

$$\Gamma_a = \langle \delta K_a \rangle = \Gamma_0 \left(\frac{\alpha}{\pi}\right)\left(-\frac{1}{4}\ln A + 2\ln 2 - \frac{5}{8}\right) , \tag{23}$$

$$\Gamma_b = \langle \delta K_b \rangle = \Gamma_0 \left(\frac{\alpha}{\pi}\right)\left(\frac{1}{4}\ln A - \ln 2 + \frac{\pi^2}{16} + \frac{1}{8}\right) , \tag{24}$$

so that

$$\Gamma_a + \Gamma_b = \langle \delta K_a \rangle + \langle \delta K_b \rangle = \Gamma_0 \left(\frac{\alpha}{\pi}\right)\left(\frac{\pi^2}{16} + \ln 2 - \frac{1}{2}\right) ; \tag{25}$$

which is finite and in full agreement with all previous calculations (note, however, that δK_a and δK_b contribute to the actual rate in a different combination).

Kernels c–d are not two–fermion irreducible; strictly speaking they enter in the calculation through the term of second order in δK of eq. (13) where $\hat{G}_c$ is approximated by the free two–fermion propagator and the two irreducible irreducible kernels are: self–mass and lowest order two–photon annihilation in δK_c, one–photon exchange and again $\delta K^{(2)}$ in δK_d. With the already discussed regularization we find

$$\langle \delta K_c \rangle = \Gamma_0 \left(\frac{\alpha}{\pi}\right)\left(-\frac{1}{4}\ln A - 2\ln\alpha - \ln 2 - \frac{9}{8}\right) . \tag{26}$$

Note the presence of the IR divergence which is naturally parametrized by $\ln\alpha$ when the proper wave function eq. (18) is used.

The last term, $\langle \delta K_d \rangle$, is also referred to as the binding diagram. It involves the one–photon exchange which, however, contains the binding kernel K_c. According to eq. (11) the proper kernel to be used, which we call δK_f, can be depicted as in Fig. 3, as the difference between the full photon exchange and K_c.
The correct contribution from δK_d is then

$$\begin{aligned}\Gamma_d = &\langle \delta K_f G_0 \delta K^{(2)} \rangle = \langle \delta K_d \rangle - \langle K_c G_0 \delta K^{(2)} \rangle \\ = &\langle \delta K_d \rangle - \langle \delta K^{(2)} \rangle\end{aligned} , \tag{27}$$

Fig. 3: δK_f expressed as the difference between the full-photon exchange and K_c.

where use has been made of eq. (10). Explicit calculation gives

(28)
$$\begin{aligned}\langle\delta K_d\rangle = \Gamma_0\cdot\left[1+a_0\left(\frac{\alpha}{\pi}\right)\right]&\left\{\frac{\sqrt{2}}{\pi^2}\gamma^2\cdot\int \mathrm{dp}\frac{\mathrm{p}}{(\mathrm{p}^2+\gamma^2)^2}\frac{\sqrt{\mathrm{E_p+W_0}}}{\mathrm{W_0E_p}}\,(2\mathrm{W_0E_p}-1)\right.\\ &\times\int\frac{\mathrm{dk}}{\mathrm{E_k}}\frac{1}{\mathrm{E_k-W_0}}\ln\frac{\mathrm{E_k+k}}{\mathrm{E_k-k}}\ln\frac{\mathrm{E_p-W_0+E_k-W_0+p+k}}{\mathrm{E_p-W_0+E_k-W_0+|p-k|}}\\ &\left.+\left(\frac{\alpha}{\pi}\right)\left(\ln 2-\frac{1}{2}\right)\right\} - \langle\delta K^{(2)}\rangle\\ =\Gamma_0\cdot&\left\{1+a_0\left(\frac{\alpha}{\pi}\right)+\left(\frac{\alpha}{\pi}\right)(3\ln\alpha+\ln 2-1)\right\} - \langle\delta K^{(2)}\rangle\,.\end{aligned}$$

Note the appearance of a_0, already introduced in eq. (21). It is to be observed that δK_d has an IR divergence as well.

Summing up these results, one finds

(29)
$$\begin{aligned}\Gamma =&\Gamma^{(2)}+2\left(\Gamma_a+2\Gamma_b+\Gamma_c+\Gamma_d\right)\\ =&\Gamma_0\left\{1+\left(\frac{\alpha}{\pi}\right)\left(2\ln\alpha+\frac{\pi^2}{4}-5\right)\right\}\,.\end{aligned}$$

Such a result is somewhat surprising since it contains an unexpected — and undesired — $\ln\alpha$ term. Eq. (28) cannot be directly compared with previous calculations, as their form depends on the way UV and IR divergences are parametrized.

When using the wave function eq. (18) the decay kernels of fig. (2) are to be evaluated for incoming fermion four momenta $(\mathrm{p_0+W_0},\vec{\mathrm{p}})$, $(\mathrm{p_0-W_0},\vec{\mathrm{p}})$; we found, however, that if we neglect the p_0-dependence in the kernels, which amounts to disregarding $(\mathrm{E_p-W_0})$ in eq. (29), our results change into

(30)
$$\langle\delta K_c\rangle =\Gamma_0\left(\frac{\alpha}{\pi}\right)\left(-\frac{1}{4}\ln A-2\ln\alpha-\frac{9}{8}\right),$$

(31)
$$\langle\delta K_d\rangle =\Gamma_0\left\{1+a_0\left(\frac{\alpha}{\pi}\right)+\left(\frac{\alpha}{\pi}\right)(2\ln\alpha-1)\right\}-\langle\delta K^{(2)}\rangle\,,$$

giving

(32)
$$\Gamma=\Gamma_0\left\{1+\left(\frac{\alpha}{\pi}\right)\left(\frac{\pi^2}{4}-5\right)\right\}\,,$$

which is the old result of Harris and Brown[7] that has been reproduced in all subsequent calculations. This result is, to $\mathrm{O}(\frac{\alpha}{\pi})$, not affected by the replacement $\sqrt{\mathrm{E_p+W_0}}\rightarrow\sqrt{2m}$ in eq. (28).

Clearly the situation is confused. The p_0-dependence is there, both in the wave function and kernels and a systematic approach cannot simply ignore it. While it is true that p_0 is in some sense of $O(\alpha^2)$ in the calculation, it can nevertheless contribute an $\alpha \ln \alpha$ as we have shown explicitly. On the other hand, if we accept that no $O(\alpha \ln \alpha)$ term is present in the final result, additional contributions from higher order terms must exist, which compensate the unwanted $\alpha \ln \alpha$ and give, presumably, additional $O(\frac{\alpha}{\pi})$ terms as well. It is indeed not difficult to find two-loop candidates for $O(\frac{\alpha}{\pi})$ corrections; the replacement of $\hat{G}_c$ with G_0 should also be considered more carefully.

As a last remark, we mention that we have used Feynman gauge throughout, although it is common wisdom that the use of Coulomb gauge usually results in faster convergence for bound state problems. It is to be observed, however, that Coulomb gauge can cause additional problems in positronium as binding photons and photons giving rise to UV divergences mix so that one has to either renormalize everything in Coulomb gauge or one needs to rely on problematic mixed gauge prescriptions. But, as the last remark, we point out that the discrepancy between eqs. (29) and (32) is not due to the use of different gauges as eq. (32) has also been obtained in ref.[7] and by subsequent authors in Feynman gauge as well.

IV. References.

[1] W.Caswell and G.P.Lepage, *Phys. Rev.* **A18** (1978) 810;

R.Barbieri and E.Remiddi, *Nucl. Phys.* **B141** (1978) 413.

[2] W.E.Caswell and G.P.Lepage, *Phys. Lett.* **167B** (1986) 437.

[3] T.Kinoshita and J.Sapirstein, *Atomic Physics 9*, R.S.Van Dyck,Jr. and E.N.Fortson, Eds., World Scientific (1984) 38.

[4] G.T.Bodwin, D.R.Yennie and M.A.Gregorio, *Rev. Mod. Phys.*, **57** (1985) 723.

[5] C.I. Westbrook, D.W.Gidley, R.S.Conti and A.Rich, *Phys. Rev. Lett.* **58** (1987) 1328.

[6] W.E.Caswell and G.P.Lepage, *Phys. Rev.* **A20** (1979) 36;

G.S.Adkins, *Ann. Phys. (N.Y.)* **146** (1983) 78.

[7] I.Harris and L.M.Brown, *Phys. Rev.* **105** (1957) 1656.

Electron $g - 2$ and High Precision Determination of α

T. Kinoshita

Newman Laboratory, Cornell University, Ithaca, NY 14853, USA

Recent progress in the Penning trap measurement of the magnetic moment anomaly a of the electron and positron has enabled Dehmelt and coworkers to determine a to a precision of 4×10^{-9}, providing the strongest challenge to date to the validity of QED. Calculation of a up to the order α^4, where α is the fine structure constant, has now reached the point where the intrinsic theoretical uncertainty is comparable to that of the measurements. Unfortunately rigorous test of QED itself must be postponed until a better value of α becomes available. Pending improved measurement of α, however, one can determine α from theory and the experimental value of a to a precision better than 1×10^{-8}, which is much more accurate than the α's determined from the ac Josephson effect, the quantized Hall effect, or the hyperfine structure of the muonium ground state.

1. Review of Previous Work

My involvement in the study of the lepton magnetic moment anomaly began in 1966 when, inspired by the beautiful muon g - 2 experiment then in progress at CERN, I developed a method for determining the coefficients of the $\ln(m_\mu/m_e)$ terms in the α^3 contribution to the muon anomaly [1]. This was the first practical application of the mass singularity theorem [2] and the renormalization group technique. Eventually this work evolved into a full scale numerical evaluation of the α^3 terms of the muon and electron anomalies. It was completed by 1974 when our results became more precise than the experimental data then available [3]. The superiority of theory over experiment, however, was short-lived. It was shattered by the remarkable breakthrough by Dehmelt and coworkers who succeeded in measuring a of an individual electron suspended in a Penning trap [4]. Their precision demanded knowledge of the α^4 term. Our work on this term started in 1977 [5]. Only now it is coming to a conclusion. In this talk I will concentrate on the electron anomaly.

Before describing the present status of the theory, let me first review the previous results. The QED prediction for a can be written as a power series in α/π:

$$a_e(\text{QED}) = C_1(\alpha/\pi) + C_2(\alpha/\pi)^2 + C_3(\alpha/\pi)^3 + C_4(\alpha/\pi)^4 + \ldots, \tag{1}$$

The Hydrogen Atom Editors: G.F. Bassani · M. Inguscio · T.W. Hänsch

where C_1 and C_2 are known analytically and given by [6]

$$C_1 = 0.5 ,$$

$$C_2 = - 0.328\,478\,965 \ \ldots \ . \tag{2}$$

Analytic evaluation of C_3 (which consists of 72 Feynman diagrams) is not yet complete [7] and its value

$$C_3 = 1.176\,5\ (13) \tag{3}$$

depends partly on numerical integration [3, 5], where the error represents the estimated accuracy (90% confidence limit) of the numerical integration.

As was mentioned already, one must also calculate C_4 to match the high precision of experimental results [4]. This is a formidable task requiring the evaluation of 891 Feynman diagrams. It is made somewhat easier by a technique developed in [3] by which Feynman diagrams having similar structure are combined into one with the help of the Ward-Takahashi identity. Together with time reversal and charge conjugation symmetries this reduces the number of integrals to be separately evaluated to 86 as is indicated below. The diagrams may naturally be classified into five groups:

Group I. Second-order vertex diagrams containing vacuum polarization loops of up to sixth order. This group consists of 25 diagrams. They are reduced to 10 independent integrals by time reversal and charge conjugation symmetries.

Group II. Fourth-order vertex diagrams containing second and fourth order vacuum polarization loops. This group contains 54 diagrams (reduced to 8 integrals by the Ward-Takahashi identity and time reversal and charge conjugation symmetries).

Group III. Sixth-order vertex diagrams containing a second order vacuum polarization loop. There are 150 diagrams (reduced to 8 integrals) in this group.

Group IV. Vertex diagrams containing a photon-photon scattering subdiagram with further radiative corrections. This group consists of 144 diagrams (reduced to 13 integrals).

Group V. Vertex diagrams containing no vacuum polarization loop. This group is comprised of 518 diagrams (reduced to 47 integrals).

Because of the enormous complexity only a handful of these diagrams have been evaluated analytically thus far [8]. We have adopted a purely numerical approach [5]. Each (combined) Feynman amplitude is represented by an integral over a multi-dimensional space. Integrands are generated by the algebraic program SCHOONSCHIP [9]. The integration, over a hypercube of up to 10 dimensions, is carried out using the adaptive Monte Carlo integration routines VEGAS [10] and RIWIAD [11].

A typical integrand of the first three groups is a rational function consisting of up to 2,000 terms, each term being a product of up to 8 or 9 factors. Its FORTRAN source code has size of up to 50 kilobytes. Numerical evaluation of these integrals is relatively straightforward and was completed in 1979. The results are [12]

$$C_4^{I} = 0.076\ 6\ (6)\ ,$$

$$C_4^{II} = -\ 0.523\ 8\ (10)\ ,$$

$$C_4^{III} = 1.419\ (16)\ . \qquad (4)$$

By 1981 we had written and debugged the FORTRAN source codes for all integrals in groups IV and V. These groups require much larger code (100 to 500 kilobytes), and numerical integration is much more difficult and time-consuming. Due to the limited computing power then available it was not possible to explore these integrals thoroughly. Nevertheless we managed to show that our program worked as expected, and obtained very crude and preliminary results [12]

$$C_4^{IV} = -\ 0.78\ (48)\ ,$$

$$C_4^{V} = -\ 1.0\ (2.4)\ . \qquad (5)$$

From (4) and (5) one finds

$$C_4 = -\ 0.8\ (2.5)\ . \qquad (6)$$

If one uses, for example, the value of the (inverse) fine structure constant α obtained by analyzing several pre-1986 quantized Hall effect measurements [13]

$$\alpha^{-1} = 137.035\ 994\ 3\ (127) \qquad (0.093\ \text{ppm})\ , \qquad (7)$$

the QED predictions (2), (3), and (6) lead to

$$a(\text{QED}) = 1\ 159\ 652\ 188\ (74)(108) \times 10^{-12}\ . \qquad (8)$$

To this one must add contributions from other known sources. They include the contributions of the muon loop, τ meson loop, and hadronic effect, as well as the effect of the weak interaction (in the standard Weinberg-Salam-Glashow model) [14]:

$$a\,(\text{muon}) = 2.8 \times 10^{-12}\,,$$

$$a\,(\tau\text{ meson}) = 0.01 \times 10^{-12}\,,$$

$$a\,(\text{hadron}) = 1.6(2) \times 10^{-12}\,,$$

$$a\,(\text{weak}) = 0.05 \times 10^{-12}\,. \tag{9}$$

Adding (9) to (8) we arrive at the theoretical prediction

$$a\,(\text{theory}) = 1\;159\;652\;192\;(74)(108) \times 10^{-12}\,, \tag{10}$$

where the first error is from theory while the second is due to the measurement error of α in (7).

The value (10) is in good agreement with the latest measurements of a for the electron and positron [15]:

$$a\,(e^-) = 1\;159\;652\;188.4\;(4.3) \times 10^{-12}\,,$$

$$a\,(e^+) = 1\;159\;652\;187.9\;(4.3) \times 10^{-12}\,, \tag{11}$$

where the experimental error arises from several sources:

$$\text{statistical error} = 0.62 \times 10^{-12}\,,$$

$$\text{error due to microwave power shift} = 1.3 \times 10^{-12}\,,$$

$$\text{error due to cavity shift} = 4 \times 10^{-12}\,. \tag{12}$$

Clearly the cavity shift correction, which results from the radiative interaction of an electron with the metallic walls of the Penning trap that surrounds it and exhibits a complicated resonance behavior, is the largest source of uncertainty at present [16].

The value of C_4 in (6) is rather crude because of very limited integrand sampling, even though it represents an outcome of more than 300 hours of computing on a CDC-7600. The primary significance of the result (6) is not in its precision but in the establishment of bounds on C_4, namely the confirmation that the renormalization of QED in fact gives a convergent result to order α^4, a nontrivial, even though expected, result in view of several thousands of divergent terms that must cancel out completely.

In spite of its crudeness the intrinsic theoretical error given in (10) is smaller than that of the α used in the calculation. This means that the study of a provides a very powerful tool for obtaining a highly accurate value of α. Indeed we find from the experimental result (11) and theory that

$$\alpha^{-1}(g - 2) = 137.035\,994\,2\,(89) \quad (0.065 \text{ ppm}) , \tag{13}$$

which is more accurate than (7).

2. Recent Developments

In order to obtain accurate and statistically reliable results it was estimated that at least 100 times more computation (more than 30000 hours on CDC-7600) was required, which was prohibitively expensive at that time. It is only in the last few years that such a calculation has become feasible in terms of time and cost. This is the primary reason why we have not been able to finish our work more quickly.

Before starting such an extensive computation, I felt that it was necessary to make sure that all of the FORTRAN code was completely free of errors. Thus in 1984-85 I generated new code from scratch in a form different from the original, and tested against each other numerically. I now have complete confidence in the entire program.

One might think that the remaining task is simply a matter of number crunching. Unfortunately it is more complicated than that for several reasons. One is the sheer size of the integrands. In order that an integration routine give a reliable result the number N of randomly chosen points where the integrand is sampled in each iteration must be sufficiently large. Otherwise the integration routine may not be able to explore the integrand closely, resulting in deceptively optimistic error estimates. In the early evaluation of the integrals we were unable to choose N larger than 120,000. Our subsequent work indicated that this N was far too small for some integrals of Group IV and most integrals of Group V. It is only when we ran our programs in vectorized form on the HITAC S-810 at Tokyo University and subsequently at KEK, with N in the range of 4- to 20 million, that we were finally able to confirm the $N^{-1/2}$ behavior expected for statistically satisfactory samplings.

Another problem we encountered is deeply rooted in our particular approach to the removal of ultraviolet and infrared divergences [17]. In our method, terms of the integrand generated by SCHOONSCHIP actually diverge on some boundaries of the integration domain and the integral is kept finite by point-by-point cancellation of divergences by carefully tailored counter terms. Because of the insufficient numerical precision of double precision arithmetic, however, round-off errors sometimes force a breakdown of divergence cancellation mechanism, causing undesirable fluctuations. Fortunately this problem can be reduced to a manageable level by switching to quadruple precision arithmetic, which slows down the (unvectorized) computation by a factor of 5 to 6. In practice quadruple precision is needed only in the neighborhood of singularities. One can evaluate the bulk of the integral in double precision resorting to quadruple precision only where it is absolutely needed.

Further complication arose when we tried to check the renormalization procedure of some diagrams of Group IV (which contain an internal light-by-light scattering subdiagram) by comparing it with a parallel scheme which, while keeping Pauli-Villars regularization to insure convergence, avoids explicit renormalization taking advantage of an identity derived from current conservation. What we found after a very extensive and time-consuming numerical experiment was that VEGAS was not able to handle this scheme adequately when regulator masses were too large. However, the results obtained using smaller regulator masses were good enough to convince us that the renormalization was correctly implemented in these diagrams.

Our calculation of C_4 is now close to completion, although further refinement and consistency check are being made on several integrals of Group V. Actually the error on C_4 has already been reduced to less than that on C_3, forcing us to do further work on C_3. The present (not yet final) values are

$$C_3 = 1.175\ 62\ (56)\ ,$$

$$C_4 = -\ 1.472\ (152)\ , \tag{14}$$

which supersede the results (3) and (6). If one uses the value in (7) for α, this leads to

$$a\,(\text{theory}) = 1\ 159\ 652\ 164\ (108) \times 10^{-12}\ . \tag{15}$$

Very recently, however, two new measurements of α have been reported [18, 19]. One is based on the equation

$$\alpha^{-1} = 2R_H/\mu_0 c \tag{16}$$

relating α to the quantized Hall resistance R_H, and the other is based on

$$\alpha^{-2} = (c/4R_\infty\gamma_p')(\mu_p'/\mu_B)(2e/h) \tag{17}$$

which relates α to the measurements of the Josephson frequency and the proton gyromagnetic ratio γ_p'. It is also possible to determine α from the formula

$$\alpha^{-3} = (R_H / 2\mu_0 R_\infty \gamma_p')(\mu_p' / \mu_B)(2e/h) \tag{18}$$

which is obtained by combining (16) and (17).

The new values of α are [18]

$$\alpha^{-1}(\text{QHE}) = 137.035\,997\,9\,(33) \quad (0.024\text{ ppm}) \,, \tag{19}$$

and [19]

$$\alpha^{-1}(\text{acJ } \& \ \gamma_p') = 137.035\,976\,9\,(77) \quad (0.056\text{ ppm}) \,. \tag{20}$$

Corresponding to (18) one finds

$$\alpha^{-1}(\text{Eq.}(18)) = 137.035\,983\,9\,(51) \quad (0.037\text{ ppm}) \,, \tag{21}$$

which is derived from (19) and (20). Although (19) and (20) are the most accurate values of the respective measurements reported thus far, it should be kept in mind that they are still preliminary having been obtained in a hurry to beat the deadline of a conference. Thus values may move and errors may change before the final reports are written. Both (19) and (20) are measured using a calculable capacitor as a reference and thus sensitive to its uncertainty. In (21), on the other hand, such a dependency cancels out and its error is mainly due to that of γ_p'.

If one accepts (19), which is the best of the three at present, as the value for α, the theoretical prediction for a becomes

$$a(\text{theory}) = 1\,159\,652\,133\,(29) \times 10^{-12} \,, \tag{22}$$

where

$$\text{error due to } C_3 = 7.1 \times 10^{-12} \,,$$

$$\text{error due to } C_4 = 4.5 \times 10^{-12} \,,$$

$$\text{error due to } \alpha = 28 \times 10^{-12} \,. \tag{23}$$

Thus the theoretical value (22) and measured value (11) of a are nearly two standard deviations apart. The values of a obtained using (20) and (21) are somewhat less accurate. As is seen from (23) the error on a is still dominated by that on α. This means that a sharper test of QED must be postponed until better value of α is found. Pending improved measurement of α, however, one can calculate α from theory and the experimental value of a :

$$\alpha^{-1}(g - 2) = 137.035\,991\,4\,(11) \quad (0.0081 \text{ ppm}) \tag{24}$$

where

$$\text{error due to experiment} = 0.003\,7 \text{ ppm} ,$$

$$\text{error due to theory} = 0.007\,2 \text{ ppm} . \tag{25}$$

Clearly the result (24) is more precise than those of (19), (20), (21), or one obtained from the muonium hfs [20, 21] :

$$\alpha^{-1}(\mu\text{hfs}) = 137.035\,992\,5\,(224) \quad (0.17 \text{ ppm}) . \tag{26}$$

In comparison with the results (19), (20), (21), and (24), the precision of $\alpha(\mu\text{hfs})$ in (26), which has not had a major improvement for some time, looks rather poor. Its error is due partly to the uncertainty in the muon mass but mostly to radiative recoil corrections involving two virtual photons which is yet to be calculated. New experiment for a more accurate measurement of the muonium hfs and muon mass is in preparation [22]. There seems to be no obstacle in improving the theoretical prediction by at least an order of magnitude although it will certainly require an extensive calculation. It must be emphasized that the comparison of α obtained from the improved muonium hfs and a is particularly important for checking the internal consistency of QED.

It is generally believed that α's determined by the ac Josephson effect and quantized Hall effect have no *theoretical* uncertainty, although this has never been proved. The difference of 21×10^{-6} between (19) and (20), which is about 3 times larger than the error in (20), might be the first indication that this belief is not above scrutiny. Before pursuing such a question, however, one must examine carefully the errors in the measurements, in particular those of the calculable capacitor and γ_p'.

An advantage of the composite result (21) is that, being independent of calibration of the calculable capacitor, its error comes mainly from the measurement of γ_p'. Thus it is really (21) that should be compared with the $g-2$ value (24) although the present error on (21) is not as good as it should be reflecting the wide separation of (19) and (20).

If the difference between (19) and (20) persists even after the experimental errors are fully understood and improved, it may become necessary to examine the theoretical basis of (16) and (17) closely and evaluate correction terms, if any, to (16) and/or (17). Strictly speaking, it is not possible to test the validity of QED at the level of 10^{-8} until the apparent internal inconsistency of condensed matter physics is resolved. The theoretical situation of QED is under better control. One may be able to compute a unambiguously at least to the level of 10^{-13}.

Sooner or later theory and experiment of a will be pushed to the point where α is determined to a precision of 10^{-9} or better. Together with improved determinations of α by other means, this will provide a far more stringent test of QED as well as the theoretical basis of condensed matter physics. At the same time it will impose strong constraints on theoretical speculations such as possible internal structure of the electron, supersymmetric theories, and superstring theories.

Acknowledgments

This work is supported in part by the U. S. National Science Foundation. The bulk of this computation was carried out on the HITAC S-810 computer at the National Laboratory for High Energy Physics (KEK), Tsukuba, Japan. The last phase of this research is being conducted using the IBM-3090 computer at the Cornell National Supercomputer Facility, which is funded in part by the U. S. National Science Foundation, New York State, and the IBM Corporation.

References

1. T. Kinoshita, Nuovo Cimento 51*B*, 140 (1967)
2. T. Kinoshita, J. Math. Phys. 3, 650 (1962)
3. P. Cvitanovic and T. Kinoshita, Phys. Rev. *D* 10, 4007 (1974)
4. P. B. Schwinberg, R. S. Van Dyck, Jr., and H. G. Dehmelt, Phys. Rev. Lett. 47, 1679 (1981)
5. T. Kinoshita and W. B. Lindquist, "Improving the theoretical prediction of the electron anomalous magnetic moment," Cornell preprint CLNS-374, 1977
6. J. Schwinger, Phys. Rev. 73, 416 (1948); C. Sommerfield, Phys.Rev. 107, 328 (1957); A. Petermann, Helv. Phys. Acta 30, 407 (1957)
7. M. J. Levine, H. Y. Park, and R. Z. Roskies, Phys. Rev. 25, 2205 (1982)
8. M. Caffo, S. Turrini, and E. Remiddi, Phys. Rev. D 30, 483 (1984); E. Remiddi and S. P. Sorella, Lett. Nuovo Cimento 44, 231 (1985)
9. H. Strubbe, Comp. Phys. Comm. 8, 1 (1974)
10. G. P. Lepage, J. Comp. Phys. 27, 192 (1978)

11. B. E. Lautrup, "An adaptive multidimensional integration technique," in Proceedings of the Second Colloquium in Advanced Computing Methods in Theoretical Physics, Marseille, 1971, A. Visconti, ed. (Univ. of Marseille, Marseille, 1971)

12. T. Kinoshita and W. B. Lindquist, Phys. Rev. Lett. 47, 1573 (1981)

13. E. R. Cohen and B. N. Taylor, Rev. Mod. Phys. 59, 1121 (1987)

14. See, for instance, T. Kinoshita in "New Frontiers in High Energy Physics", B. Kursunoglu *et al*., eds. (Plenum, 1978), pp.127-143

15. R. S. Van Dyck, Jr., P. B. Schwinberg, and H. G. Dehmelt, Phys. Rev. Lett. 59, 26 (1987)

16. L. S. Brown and G. Gabrielse, Rev. Mod. Phys. 58, 233 (1986)

17. T. Kinoshita and W. B. Lindquist, Phys. Rev. *D* 27, 886 (1983)

18. M. E. Cage *et al*, presented at the 1988 Conference on Precision Electromagnetic Measurements, Tsukuba, Japan, June 7-10, 1988

19. E. R. Williams *et al*, presented at the 1988 Conference on Precision Electromagnetic Measurements, Tsukuba, Japan, June 7-10, 1988

20. F. G. Mariam *et al*., Phys. Rev. Lett. 49, 993 (1982)

21. M. I. Eides, S. G. Karshenboim, and V. A. Shelyuto, Phys. Lett. B 177, 425 (1986); 202, 572 (1988)

22. V. W. Hughes and G. Z. Putlitz, Comm. Nucl. Part. Phys. 12, 259 (1984)

General QED/QCD Aspects of Simple Systems

V.L. Telegdi[1] *and S.J. Brodsky*[2*]

[1]Institute for High Energy Physics, ETH, CH-8092 Zürich, Switzerland
[2]Stanford Linear Accelerator Center (SLAC), Stanford University, Stanford, CA 94305, USA
*Currently Humboldt Fellow, MPI Heidelberg

1. Bringing you up to date

The honor of addressing this gathering of distinguished atomic physicists came to one of us (VLT) as a shocking surprise. It is true that quite some time in the past VLT too was a member of the "Inverse Millionaires Club" - that circle of people who measure things to a fraction of a ppm - but that was so long ago that it could hardly justify my talking to you now. For a while VLT thought that the invitation was prompted by his fluency in Italian, but that turned out to be wrong, since the talks are to be given in English (presumably largely broken).

The shock of the invitation became even greater when VLT saw the title proposed for his talk: "General Quantum Electrodynamic Aspects Related to the Spectroscopy of Simple Atomic Systems". Only a committee of seasoned sadists could assign such a subject to an experimental physicist, and only an inveterate masochist could volunteer to accept it! Very fortunately the printed program had a vague title: "General Quantum Electrodynamic Aspects", but even that sounded like an impossible challenge.

Under these circumstances, after having foolishly accepted (who can resist a chance to see Pisa again?), VLT decided on the following strategem: a) change the title so as to bring this audience up to date on some modern topics less familiar to this audience than the one proposed, b) get himself a collaborator with impeccable credentials. Stan Brodsky has kindly agreed to assist VLT in an otherwise impossible task.

Paraphrasing what has been said of the famous treatise by Landau and Lifshitz, one could say "This talk will not contain a single formula by Telegdi, and not a single word by Brodsky".

This Conference is devoted to the Hydrogen Atom and its younger relatives like positronium. The latter, composed of (presumably) point-like objects, is the ideal testing ground for QED. It should hence be of interest to this audience to be reminded of the fact that the last decade has led to the discovery and detailed study of new bound particle-antiparticle systems, which we call <u>quarkonia</u>, since they consist of bound quark-antiquark pairs. There can in principle be as many

---

*Currently Humboldt Fellow, MPI Heidelberg

The Hydrogen Atom Editors: G.F. Bassani · M. Inguscio · T.W. Hänsch

such systems as there are "flavors" of quarks (e.g. s, c, b ...) in increasing order of heavyness). The most interesting ones of these are "charmonium" ($c\bar{c}$) and "bottomium" ($b\bar{b}$), since for these heavy quarks a non-relativistic description is quite adequate, ($m_c \cong 1.5$ GeV, $m_b \cong 5$ GeV ; it is amusing to note that the ground state of bottomium has about 10^4 times the mass of positronium!).

Figures 1 and 2 show, respectively, the presently well established levels of charmonium and bottomium. Today more levels are known for these systems than for positronium, and more "spectral lines" (transitions) have been identified than were known for hydrogen in Balmer's days!

What is most remarkable about these levels? Probably two facts: first, although they are hadronic states, they are long-lived; electromagnetic transitions (E1) compete in general appreciably with the emission of mesons. Second, there is really no "series limit" in the sense of ionization into $Q + \bar{Q}$ (Q = c or b).

The Ψ and Υ states are formed as sharp resonances in $e\bar{e}$ collisions. This identifies their spin (J), parity (P) and charge conjugation (C) quantum numbers readily as those of the photon: $J^{PC} = 1^{--}$. The quantum numbers of the states are readily assigned by using well-known (e.m. and hadronic) selection rules. This results in the J^{PC} values given at the bottom of Figs 1 and 2.

From a certain excitation on, the Ψ (and Υ) states can dissociate into two charge-conjugate mesons M, $\bar{M}$ according to the scheme

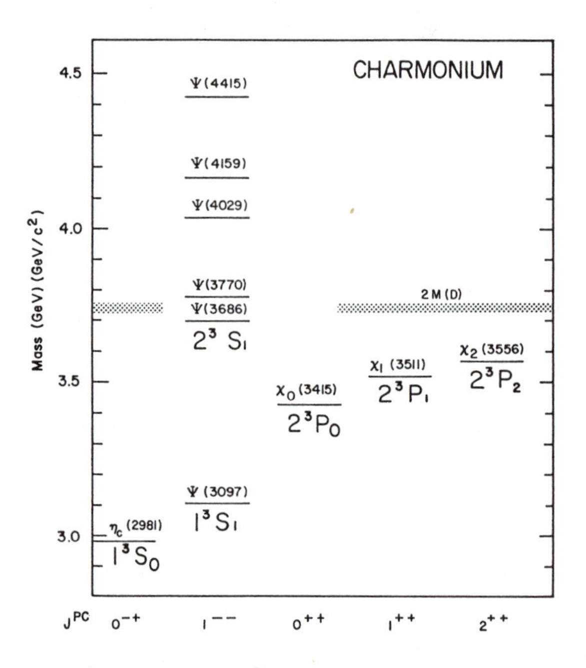

Fig. 1 Charmonium ($c\bar{c}$) spectrum. The band at mass = 2M(D) denotes the flavor threshold, above which levels are broader than those below it.

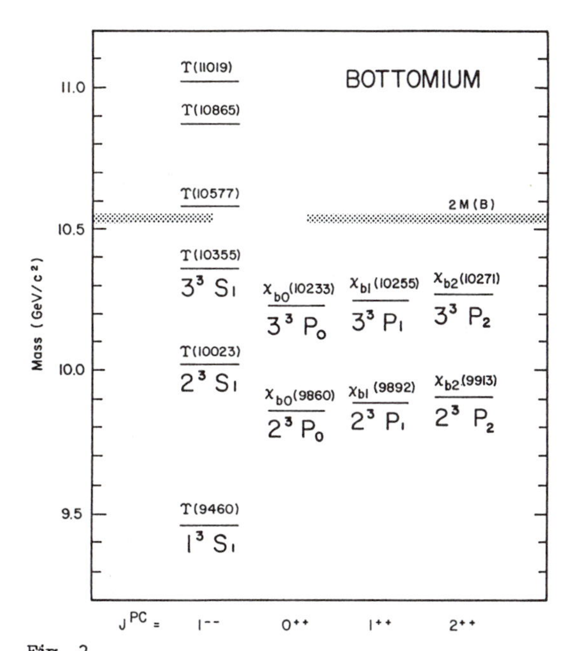

Fig. 2

Spectrum of the upsilon ($b\bar{b}$) family. Levels above flavor threshold [band at mass = 2M(B)] are broader than levels below it.

$$Q\bar{Q} \rightarrow Q\bar{q} + \bar{Q}q = M + \bar{M},$$

where q is a very light quark (d or u). The combination ($c\bar{q}$) is called a D-meson, the combination ($b\bar{q}$) a B-meson. The corresponding thresholds are indicated in the Figs. by shaded bands. Above these, "hidden charm" turns into "open charm", "hidden beauty" into "open beauty". (The reason for a new name for the flavor "b" should be obvious.) After all the J^{PC} assignments are made, one can - within the framework of the "naive" quarkonium model - assign the standard spectroscopic labels to the levels. This is shown in the overlay. The standard n = 1 and n = 2 positronium levels appear, but in addition many excited ${}^3S_1$ states. The spacing of the latter indicates that the effective potential (if there is one!) is much softer than the familiar $1/r$.

Many authors have proposed phenomenological potentials which yield all the observed states, and predict new ones (e.g. D states) yet to be discovered. The corresponding wave functions yield E1 matrix elements in reasonable agreement with experiment.

The task is to predict the "observed" potentials from first principles. The current theory of strong interactions, quantum chromo-dynamics (QCD), qualitatively succeeds in achieving this. This gauge theory patterned after QED is believed to explain why there is no series limit for quarkonia: quarks are forever "confined" within any hadron. It also explains why the quarkonium states are so narrow. In strict analogy with positronium, the C = -1 states can go only into three, the C = +1 states only into two C = -1 field quanta (called gluons). Indeed the $\chi(={}^3P)$ states are observed to be wider than the Ψ or Υ $({}^3S_1)$ states. We shall return to the QCD-QED analogies later.

Another novelty which deserves your attention is the nature of the beloved fine-structure constant α. It is, as we shall discuss later in more detail, a "constant" only in processes involving very small momentum transfers.

Next, and more importantly, there is the fact that QED has become but part of a broader gauge theory which includes "weak" interactions. Through the discovery of the heavy vector bosons Z^0 and $W^{\pm}$ at CERN this theory has been brilliantly confirmed. The photon's heavy partner, the Z^0, is exchanged between essentially <u>all</u> particles, not only the charged ones. Atomic parity violation experiments have confirmed this: Laporte's rule is dead. The "weak" analogs of α are also energy dependent, so that at some point the "weak" and electromagnetic forces become comparable, whereby the term "weak" loses its meaning. This is illustrated in Fig. 3.

The coupling constant of the strong interaction (QCD), α_s, decreases with increasing momentum transfer - a point we shall discuss in detail later. There have been proposals for a Grand Unified (gauge) Theory, GUT, where all three interactions become equally "strong" at some very high energy. This is also indicated in Fig. 3.

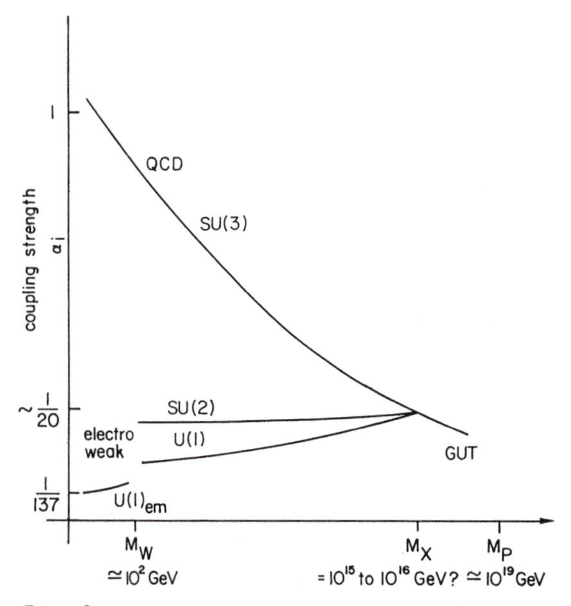

Fig. 3

QED is the model gauge theory after which all others are patterned. We shall divide the discussion of its current status into two parts: Closed subjects, and open subjects. To these one may refer respectively as the "rug" and the "dirt", recalling Feynman's famous statement that he got the Nobel Prize for being better than others in sweeping the dirt under the rug.

2. "Closed" subjects (the "rug")

QED, which is supposed to provide finite answers to all orders of perturbation theory (AOPT), can be represented as resting on a foundation (local gauge invariance) and on three pillars (see Fig. 4). Local gauge invariance implies that the theory is invariant under arbitrary phase transformations of the electron field at each point in space and time. The generalization of this principle to invariance under unitary matrix transformations of the fermion fields leads to the concept of non-Abelian gauge theories which include quantum chromodynamics and the unified electroweak theory. The three pillars are:

2.1 Renormalization theory, and in particular the treatment in terms of the renormalization group. The latter goes back to an idea of Petermann and Stueckelberg, and was formulated quantitatively by Gell-Mann and Low. The essence is that only the observed mass m and the observed charge e of the election (and/or its heavier brother leptons μ and τ) enter into the final results. Ultraviolet infinities ($k \rightarrow \infty$) are consistently eliminated to AOPT. The coupling is characterized by a "running" coupling constant which incorporates vacuum polarization to all orders, viz.

$$\alpha_r(Q^2) = \frac{\alpha(Q_0^2)}{1-\Pi\ (Q^2/Q_0^2)} \tag{1}$$

where Q = (4-momentum) of interest, and Q_0 a "reference" 4-momentum. The function Π is given by

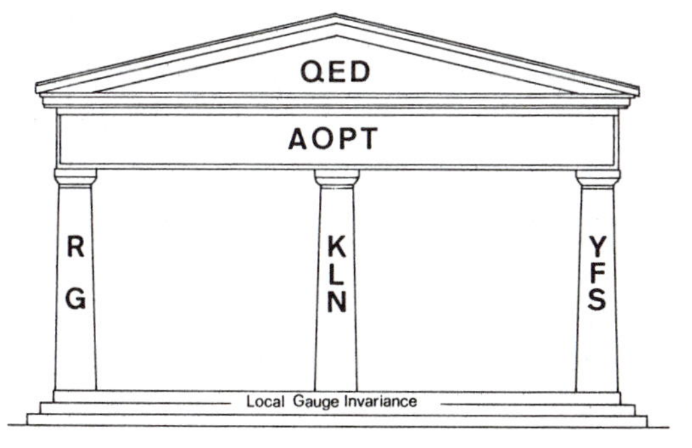

Fig. 4

$$\Pi = \frac{\alpha(Q_0^2)}{3\pi} \ln(Q^2/Q_0^2) + \ldots \qquad (1a)$$

where both Q_0^2 and $Q^2 >> m_l^2$ (lepton mass). Reinterpreting things in coordinate space, (1) simply means that the effective coupling decreases with increasing distance: one observes the shielding due to virtual pairs. At extremely small distances where $\Pi(Q^2)$ is of order 1, i.e., $\sim 10^{-291}$ cm, one could have a blow-up ("Landau singularity") where the theory becomes undefined; this may however be "cured" by the unification of QED with other interactions.

The current, rather successful, theory of strong interactions, QCD, is patterned after QED. It is a scenario where quarks play the rôle of leptons, massless vector gluons the part of the photon (gauge bosons), and "color" that of the charge. The big difference with electromagnetism is that both the sources (quarks) and the fields (gluons) carry color, i.e. charge. One is again led to running coupling constant analogous to α_r, viz.

$$\alpha_s(Q^2) = \frac{\alpha_s(Q_0^2)}{1-\Pi(Q^2/Q_0^2)}, \quad \Pi = - [(11 - \frac{2}{3} n) \frac{\alpha_s(Q_0^2)}{4\Pi} \ln (Q^2/Q_0^2)] \qquad (2)$$

n = number of flavors

with, however, effectively a plus sign in the denominator. As $Q \to \infty$, i.e. $r \to 0$, the coupling becomes weaker, one has antishielding (in current slang, this plenomenon is called "asymptotic freedom"). It makes it possible to justify the soft potentials corresponding to the observed levels (Figs 1, 2) of the quarkonia. We mention in passing that in virtue of the quark spins and of the vector nature of the gluons one has the fine structures so dear to atomic physicists.

2.2 The Kinoshita-Lee-Nauenberg (KLN) theorem

This theorem, of rather formal character, guarantees that one may (summing over the final states of any inclusive e.m. process) let the lepton mass m_l tend to zero without creating terrible havoc.

2.3 The Yennie - Frautschi - Suura relation

This relation, similar in essence to the old Bloch-Nordsieck theory, guarantees the absence of catastrophes (infra-red divergencies) in the limit $k \to 0$. Such a catastrophe could be anticipated, but obviously does not happen in, say, elastic electron scattering where the final state electron could radiate an infinite number of softer and softer photons.

From these three "pillars" and the "foundation" of local gauge invariance, one can derive - besides the innumerable atomic properties you are all familiar with - many important consequences. These are either interesting in themselves, or through the fact that they are readily generalized to strong interactions (QCD). We discuss a few:

2.3.1 Scale invariance at large momentum transfer

This means that in an inclusive reaction like

$$e + \bar{e} \rightarrow \gamma^* \rightarrow \mu + \bar{\mu} + X \tag{3}$$

where X = any neutral state composed of leptons and photons

the cross section exhibits, to AOPT, a pointlike behaviour (thus scale invariance meaning that no lengths appear in the formulae):

$$\sigma(e + \bar{e} \rightarrow X) = \frac{4\pi\alpha(Q^2)}{3Q^2}\left(1 + \frac{3}{4}\frac{\alpha(Q^2)}{\pi} + C_2\left(\frac{\alpha(Q^2)}{\pi}\right) + C_3\left(\frac{\alpha(Q^2)}{\pi}\right)^3 + .. \tag{4}$$

(valid for $Q^2 \gg 4m^2_\mu$).

Note the absence of terms in $\ln m_l$, a consequence of the KLN theorem. The reaction (3) is not one of purely academic interest. In fact, in $e\bar{e}$ colliders the muon pair production is used in practice to monitor the luminosity of the machine, i.e. for normalization purposes. We shall come back to the term in C_3 at a later point.

It is interesting to replace the leptons in (3), either in the initial or the final state, by quarks. We thus consider

$$e + \bar{e} \rightarrow (q + \bar{q}) + X \tag{5}$$

and

$$q + \bar{q} \rightarrow \mu + \bar{\mu} + X \tag{6}$$

The brackets in the first reaction represent the fact that the quark and antiquark never appear as isolated physical particles in the final state. They can be produced in a bound state (of spin-parity 1^- equaling that of the γ^*). Such pairs are precisely the ^3S quarkonia shown in Figs 1, 2. Their production cross sections contain factors allowing for the fractional charges of the quarks and for their "color". Process (5) represents muon pair production in the collision of any two hadrons, to the extent that these contain (real or virtual) $\bar{q}$'s. In the jargon it is called the "Drell-Yan" process; it has been the subject of much experimental investigation, and is one of our major sources of information about the quark "wavefunction" of hadrons.

Finally, one may replace the leptons on both sides of Eq.(3) by quarks. Electromagnetism then plays a subordinate rôle, so that the virtual photon γ^* has to be replaced by a virtual gluon g^*. Thanks to the gauge structure common to QED and QCD, the essential results remain valid in the latter, with $\alpha_s(Q^2)$ replacing $\alpha(Q^2)$.

2.3.2 Scaling and scaling violation at large momentum transfers ("deeply inelastic" scattering)

Consider (for pedagogical reasons!) the process

$$\mu + e \rightarrow \mu + e \,. \tag{7}$$

One has for the differential cross section without radiation

$$\frac{d\sigma}{d\Omega} = \frac{\pi\alpha_r^2}{s} f(\vartheta) \,, \qquad (\sqrt{s} = \text{c.m. energy}) \tag{8}$$

which can be generalized to AOPT and to QCD processes. Next consider, to please the tastes of atomic physicists, the inelastic scattering of electrons by muonium

$$e + (\mu\,\bar{\mu}) \rightarrow e' + X \,. \tag{9}$$

Because of the inelasticity, one has now a doubly differential cross section, which can be written as

$$\frac{d\sigma^2}{dQ^2\,dx} = \left(\frac{d\sigma}{dQ^2}\right)_{e\mu} F(x) \tag{10}$$

where x is the dimensionless scaling variable

$$x \equiv \frac{Q^2}{2P\cdot q} \cong \frac{(\text{momentum transfer})^2}{M\,(\text{energy transfer})} \tag{11}$$

with P the 4-momentum and M the mass of the "incident" muonium. Equ. (10) is the basis of the parton model of deeply inelastic scattering of leptons, where the rôle of the muons in our "pedagogical" example is played by the quarks. The elastic collision between quark and lepton is turned into a (deeply) inelastic scattering of the lepton by the hadron, the final state X consisting of real hadrons rather than free partons.

Because of the gauge nature of QCD, entirely similar arguments hold for parton-parton collisions. Radiative corrections are, however, generally more important here, because α_s (s for strong!) is, at given Q^2, larger than $\alpha(Q^2)$: gluons are more easily radiated then photons! Consider reaction (7) with photon radiation by the incident muon. The differential cross section (8) is modified as

$$\frac{d\sigma}{d\Omega} = \frac{\pi \alpha_r^2}{s} f(\vartheta) \left[1 + \frac{\alpha}{\pi} \ln \frac{Q^2}{m_\mu^2} \ln \Delta E/E\right] , \tag{12}$$

where $\Delta E/E$ is an experimental resolution. Similarly, the "structure function" $F(x)$ of the $\mu^+\mu^-$ atom in Equ. (10) becomes

$$F(x, \ln Q^2/Q_0^2) \tag{13}$$

Thereby scale invariance is broken, although no explicit dependence on a length enters. Again, a logarithmic dependence as in (13) is taken over into QCD. All structure functions "evolve", as was shown by Gribov and Lipatov, and by Altarelli and Parisi.

2.3.3 Low-energy theorem in Compton scattering

One can show that the forward scattering anglitude is given, as $\omega \to 0$, to AOPT for any spins by

$$f(0) = -\frac{e^2}{m} \vec{\epsilon}' \cdot \vec{\epsilon} - i\omega\mu_a^2 (\vec{S}/S) \cdot \vec{\epsilon}' \times \vec{\epsilon} + O(\omega^2), \tag{14}$$

where $\mu_a = \mu - eS/m$ defines the anomalous moment for any spin. This relation, in combination with the optical theorem, enables one to set limits on the composite scale of leptons. It also implies that the normal g-factor $g = (\mu/S)/(e/2m)$ of any pointlike particle is 2. Indeed if the electron or muon were composite, i.e. if they had internal excitations at the mass scale Λ, their anomaly $a = \frac{g-2}{2}$ would be of order (m_e/Λ) or $(m_e/\Lambda)^2$.

The two cases depend whether or not the interactions of the underlying theory resemble gauge theories and conserve chiral invariance. In either case, the present agreement between theory and experiment for the electron and muon anomalous moments rules out an internal scale below 1 TeV, [see e.g. S. J. Brodsky and J. Primack, Ann. Phys. 52, 315 (1969). S. J. Brodsky and S. D. Drell, Phys. Rev. D22, 2236 (1980).]

2.3.4 Renormalization of the weak angle Θ_W.

The standard theory of electroweak interactions contains two coupling constants but only one free parameter, the Weinberg angle Θ_W. The latter fixes the e.m. - weak connection:

$$e = g \sin\Theta_W = g' \cos\Theta_W \tag{15}$$

as well as the mass ratio of the two heavy gauge bosons:

$$m_W/m_Z = \cos\Theta_W. \tag{16}$$

Since e, i.e. α, is a "running" coupling constant (see above), it is clear that Θ_W itself must be "running". These considerations are of interest for two reasons: (i) they will tell us at which energy e.m. and "weak" interactions will become equally "strong", (ii) by determining Θ_W at two energies, one can experimentally verify the gauge nature of the theory.

2.3.5 The Nambu-Bethe-Salpeter (NBS) equation

This covariant two-body equation, with which this audience is certainly familiar, allows to solve everything in principle, but little in actual practice. This is for two reasons: (i) one needs an infinite number of kernels, (ii) even in the ladder approximation no analytic solution for QED has been produced.

One interesting consequence of the NBS equation is that by its reduction (in the case of two quarks) a Schrödinger equation with a non-local potential emerges.

See also comments below under "open problems".

3. Open problems ("the dirt")

3.1 Does the perturbation series in QED converge?

Nobody knows the answer, but perhaps there is no answer within the old classical framework, i.e. in a world made of leptons and photons alone. Indeed charged leptons interact with each other by both γ and Z^0 exchange, a fact already verified by experiment (μ-pair asymmetry in $e\bar{e}$ collisions). There are "grand" schemes to unify electroweak and strong (QCD) forces, giving them equal strength at some very high (say 10^{14} GeV) energy. In such schemes the "Landau singularity" might be cured.

There exist some exciting warnings from PT that the PT series may not converge. Let us mention two:

3.1.1 The decay rate of $^3S_1$ positronium

The current theoretical prediction is

$$\Gamma = \Gamma_0[1-10.282(\alpha/\pi) + \frac{1}{3}(\alpha^2\ln\alpha)+(300\pm30)(\alpha/\pi)^2]. \tag{17}$$

The unexpectedly large coefficient of the last, experimentally determined term might well be the presage of worse things to come! A similar behavior in QCD, say in the analogous 3-gluon annihilation of $^3S_1$ charmonium, would be a real disaster, since α_s is larger than α.

3.1.2 Radiative corrections to QCD Born cross section

The inclusive cross-section for e+e $\longrightarrow$ hadrons is given by

$$\sigma = \sigma_0 \; [1 + \left(\frac{\alpha_s}{\pi}\right) + 1.41 \left(\frac{\alpha_s}{\pi}\right)^2 f - 64.809 \left(\frac{\alpha_s}{\pi}\right)^3 + ..]$$

as reported by Gorishny, Kataev and Larin (Dubna). This may be, if confirmed by independent calculations, an indication of the breakdown of the PT series in gauge theories.

3.2 Progress on the relativistic 2-body equation

There are three methods other than NBS. In the approach of Grotch and Yennie one uses an effective Dirac equation with non-local potentials derived from $e\bar{e}$ scattering. In a more recent method, that of Caswell and Lepage, one starts from an effective Schrödinger equation, again with non-local potentials. Both methods have been used to calculate higher order terms for ep, $e\bar{e}$ and $e\bar{\mu}$ atoms. A third approach, currently being used by S. Brodsky, T. Eller, H.C. Pauli and A. Tang, is that of "discretized light-cone quantization". These authors directly (i.e. numerically) diagonalize the light-cone Hamiltonian, of course with a truncated basis of Fock states. This yields both the mass spectrum (levels) and the wave functions. The method works for any α, but results have only been reported to date for 1 + 1 dimensions.

Search for Parity Nonconservation in Hydrogen

E.A. Hinds

Physics Department, Yale University,
New Haven, CT 06520, USA

1. Theory

Weak interactions are mediated by the exchange of virtual W^+, W^- and Z^0 particles, which together with the photon are the gauge bosons of the standard electroweak theory of Glashow, Weinberg and Salam [1]. In stable atoms the electrons have no charged interactions (exchange of W^+ or W^-) in first order (that would constitute β-decay) and the second order effects are, of course, very feeble. On the other hand Z^0 exchange does not affect the charges of the constituent particles and contributes in first order to atomic structure. The range of the virtual Z^0 is very short ($\sim 10^{-18}$ m) because it is massive (~ 100 GeV/c^2) and for our purposes it is adequate to consider the interaction as a simple current-current interaction at a point. Assume, as in the standard theory, that the currents contain only vector (V) and axial vector (A) components. Then there are four main terms to consider in the effective Hamiltonian

$$H_{eff} = \frac{G_F}{\sqrt{2}} \sum_N (C_{VV}^{eN} \bar{\psi}_e \gamma^\mu \psi_e \bar{\psi}_N \gamma_\mu \psi_N + C_{AA}^{eN} \bar{\psi}_e \gamma^\mu \gamma_5 \psi_e \bar{\psi}_N \gamma_\mu \gamma^5 \psi_N$$

$$+ C_{AV}^{eN} \bar{\psi}_e \gamma^\mu \gamma_5 \psi_e \bar{\psi}_N \gamma_\mu \psi_N + C_{VA}^{eN} \bar{\psi}_e \gamma^\mu \psi_e \bar{\psi}_N \gamma_\mu \gamma^5 \psi_N). \quad (1)$$

ψ_N and ψ_e are the nucleon and electron field operators, the C's are coupling constants, γ_μ and γ_5 are the usual Dirac matrices and G_F is the Fermi constant. Of course the fundamental couplings are to the quarks but at low energy it is convenient to consider the nucleons as fundamental. There are also terms proportional to the momentum transfer but they are small in atoms.

The first two terms in (1) are of even parity because they involve the product of two odd currents (the C_{VV} term) and of two even currents (the C_{AA} term). These interactions cause shifts of the energy levels of an atom. The last two terms are of odd parity and according to first order perturbation theory they do not shift the levels. These are the terms responsible for parity nonconservation in hydrogen. They may be replaced to a good approximation by a non-relativistic potential H_{pv} for an electron interacting with a point proton at $r = 0$. In atomic units [2]

$$H_{pv} = \frac{\alpha G_F}{2\sqrt{2}} \{-C_{AV}^{ep} (\vec{\sigma}_e \cdot \vec{p}) \delta(\vec{r}) + C_{VA}^{ep} (\vec{\sigma}_e \cdot \vec{p})(\vec{\sigma}_e \cdot \vec{\sigma}_N) \delta(\vec{r}) + \text{h.c.}\}. \quad (2)$$

The matrix of H_{pv} between the $nS_{1/2}$ and $nP_{1/2}$ states of hydrogen is diagonal in the hyperfine quantum numbers F and M and the elements are

The Hydrogen Atom Editors: G.F. Bassani · M. Inguscio · T.W. Hänsch

$$<nP_{1/2}FM|H_{pv}|nS_{1/2}FM> = iV(C^{ep}_{AV} - C^{ep}_{VA}[2F(F+1)-3]) \quad (3a)$$

where

$$V = \frac{\alpha G_F}{2\sqrt{2}} \cdot \frac{Z^4}{\pi n^4}\sqrt{n^2-1}. \quad (3b)$$

In the case of hydrogen with $n = 2$, V/h is 0.0128 Hz.

Parity nonconservation in hydrogen was discussed as early as 1959 by Zel'dovich [3]. Later many practical aspects of making a measurement were explored by Michel [4], but the subject really came to life with the discovery of weak neutral currents in 1973/74 [5,6]. Almost immediately Feinberg published two important papers on the weak circular polarization of spontaneous decay radiation from hydrogenic atoms [7] and Lewis and Williams discussed the rate of excitation of hydrogen atoms by polarized light [8]. It was subsequently suggested that the best hope of detecting parity nonconservation in hydrogen lay in microwave transitions among the $2S_{1/2}$ levels [9] and the three experiments tried on hydrogen (at Seattle, Michigan and Yale) are all based on such transitions. Dunford [10] has proposed a similar scheme with He^+. As yet there are no results of sufficient accuracy to detect the parity nonconservation. For a fuller and more general account see ref. [11].

2. Principle of the Yale Experiment

The Yale experiment involves the $2S_{1/2}$ and $2P_{1/2}$ states in zero magnetic field. The strength of the parity nonconserving interaction is measured by a study of hyperfine transition from $(2S_{1/2}F = 0)$ to $(2S_{1/2}F = 1)$ at 178 MHz using an atomic beam of hydrogen.

According to first order perturbation theory the usual $(2S_{1/2}F)$ state $|SF>$ is modified by an admixture of the $(2P_{1/2}F)$ state $|PF>$ to become

$$|S'F> = |SF> + \frac{<PF|H_{pv}|SF>}{E_{SF} - E_{PF}}|PF> \quad (4)$$

where E_{SF} and E_{PF} are the eigenvalues of those two states and we have neglected the smaller effect of $2P_{3/2}$ and other more distant states. Hence the electric dipole matrix element between the states normally labeled (2S, F = 0) and (2S, F = 1) is

$$<S'1|z|S'0> = <S1|z|P0>\frac{<P0|H_{pv}|S0>}{E_{S0} - E_{P0}} + \frac{<S1|H_{pv}|P1>}{E_{S1} - E_{P1}}<P1|z|S0>. \quad (5)$$

The weak interaction matrix elements are given by (3a) and (3b). The energy denominators are both approximately equal to the Lamb shift S; 969 MHz for $F = 0$ and 1088 MHz for $F = 1$. Consequently the terms involving C^{ep}_{AV} almost cancel and the electric dipole matrix element is approximately

$$<S'1|z|S'0> = \frac{4\sqrt{3}iV}{S}C^{ep}_{VA}. \quad (6)$$

Of course the main point is that this matrix element is not zero and that an electric dipole hyperfine transition can therefore be excited.

Fig. 1 indicates the main features of the apparatus. A high intensity $2S$ atomic beam is produced by passing a 500 eV beam of protons through a cesium target. The atoms are prepared in the $F = 0$ state using a microwave state selector [12] to stimulate decay of the $F = 1$ atoms through transition to the short lived $2P_{1/2}$ levels. The beam now enters the main interaction region in which it passes sequentially through two separated parallel, coherent oscillating fields. The first field $\vec{\beta}$ is magnetic and drives the normal $F = 0 \rightarrow F = 1$ magnetic dipole transition. The second field $\vec{\epsilon}$ is electric and drives the parity-forbidden electric dipole transition $F = 0 \rightarrow F = 1$ through the transition moment discussed above. The $F = 1$ amplitudes induced at resonance in the two regions may be written

$$\vec{A}_{M1} = \mu\vec{\beta} \qquad \vec{A}_{E1} = id\vec{\epsilon}e^{i(\phi_0+\phi)} \tag{7}$$

where μ and d are real and proportional respectively to the magnetic dipole and (parity-forbidden) electric dipole transition moments. The angle ϕ_0 is a quantum mechanical phase (equal to zero if we ignore the decay of the P states) and ϕ is the phase angle between $\vec{\epsilon}$ and $\vec{\beta}$. The state amplitudes are polarized parallel to the fields because the initial $F = 0$ state is spherically symmetric. Thus the probability of transitions to the $F = 1$ state is

$$\begin{aligned} P &= (\vec{A}_{M1} + \vec{A}_{E1})^* \cdot (\vec{A}_{M1} + \vec{A}_{E1}) \\ &= \mu^2\beta^2 + d^2\epsilon^2 + 2\mu d\vec{\beta}\cdot\vec{\epsilon}\sin(\phi_0 + \phi). \end{aligned} \tag{8}$$

Here we see first the allowed $M1$ term, then the negligible forbidden $E1$ term and last the interference term to be measured, which is proportional to the weak interaction through d and depends on the odd parity combination of fields $\vec{\beta}\cdot\vec{\epsilon}$.

Atoms driven into the $F = 1$ state are picked out by a second state selector (Fig. 1) which stimulates $F = 0$ atoms to decay, but transmits $F = 1$ atoms. The latter are then detected by measuring the flux of emitted Lyman-alpha photons when the beam enters a region of strong static electric field. Thus the number of detected atoms is proportional to P (8). The interference term of interest is modulated by chopping the phase of the *rf* electric field between ϕ and $\phi + \pi$ and is extracted from the total detector signal by phase sensitive detection.

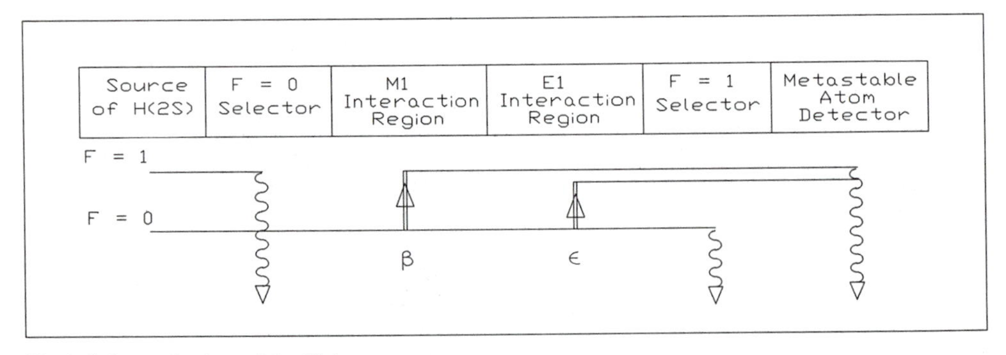

Fig 1. Schematic view of the Yale apparatus

3. Status of the Yale Experiment

At present the apparatus runs with sufficient statistical sensitivity to determine C_{VA}^{ep} with an error of order unity in one day. Unfortunately, there is a relatively large and unexpected transition amplitude that appears to be both proportional and parallel to the electric field $\vec{\epsilon}$. It is evident from our measurements of the phase and frequency dependence of this amplitude that it is not the interesting $\vec{A}_{E1}$ of (7). We believe therefore that it is the result of some imperfection of our apparatus which so far we have been unable to identify in spite of several years of effort.

The hypotheses that we have tested include the obvious ones such as a small rf magnetic field or a stray static electric field in the $E1$ region. We have also excluded a large class of possible effects that could occur in the fringes of the electric field $\vec{\epsilon}$ and we have demonstrated that stray magnetic fields are not responsible. In fact we have not found any experimental variable apart from $\vec{\epsilon}$ that can influence this transition amplitude.

Although this problem prevents us from observing parity nonconservation in hydrogen we are able to determine an experimental upper bound $C_{VA}^{ep} < 300$ which is the most accurate result so far from hydrogen. Of course we hope that the mechanism responsible for the systematic error will come to light, but at present we are not optimistic about the prospects for a large improvement. Similar problems have been encountered in the Seattle and Michigan hydrogen experiments at a similar level of precision.

At present several experiments have determined C_{AV} in various heavy atoms at the 10% level of accuracy or better but C_{VA} has not been detected in any atom. For a review of parity nonconservation in atoms, see [13] or [14].

References

1. E.D. Commins and P.H. Bucksbaum, *Weak Interactions of Leptons and Quarks* (Cambridge University Press, 1983).
2. R.W. Dunford, R.R. Lewis and W.L. Williams, *Phys. Rev.* **A18** (1978) 2421.
3. Ya. B. Zel'dovich, *Zh. Eksp. Teor. Fiz.* **36** (1959) 964.
4. F.C. Michel, *Phys. Rev.* **B138** (1965) 408.
5. F.J. Hasert, *et al., Phys. Lett.* **46B** (1973) 121; *Phys. Lett.* **46B** (1973) 138.
6. A. Benvenutti *et al., Phys. Lett.* **32** (1974) 800.
7. G. Feinberg and M.Y. Chen, *Phys. Rev.* **D10** (1974) 190; Errata **10** (1974) 3145; G. Feinberg and M.Y. Chen, *Phys. Rev.* **D10** (1974) 3789.
8. R.R. Lewis and W.L. Williams, *Phys. Lett.* **59B** (1975) 70.
9. E.A. Hinds and V.W. Hughes, *Phys. Lett.* **B67** (1977) 487.
10. R.W. Dunford, *Phys. Lett.* **99B** (1981) 58.
11. E.A. Hinds, in *The Spectrum of Atomic Hydrogen Advances*, ed. G.W. Series, World Scientific, 1988, p. 245.
12. J. Wm. Edwards, G.L. Greene and E.A. Hinds, *Nucl. Instrum. Methods* **197** (1981) 581.
13. E.N. Fortson and L.L. Lewis, *Phys. Reports* **113** (1984) 290.
14. E.A. Hinds, in *Atomic Physics 11*, eds. J.C. Gay, G. Grynberg and S. Haroche, World Scientific, 1989.

Part IV

Hydrogen in Strong Fields and Chaos

Multiphoton Transition to the Continuum of Atomic Hydrogen

N.K. Rahman

Dipartimento di Scienze Chimiche, Università di Trieste,
Piazzale Europa, 1, I-34100 Trieste, Italy

The current experimental and theoretical status of multiphoton transition to the hydrogenic continuum, its prospects in the near future and the relation of these studies vis-a-vis standard QED are described.

1. Introduction

Non-linear processes involving transitions to the continuum brought about by high intensity pulsed lasers are the subjects of numerous investigations in the area of atomic, molecular and optical physics. New processes are being experimentally uncovered at a rapid rate, and the field promises to be of continued vigour due to the remarkable progress that we are witnessing in the development of high power lasers. Almost all of these studies have been done with non-hydrogenic atoms (or molecules). These processes are treated theoretically with necessarily approximate inputs and at some stage of the calculations, wave functions, matrix elements, summations over the entire spectra, what you will, that are utilized give us results that can hardly be considered as exact. Often the agreement between theory and experiment is either fortuitous or non-convincing for many reasons and more often, the agreement between theory and expriment is simply lacking.

The mathematical structure of the quantum mechanical amplitudes of non-linear processes involving continuum for system is rather complicated and for non-hydrogenic systems can be the theoreticians nightmare. To cite an example, a large number of experimental results have been reported in the last few years with the rare gases, but no reliable calculations for these experiments exist. Essentially the detailed confrontation between theory and experiments have been given up in this area and very often one is satisfied with the order of magnitude estimates.

Hydrogen atom is in the pleasant exception to this rule: the theoreticians are expected to produce accurate estimates of these processes. The experiments then serve to test these benchmark calculations and the satisfactory conclusion of this dialogue serves as the foundation of the atomic physics of non-linear processes involving continuum just as the photoionization process of the hydrogen atom [1] serve as the cornerstone of all photoionization processes. In what follows, I shall explain what three recent experiments have measured in this area and how theoretical calculations have

The Hydrogen Atom Editors: G.F. Bassani · M. Inguscio · T.W. Hänsch

helped to explain quantitatively these measurements of three highly non-linear processes involving multiphoton absorption in the continuum of the hydrogen atom.

Further on, I shall discuss a few other such processes which have not yet been experimentally measured for the hydrogen atom. These could be the next set of experiments with the hydrogen atom and intense lasers. The future of this class of processes vis-a-vis the next generation of high power lasers useful for basic atomic physics experiments, and the non-linear processes viewed from the perspective of QED theorists are the final comments in this paper.

2. REMPI, ATI and REATI

There is a large number of processes that one can envisage being induced by intense lasers with the hydrogen atom. I have selected here three of them, the reasons being that in the last three years, three experiments [2,3,4] have been performed to study them. These processes may be called with the acronyms REMPI (Resonance enhanced multiphoton ionization), ATI (Above threshold ionization) and REATI (Resonance enhanced above threshold ionization) [Fig. 1]. All of these three processes are non-linear, i.e. the probabilities are not linear functions of the intensity of the lasers.

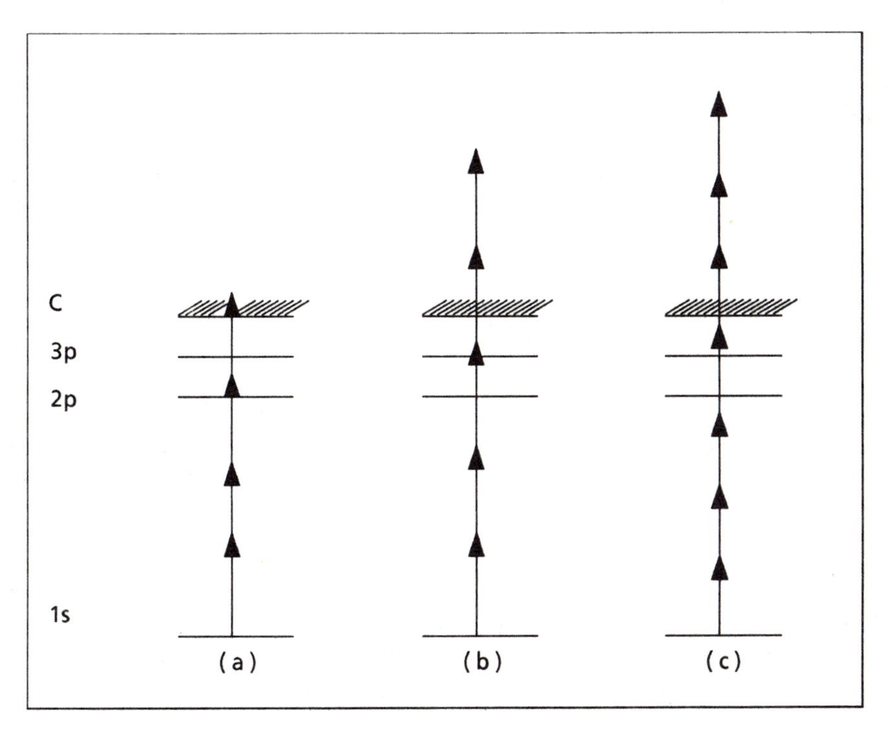

Fig.1. Examples of (a) REMPI, (b) REATI and (c) ATI for the hydrogen atom. Each arrow represents a photon absorption.

REMPI is the bound-free, i.e. ionization, process in which N (N>1) photons are needed to ionize the atom and the frequency of the photons is such that M(M<N) photons can cause a bound-bound transition from the initial state of the atom. This latter condition causes an enhancement of the probability of the usual multiphoton ionization with no frequency condition.

ATI is the ionization by the absorption of N (N>1) photons by the atoms while in order to ionize the atom, one needs M (M<N) photons. Thus the electrons that emerge in this processes acquire energy due to the absoption of additional photons in the continuum. The process involves an admixture bound-continuum and continuum-continuum transitions. The electron spectra is the tell-tale sign of such processes showing how many photons have been absorbed beyond the minimum number M.

REATI is the above threshold ionization in which P (P<M,N) photons can cause a bound-bound transition in the atom, M photons are needed to ionize and N photons are number of photons absorbed for the overall detected signal. In contrast to (P,M,N) which specifies the REATI, the duo (M,N) that specifies ATI and REMPI can be in principle any set of integers with N>I. For the three experiments that have been done, they are (P = 3, M = 4, N = 5) for REATI, (M = 6, N = 9) for ATI and (M = 3, N = 4) for REMPI.

3. Current Status

The theory of these processes for a non-relativistic hydrogen atom involves utilizing perturbation theory of the N-th order as the basic starting point. The amplitude is given by:

$$T_{fi}^{(N)} = \langle i \left| V^- G_o V^- G_o \ldots\ldots G_o V^- \right| f \rangle$$

where G_0 is the Coulomb Green's function and V^- is the photon absorption part of the interaction hamiltonian.

The above expression contains for 4 photon processes (for example provoked by radiation at 355 nm) 3-fold infinite summation (over all the discrete states of H°) and integration over the entire Coulomb continuum. This can be done for the hydrogen atom with a variety of methods, only one of which will be relevant for our purpose and we shall discuss that later. The ATI process starts with utilizing the same amplitude as in Eq.(1) with N = 5 (at the same frequency 355 nm). The remarkable feature of the amplitude now is that while previously the basic radial part of the amplitudes are real, the corresponding ones are now complex. This happens due to the fact that in ATI the last intermediate integration has a pole in the denominator, producing thereby a complex amplitude. This fact recurs for all the higher order ATI processes. For example, at 532 nm, we need to consider N = 7, 8 and 9 for ATI [5] . The integration through the pole occurs one, two and three times respectively. The resulting amplitudes are then utilized to write down the so-called differential cross-section, which gives us the angular distribution of the ejected electrons. These angular distributions of the ejected electrons once measured, give us an excellent check on the validity of the calculation.

How are these calculations done? The work requiring the N-fold infinite summations and integrations is a non-trivial task. A large amount of computational effort has been devoted to their calculation. The methods range from brutal computation, to special representation of the Coulomb Green's function (the notheworthy one being that in terms of the Strumian functions) and the implicit method utilizing a bypass by solving coupled differential equations. This latter method has turned out to be the most effective for the calculations regarding the experimental data in [3]. The method has been utilized to calculate ATI at 355 nm (N,M = 5,4 and 532 nm (N,M = 9,6) and have reproduced rather strikingly all the experimental angular distributions [5]. One may conclude that the lowest order perturbation theory is satisfactory for ATI under the given experimental condition of the intensity of the laser and the photon frequencies.

The REMPI experiment involves coupling 1S and 2P or 3P states by a 3 photon bound-bound transition and then bound-continuum transitions to produce ionization. The process is both non-linear as well as non-perturbative, the latter being manifested by the non-observance of the I^N power law. The calculation essentially uses a formalism that ensures the strong coupling of the two relevant states while the rest are treated perturbatively. The shifts and these states as well as the width of the resonant state appear naturally. The hydrogenic energy formula $E_n = 1/(2n^2)$, also creates the curiosity that the exact resonance with one state is also the threshold of ionization, i.e. the electron should be ejected with zero velocity. This produces certain asymmetry to the probability of ionization as a function of frequency. Taking the experimental pulse structure as well as coherence properties of the laser into consideration, the calculation has been completed to the extent that the results can be directly confronted to the experiment. The agreement is indeed very satisfactory.

The third experiment which concerns REATI is of interest because it measures the branching ratio between the electrons issued with the minimum energy and the ones that had acquired the energy of one photon, under the condition of REMPI. A minor modification to the theory of REMPI allows one to calculate REATI and therefore the branching ratio. The agreement between the experiment and the theory is within 8%.

4. Other Processes

There is a large variety of non-linear processes for hydrogen that have not been studied experimentally, but have been theoretically investigated. It would not be possible for me to discuss them all here, but let me briefly mention a few.

Probabilities of multiphoton ionization have been calculated for the hydrogen atom for N quite large [8,9,10], up to even N = 16. These are rather formidable calculations and no experiments of these have still been made. (There exists a remarkably early experiment [11] with hydrogen atom). The theoretical values of these probabilities arise from large order of the perturbation series. It is noteworthy that some of these processes have been calculated utilizing independent methods (Sturmian function representation of the Green's function and coupled hierarchical set of differential equations) and the results agree with one other.

Another class of processes involves free-free multiphoton transition. These are the absorption or emission of N photons entirely in the hydrogenic (i.e. Coulombic) continuum. These processes could be multiphoton bremsstrahlung or inverse bremsstrahlung. Exact calculation [12] for 2 photons have been recently reported, which is the natural extension of the classic Sommerfeld calculation for the single photon. Measurement of such processes for hydrogen i.e. electron absorbing N photons in the Coulomb potential have still not been attempted. Another class of processes that have drawn the attention of theorists is in the Raman-like processes involving N Photons from an external source. Laborious calculations and accurate predictions have been made: what are now lacking are the corresponding experiments.

5. Discussion

The agreement between the three recent experiments and the corresponding calculations augers well for the future of the physics of multiphoton processes in which a wide variety of newer processes are to be observed experimentally. To be able to observe them for the hydrogen atom means that one can confront them with exact calculations and the basic understanding of these processes leaves no room of uncertainty.

Hydrogen atom may now prove to be an extremely valuable system for understanding the dynamics of the absorption of photons from extremely intense fields. Intensities of the order of 10^{16-18} Watt/cm^2 with pulse duration in the second region are beginning to be utilized for a newer class of experiments. The interaction term in the hamiltonian $H_{int} = - e/mc\ A \cdot P + e^2A^2/2mc^2$ for these field strengths are on the average comparable or higher than the Coulomb potential. The perturbation theory, at least the lowest order calculation, then become inoperative and one has to take recourse to resummation of the higher order terms of the perturbation series [13,14].

Other techniques such as Floquet theory then may become of much importance. One starts to wander into the uncertain regions of theory and it behooves us to consider above all the hydrogen atom (whose every eigenfunction we know to the T) as our theoretical guinea pig for these calculations. The experiments can then be done to confront these predictions and the strong field problem will then progress with a logical foundation. Technical progress in this direction appears to be really exciting.

Finally it is not unreasonable to place some of these processes and the corresponding experiments in the perspective of quantum electrodynamics. In QED, one nowadays, with large computation effort, calculates the higher order terms which increase the accuracy of a prediction. One thus adds more and more digits to measurable number and test how accurate is QED. The accuracy of the calculation and the corresponding measurement of Lamb shift of the hydrogen atom is truly remarkable. In a broad sense, the multiphoton transition into the continuum of the hydrogen atom is adding another dimension to the picture. The effects are due to the external field and the processes that are being investigated are the physical manifestation of rather high order terms of the perturbation series. Indeed, by means of the external field, one has been able to calculate 16th order term of the perturbation series (no experiment yet) and measure a 9th order term, (the experiment in Ref. 3 and the calculation in Ref.5), essentially exactly. Even without neglecting the fact that the calculations are for the non-relativistic hydrogen atom, the order of these processes certainly is a remarkable testimony of how much progress has been made in the quantum electrodynamics of the external electromagnetic field.

Acknowledgements
Much of the work that I have discussed here, happens to be partly the result of my own involvement along with Dr.Y.Gontier and Dr.M.Trahin of the Institut Recherche Fondamental, Service de Physique Atomique et des Surfaces at Saclay Laboratories. I must thank them for their friendship and cooperation. Thanks are due to Ms. Stefania Grassini for an excellent job of preparing this manuscript with the shortest of notice.

References

1. H.A.Bethe, E.E.Salpeter: Quantum Mechanics of One- and Two- Electron Atoms, (Springer Verlag, Berlin, 1957)
2. D.E.Kelleher, M.Ligare, L.R.Brewer: Phys.Rev. A31, 2747 (1985)
3. D.Feldmann, B.Wolff, M.Wenhöner, K.H.Welge: Z.Phys.D. 6, 293 (1987)
4. M.J.Muller , H.B. van Linden van der Heuvell, M.J. van der Wiel: Phys.Rev.A34, 236 (1986)
5. Y.Gontier, N.K.Rahman, M.Trahin: Europhys.Lett. 5, 595 (1988)
6. E.Karule: J.Phys.B: At.Molec.Phys. 21, 1997 (1988)
7. Y.Gontier, N.K.Rahman, M.Trahin: Phys.Rev. A37, 4694 (1988)
8. Y.Gontier, M.Trahin: Phys.Rev. A4, 1896 (1971)
9. E.Karule: J.Phys.B: At.Molec.Phys. 4, L 67 (1971)
10. A.Maquet: Phys.Rev. A15, 1088 (1977)
11. M.Lu Van, G.Mainfray, C.Manus, I.Tugov: Phys.Rev.A7, 91 (1973)
12. V.Veniard, M.Gavrila, A.Maquet: Phys.Rev. A 35, 448 (1988)
13. Y.Gontier, N.K.Rahman, M.Trahin: Phys.Rev. A14, 2109 (1976)
14. Y.Gontier, N.K.Rahman, M.Trahin: Nuovo Cimento D4, 1 (1984)

Hydrogen in Strong DC and Low Frequency Electric Fields – One Dimensional Atoms

M.H. Nayfeh, D. Humm, and M. Peercy

Department of Physics, University of Illinois at Urbana-Champaign, 1110 West Green Street, Urbana, IL61801, USA

Introduction

Recently there has been a resurgence of interest in the effects of externally applied electric, magnetic, and radiation fields upon atomic spectra and collision processes, due to the development of the ability to produce, in the laboratory, fields which are comparable in strength to internal atomic fields[1-3]. But it is almost impossible to create fields in the laboratory (10^9 V/cm) which are strong enough to disrupt atoms in their normal states. On the other hand, a Rydberg atom, which can be thousands of angstroms in size, is very sensitive to external perturbations, and the Coulomb field of the nucleus can be overcome by an external electric field of only 5 kV/cm. Thus entry of atomic physics into the strong-field regime has been accomplished by dealing with highly excited states, rather than by generating enormous laboratory fields. Diverse strong field effects are now studied in the laboratory under easily controlled conditions. These studies have extended the scope of research in areas which once used only weak field effects, and they have also opened up new avenues.

The treatment of the problem is not straightforward because of the different symmetries involved and because the Rydberg electron does penetrate the core; thus the many-particle nature cannot be avoided. A strong candidate for a unified treatment of Rydberg atoms in fields is the frame transformation approach.[4] Its basic idea is that configuration space can be separated into two regions where different dynamics prevail: an inner region, where the Rydberg electron penetrates the ionic core, and where the many-particle nature of the system must be considered; and an outer region, where the interaction between the Rydberg electron and the residual ion can be adequately represented by long-range potentials. The equations of motion are solved separately in each region and matched to a common boundary condition. This approach has been extensively developed in field-free atomic physics, under the general heading of R-matrix and quantum defect methods. It offers the possibility for dealing with the interplay between effects of particular atomic structures and of electron motion in Coulomb plus external fields, since these are associated respectively with inner and outer region dynamics.

An important implication of this point of view is that the physics of highly excited atoms in external fields should be described in terms of hydrogen, and that specific knowledge of the behavior of atomic hydrogen in external fields should provide the key to general understanding. In fact, theory has tended to focus almost exclusively on hydrogen. Because applications to complex atoms have been in mind, these calculations focused on excitation from spherical low-lying states of good orbital angular momentum. But under these field values this "hydrogenic theory" does not really apply to hydrogen itself. This is due to the fact that the Stark splitting at these fields are larger than the Lamb shifts, fine and hyperfine splittings for all levels of hydrogen including n=2 for which these are the largest.

The Hydrogen Atom Editors: G.F. Bassani · M. Inguscio · T.W. Hänsch

Thus as a result of the near ℓ degeneracy, the field mixes fully the ℓ quantum numbers, causing ℓ not to be a good quantum number and the states become pure parabolic with permanent dipole moments.

Our experimental and theoretical work has indeed shown that hydrogen in strong electric fields is unique among all atoms even in the energy region close to E = 0 where complex atoms are believed to become practically hydrogenic. In fact hydrogen in strong electric fields offers the only atomic system of combined true spherical symmetry (non-relativistic interaction) and cylindrical symmetry (dc field).

One of the novel features of the combined pure symmetries is the ability to drastically manipulate and even design and construct specific atomic structures.[5-6] Although the first generation of experiments used alkali and earth alkali and metastable excited rare gas atoms because their excitation is within the reach of the visible tunable dye lasers,[1,7-10] however, it was soon realized that the noncoulombic interaction with the ion core in complex atoms hampers one's control of the manipulation and design. Therefore the attention turned to atomic hydrogen in spite of the fact that it is not readily amenable to experimentation.[11-12] However this difficulty is offset by much theoretical interest in this system.[13-16]

In our studies, we use strong external dc electric fields to manipulate and control the atomic structure of highly excited atomic hydrogen. We can construct nearly one-dimensional atoms whose electronic distribution are highly extended along the field, and which may have enormous dipole moments ("giant dipole" atoms).[17-18] The nuclear charge Z_1 that defines the energy and other properties of the "new" atom is a fraction of the proton charge. We can construct orbits that are aligned ($m = 0$) or at an angle ($m \neq 0$) with the field. For $m = 0$, the fractions 0, 1/4, 1/2, 3/4 and 1 define four quarters that classify the atoms.[18] The dipole moment is found to be opposite to the field in the first and third, and in its direction in the second and fourth quarters, and zero at the boundaries. For $|m| = 1$, on the other hand, only the fractional charges, 0, 1/2, and 1 are of importance; they give two halves that classify the atoms with the dipole moment being in the direction and opposite to the field in the first and second halves respectively.[18]

These one-dimensional atoms can be prepared with total positive or negative energy. Although these are unstable against ionization, however at $E < 0$ F but $E > -2\sqrt{F}$ one can populate "giant dipole" atoms whose potential barriers are large enough (tunneling small enough) to render their lifetimes quite long, important for further experimentation and application.

Another unconventional property of these one-dimensional atoms is their radiative interaction with strong fields (EP of low frequency ω).[19] Their huge permanent electric dipole moment d couples very strongly to the oscillating field, dressing the atom frequency modulations and producing sidebands with spacing ω and whose strength is governed by dE/ω. A red one of these bodies can then interact near resonantly with the oscillating field effectively inducing transitions between the original undressed levels. Moreover our analysis indicates that as the excitations proceed from an initial state, the dressing might become strong enough at higher levels to cause the off ladder transitions to become comparable in strength to the ladder transitions causing a loss of the one-dimensionality.

Experimental Technique

Recently, a number of schemes have been employed for the preparation of highly excited atomic hydrogen.[11,20-22] Here simultaneous absorption of two photons from a single tunable pulsed laser beam at 243 nm results in excitation from 1s to n-2, and some photoionization of the resulting n=2 population. A second pulsed beam excites states near the continuum from the n=2 states as shown in Fig. 1. For properly chosen 243 nm energy densities, a large population at n=2 atoms can be produced with only a few percent of them being photoionized. The second beam is then capable of promoting a large portion of the remaining excited atoms to a well-defined high-lying state without saturating the process. Previously the two photon Doppler-free spectroscopy of the n=2 state has been performed using radiation at 243 nm; however, no attempt to excite the hydrogen atoms further was reported.[25]

Figure 2 shows a block diagram of the experimental setup.[12] The atomic hydrogen source is a modified Wood discharge tube. An atomic beam is formed by effusion from the discharge region through a multicollimator assembly composed of 25 small glass capillaries. The thin-walled capillaries are 4 mm long, with an inside diameter of about .2 mm. The resulting atomic beam is directed into the diffusion pumped cell which contains the field plates. The beam is loosely collimated, but produces a density of about 10^{11} H°/cm^3; the background gas density is on the order of 10^{12}/cm^3. One of the plates has a 3 mm × 10 mm slot cut into it to allow the passage of ions. Since the presence of the open slot would lead to an unacceptably non-uniform field, a .5 mm spacing copper mesh was soldered over the whole surface of the plate. The electric field between the plates is determined to an accuracy of 0.5% by measurement of the applied voltage and the separation of the field plates. Ions produced by the laser radiations are driven by the electric field through the slot in the grounded plate. They travel through a 100 cm long, field-free drift tube which provides mass analysis. This is necessary since molecular impurities are easily ionized by the ultra-violet wavelengths in use. At low electric fields, the mass resolution is sufficient to verify that the signal under study is indeed due to atomic hydrogen.

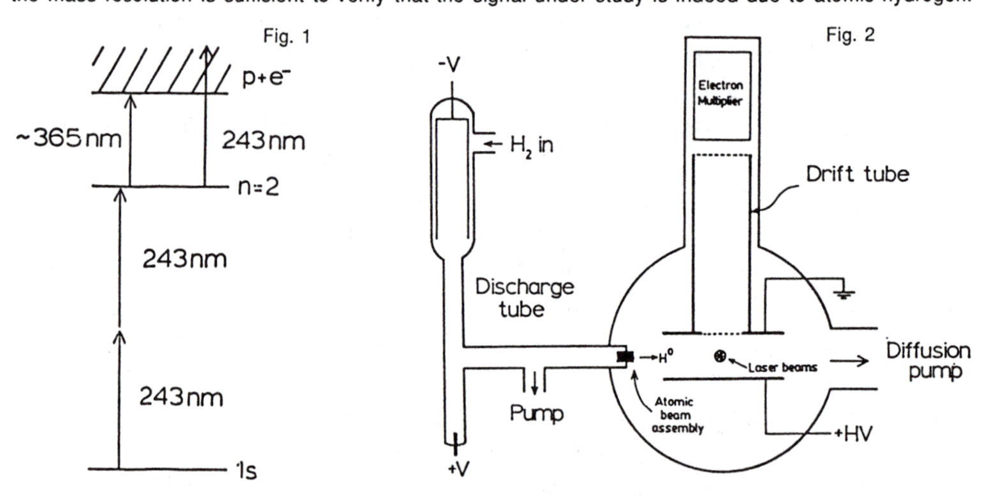

Fig. 1 Excitation of highly excited states of atomic hydrogen. Absorption of two photons each at 2430 Å excites the ground state to 2s followed by one photon excitation at 3660 Å to a highly excited state.

Fig. 2 A simplified block diagram of the experimental set up.

The resolution at kilovolt/cm range fields is sufficient to separate the hydrogen signal from those due to the impurities. Ions are detected using an 18 stage venetian blind electron multiplier capable of single ion detection. Under typical experimental conditions, several hundred ions are detected per pulse.

The transit time t_S of thermal hydrogen across the effective observation region (1.5 mm - half of the width of the slot in the upper plate) will cutout the ionization produced by states that ionize very slowly such that $\tau >> t_S$ where τ is the ionization lifetime of the atom. On the other hand, ionization from states that ionize in time much shorter than t_S will not effectively be affected by the slot. For intermediate cases, the detection factor is just $1\text{-}e^{-t_S/\tau}$. Using 1.5×10^3 m/s for the thermal velocity of hydrogen in the direction parallel to the width of the slot gives 10^{-6} s for t_S.

The optical beams needed for the excitation of atomic hydrogen are produced using a pulsed laser system: an Nd^{+3}:YAG laser and two dye lasers. A fraction of the second harmonic of the YAG laser at 532 nm is used to pump one of the dye lasers producing a beam at 630 nm, which is frequency doubled to 315 nm by a KDP crystal and then summed with the residual YAG fundamental by a KDP crystal resulting in a beam at 243 nm of pulse length of about 10 ns, a bandwidth of about 1.5 cm^{-1} and pulse energies on the order of 10 microjoules. The second dye laser produces a beam at about 555 nm which is summed with part of the YAG fundamental to produce a beam with pulse length near 10 ns, bandwidth of .8 cm^{-1}, pulse energies of a few tenths of a millijoule and a wavelength near 365 nm. The data are collected and analyzed using an LSI-11 computer system.

Thus, for this test both field plates were grounded until one microsecond after the passage of the laser pulses. At this time a high voltage pulse resulting in a 1 kV/cm field between the plates was applied, thereby field-ionizing the highly excited atoms and collecting the ions. Figure 3 shows the results of averaging 8 scans of the ionization in the region of some high Rydberg states.[23,24] Resonances are seen corresponding to the excitation of states up to about 60p. The loss of the series at

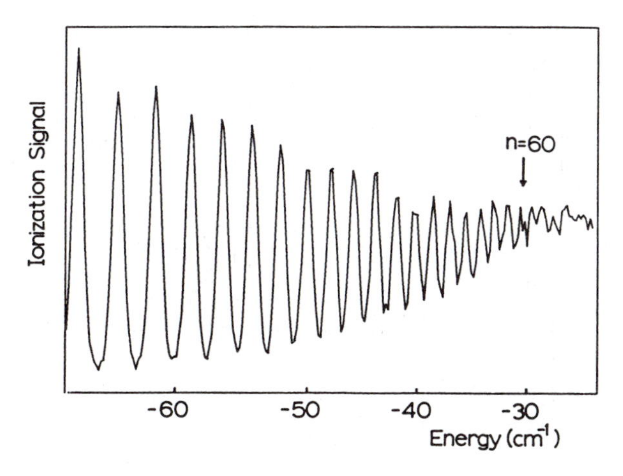

Fig. 3 Highly excited hydrogen spectrum produced by pulsing the electric field after the excitation which lasts 10ns has taken place. The loss of resolution beyond n = 66 is due to the fairly wide bandwidth of the 3660 Å laser (1 cm^{-1}).

this principle quantum number indicates an instrumental resolution of about 1 cm^{-1}, due mostly to the laser linewidth and some residual Doppler broadening (.1 cm^{-1}) due to the relatively large angular spread of the dense atomic beam.

2. Effect of field on n = 2

We first discuss the effect of the electric field on the n = 2 state, the lowest excited state.[26] For one reason, the response of this state is known and allows us to discuss the weak and intermediate field limit. More importantly, this state acts as an intermediate resonance state in our excitation of the highly excited states and therefore it is necessary to understand its structure. Since we are interested in applying fields larger than 2 kV/cm, we will not include the Lamb shift in our analysis. In the region of 2-5 kV/cm, only n and m_j are good quantum numbers, but neither j and ℓ, nor the parabolic quantum numbers (n_1, n_2, m) are. At fields higher than 5 kV/cm, the interaction with the Stark field dominates over the fine structure, thus leading to a linear Stark splitting; consequently, the states can have good parabolic quantum numbers: the state (1,0,0) originating from $p_{3/2}$ is what we call the m=0 blue state, whereas the state (0,1,0) originating from $p_{1/2}$, $s_{1/2}$ is what we call the m=0 red state. The |m|=1 state is the least shifted state. Such calculations were previously done for n=2, 3, and 4.[27] Numerical results based on these calculations are given in Fig. 4.

We will now discuss the efficiency of populating the various Stark states of n=2 using the two-photon (π polarization) process. We calculated the percentage of population of the |m|=1,m=0 blue state and m=0 red state as a function of the electric field ($F \geq 2$ kV/cm) assuming the states are not resolved. It is known that in the zero field limit |m|=1 state is not excited in the π-π excitation. The presence of a field of 3 kV/cm produces about 30%, however, at high fields (> 10 kV/cm) this population drops to less than a few percent. We also note that the two m=0 states approach 50 percent populations at higher fields with the red one approaching this value faster than the blue one.

We calculated the percentage of purity of the various states if each state is selectively excited by radiation whose effective bandwidth is less than the splittings (those are shown in Fig. 5). Above 10 kV/cm both of the m=0 states can be purely excited (> 97%) whereas the |m| =1 state is not excitable. But, because our laser bandwidth is ~1.5 cm^{-1}, then in practice we can only excite pure parabolic states using fields larger than 10 kV/cm such that the Stark splitting is larger than 3 cm^{-1}. Since we use quite low atomic hydrogen density, we find no problem in dropping up to 18 kV/cm across our interaction region; thus making these kinds of studies feasible.

Figure 6 gives the charge distribution of the 100, 010, and 001 parabolic states of n=2. The figure shows the x-z cross section through the atom with the nucleus being at the center of the system. The curves are lines of constant charge density ($\rho = \psi\psi^*$ = constant) where the total charge is normalized to unity. The large eccentricity in the charge distribution of the m_ℓ=0 blue and red states is quite evident, the concentration being up and down field respectively. On the other hand, the distribution of the image 001 state is symmetric with respect to the field. This figure also shows that the charge distribution of the red and blue states are mirror images of each other when reflected in the z = 0 plane. This property is utilized in the present study as a means of probing the shape of the electronic distributions of highly excited Stark states.

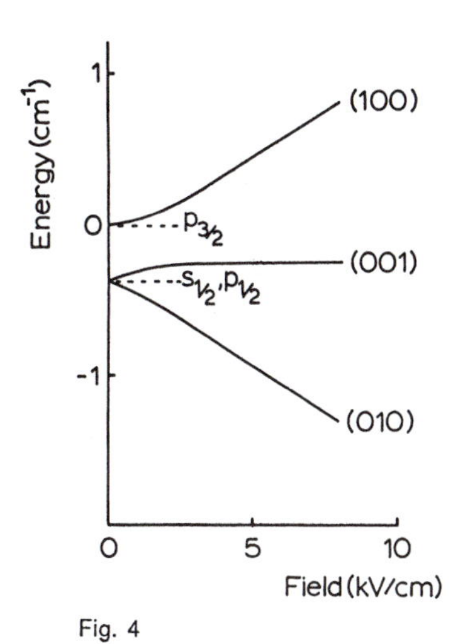

Fig. 4

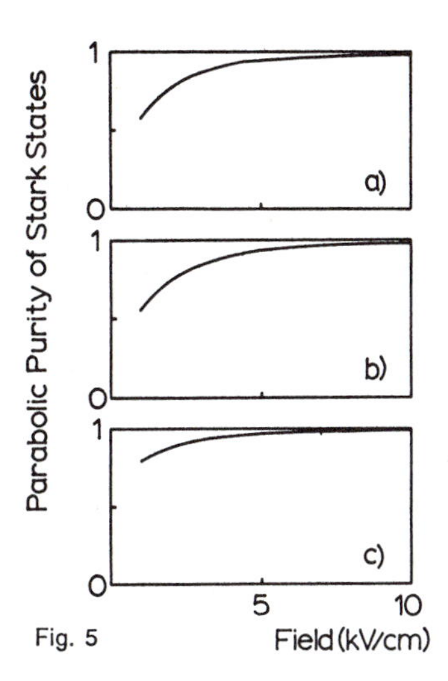

Fig. 5

Fig. 4 The splitting of the n = 2 states vs. field including the effects of the fine structure (but ignoring the Lamb shift). The Stark states are labeled at high field by the parabolic states to which they tend.

Fig. 5 Purity of the Stark states vs. field in terms of their high-field limit parabolic states, (a) is for the blue-shifted 100 state, (b) is for the unshifted 001 state, and (c) is for the red-shifted 010 state.

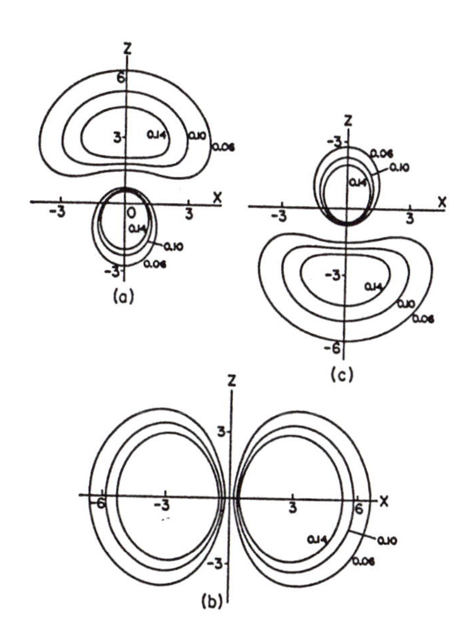

Fig. 6 Charge distribution of the parabolic states of n = 2 state of hydrogen, (a) is for the blue-shifted (1,0,0) state, (b) is for the unshifted (0,0,1) state, and (c) is for the red-shifted (0,1,0) state.

3. <u>Effect of dc field on highly excited hydrogen--Atomic Engineering</u>

Although the potentials represented using spherical coordinates in the z = 0 plane (shown in Fig. 7a) can be very useful in bringing out some features of the interaction, they are not very useful for quantitative calculation. This is because the nonspherical symmetry of the potential makes the interaction non-separable: that is, it cannot be separated into three independent one dimensional motions in spherical coordinates. The interaction, however, is separable in parabolic coordinates $\xi = r + z$, $\eta = r - z$, and ϕ the azimuthal angle, with quantum numbers n_1, n_2, and m respectively. The effective potentials for the ξ and η motions shown in Figs. 7 b,c in fact have good resemblance to that of the z cut in the $z < 0$ and $z > 0$ regions respectively, and hence govern the energy of the system (location of the energy) and the ionization lifetimes of these levels, respectively. The quantum number m is common to both parabolic and spherical descriptions, and the principle quantum number $n = n_1 + |m| + 1$. The spherical 1 and parabolic n_1, n_2 quantum numbers do not have a one-to-one correspondence: a state with definite values of n_1 and n_2 is composed of many different values of ℓ.

One important property of the atom that comes out of this procedure is the fact that only a fraction of the nuclear charge $Z_1 < 1$ drives the ξ motion and hence dictates the energy of the system, while the rest of the charge $Z_2 = 1 - Z_1$ drives the free η motion and hence dictates its lifetime. Thus the presence of an external electric field provides us with a situation where the nuclear charge that drives the bounded motion can be varied, in a near continuous fashion. Considering the fact that the physical and chemical identity of isolated atoms is defined by the nuclear charge, then it is clear that we have at our hand a means for creating new "types" of atoms.

We will now discuss the preparation and nature of the new types of atoms by discussing their spectroscopic properties such as ionization lifetimes, charge distributions (or excitation dipole moments), and branching ratios (or excitation strengths). To do so we will consider the positive and negative energy regimes separately, starting with the former.

Let us assume that atomic hydrogen is immersed in laser radiation of energy just larger than 13.6 eV, the ionization potential of hydrogen, and whose polarization is along the external dc electric field in

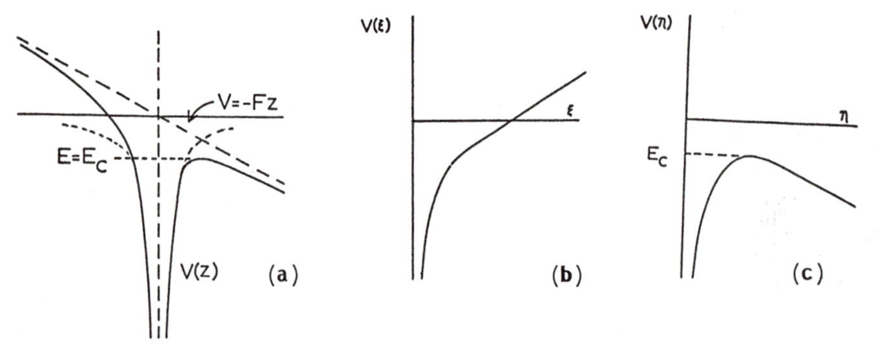

Fig. 7 Schematic of the potential of hydrogen in external dc electric. (a) Potential for a cut along the z axis in the presence of an electric field along the z axis showing a rise and a depression of the Coulomb potential in the $z \leq 0$ and $z \geq 0$ regions respectively. (b) and (c) are the same potential of (a) plotted as two one dimensional potentials in the $\xi = r + z$ and $\eta = r - z$ parabolic coordinates.

which the atom is immersed. Because of this choice of polarization, the electron gets an initial kick along the dc field, and the energy of the system is raised by 13.6 eV, thus rising to zero total energy. The electron can now execute bound motion even for this positive energy. The motion of the electron is nearly a one-dimensional motion with the orbit resembling a cigar whose axis is along the external field, the nucleus being located inside it near its lower tip (Fig. 8a). This specifically tailored atom lives on the order of 5×10^{-13}s (giving very broad widths), and the electron executes on the average about 5 rounds before it breaks away from the proton on its own, and it is found to spend most of its time away from the nucleus, near the upper tip of the cigar. If the electron were initially kicked perpendicular to the field (laser polarization perpendicular to the external field), the cigar would have been created at an angle with the field. (See Fig. 8b.).

The extraordinary thing about this cigar atom is that such a "separated charge" distribution gives a dipole moment P which points opposite to the external field. Moreover, the dipole is very large since the separation of the charge (length of the cigar) is about 1600 Å, hence giving dipole moments that are 3000 times larger than those of normal atoms. For this reason we call these atoms "giant dipole" atoms. However, in general, one cannot exclusively prepare these types of atoms without preparing the highly excited normal atom since, first of all, the excitation has to start from the ground state of the normal atom which is only weakly affected by electric fields and secondly both Coulomb and Stark fields will have to compete. Therefore, after the excitation process we always have a superposition of these two types with the branching ratio depending on a number of parameters including the total energy of the system, field strength, and the properties of the exciting laser radiation.[13-17] For example, the "visibility" of the giant dipoles, which is a measure of how much they rise above the accumulated smooth continuum, tends to be very small (4% at 5 kV/cm). This visibility gets worse at higher energy because these states get closer to each other in energy as the potential opens up. Those states were first seen in complex atoms such as rubidium, sodium, barium, krypton, and yttrium during 1978-1983,[1,7-10] but were found to have strengths that are smaller than is predicted for hydrogen. Theories that included the effect of core electrons explained the reduction of the strength.[15] The first observation of the giant dipoles in hydrogen was made in 1984 in our laboratory at the University of Illinois[12] (see Fig. 9). Similar observation was also achieved at the University of Bielefeld.[11]

Considering the shortness of their lifetimes, and the low efficiency of excitation, it is clear that experimentation with these "new atoms" will not be easy unless these two properties are enhanced. Recently we have been able to improve the efficiency[26] and to produce giant dipoles that live much longer than 10^{-12} s[18,19] in atomic hydrogen. We will discuss the efficiency first. The scheme we devised for this purpose relies on a process we call multistage shaping or charge shape tuning of the charge of the atom. In one-photon excitation from the ground state one effectively starts from a spherically symmetric charge distribution (zero dipole moment), and tries to mold it by a single operation into a giant dipole whose charge is highly focused along the field. On the other hand in multistage shaping one uses one photon to create from a ground state a not too large dipole of charge distribution that is focused along the field at an intermediate state followed by another photon absorption from this intermediate state that produces larger dipole whose charge is even more focused along and so on till one excites the giant dipole in a highly focused distribution along the field.

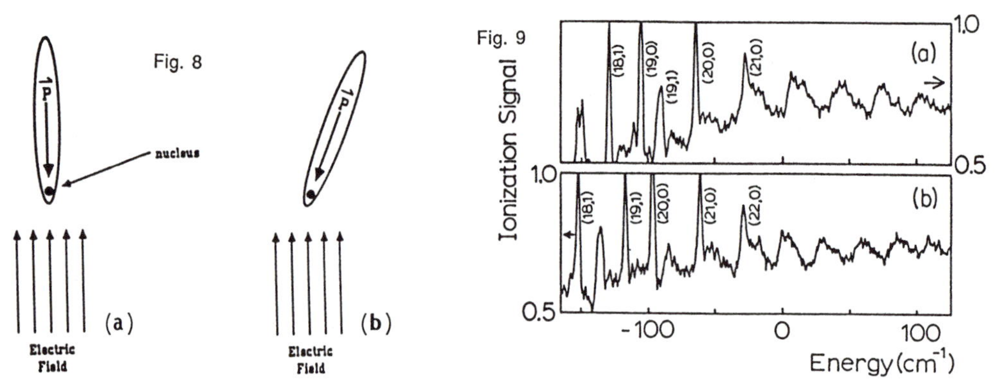

Fig. 8 Schematic of the elliptical orbit (cigar shape) of a giant dipole (a) aligned along the field, and (b) of a giant dipole tilted with respect to the field showing the proton being located near the lower tip.

Fig. 9 The spectrum of hydrogen near E = 0 in the presence of 5 kV/cm (top) and 3 kV/cm (bottom) showing the broad giant dipole states in the positive energy region.

The ability to create moderately large focused dipoles as intermediates is the key to the success of the multistage shaping operation.[28,4,6] This is explained in Fig. 10 for a two-stage process using as an intermediate n = 2 of hydrogen. Because the level splittings in n = 2 of hydrogen are small enough (0.3 cm^{-1}) such that an electric field imposed on the atom which is larger than 5 kV/cm will be able to mix all of these sublevels and hence their charge distributions (each has a zero dipole) to produce distinct dipole distributions needed for the shaping process. Our calculations show that by utilizing the up-field extended dipole of n = 2 as an intermediate, the efficiency can be increased from 10 to 30%, whereas by utilizing the down-field extended dipole the efficiency is reduced to 1%. These were confirmed in our hydrogen experiment as shown in Fig. 11. Our further calculations using higher n states whose charge can be focused along the field more easily, as shown in Fig. 12a, and hence can be matched or tuned more closely to the charge of the giant dipoles, showed dramatic effects on the efficiencies.[28] The use of, for example, n = 9 as the intermediate step in the process, rejects almost completely the excitation of the spectrum of the normal atom in favor of the one-dimensional atom. Results for n = 1, n = 2, and n = 9 are shown in Fig. 12b, along with a schematic of the focusing effect on the intermediates.

The enhanced efficiency is very nice, but it is found that it is practically not possible to increase the lifetime of these giant dipoles in this positive energy region by too much. Such inability is related to the fact that the bound motion of all these giant dipoles in this region are driven by nearly the same charge, most of the nuclear charge $Z_1 \sim 1$, which also dictates very similar orbits where the nucleus is located at the lower tip of the cigar. However, it is found that such enhanced efficiency can be extended to the negative energy region where it is also possible to produce giant dipoles that live quite long. Therefore we will now discuss such promising negative energy regions between E = 0 and $E = E_C$.

In this region the giant dipole atoms take on different properties than the one in the positive energy region. Firstly, the fraction of the charge that drives the bound motion can be varied from 0 to 1 by varying the energy of the system, and consequently the position of the nucleus inside the cigar can also

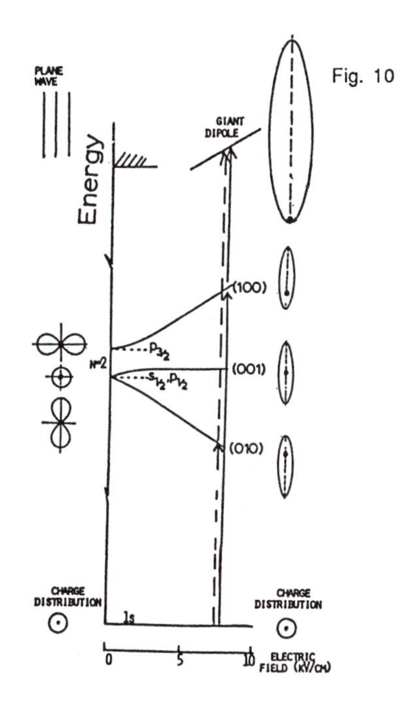

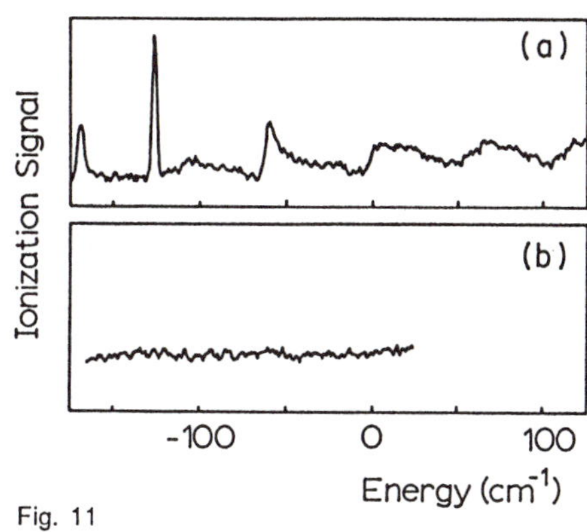

Fig. 11

Fig. 10 Schematic of energy level diagram of hydrogen as a function of the magnitude of the electric field, showing also the charge distribution of the atom at zero field (to the left) and at high fields (to the right). The figure also shows the two stage excitations via the parabolic states of n = 2.

Fig. 11 Absolute experimental spectra in arbitrary units for the two stage process (described in Fig. 10) utilizing the processes shown as solid line (a) and dashed lines (b).

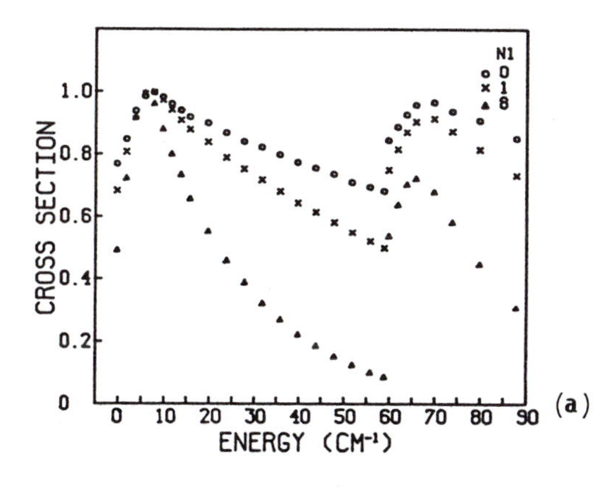

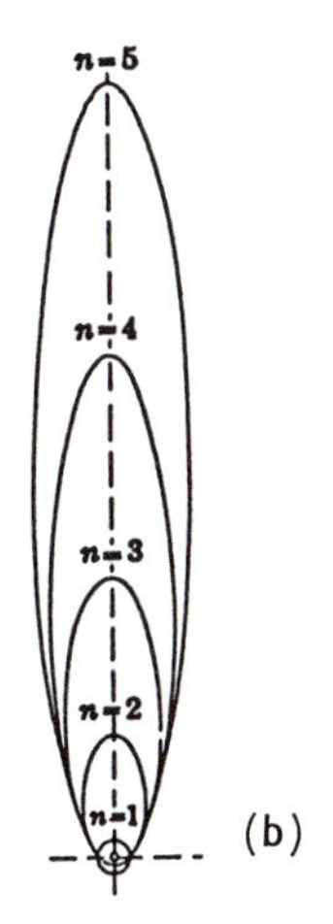

Fig. 12 The two-stage excitation of the giant dipoles utilizing higher states as intermediates. (a) The charge of high n states can be focused along the field more easily and hence can be matched or tuned more closely to the charge of the giant dipoles. (b) Theoretical results for n = 1, n = 2, and n = 9 are shown.

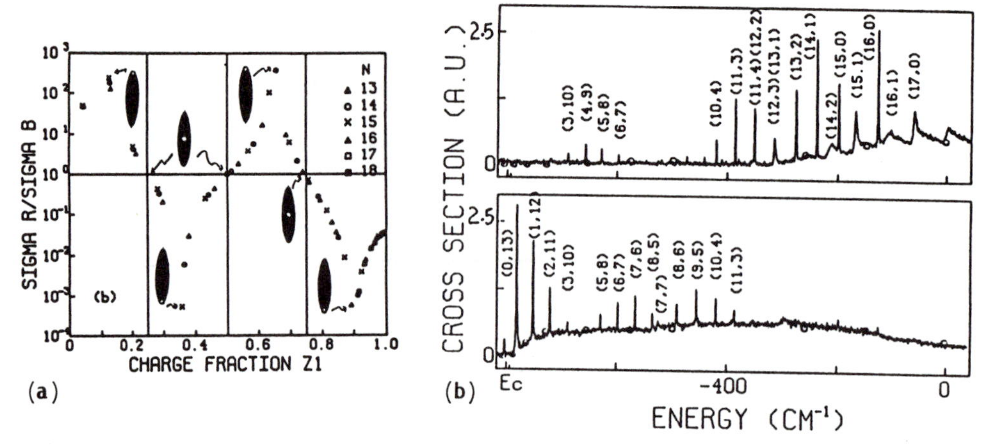

Fig. 13 Long-lived giant dipole atoms in the negative energy region. (a) We plotted an indicator of the location of the nucleus inside the cigar as a function of Z_1 along with sketches of some possible orbits. This indicator is related to the ratio of the area of the orbit below the nucleus to that above the nucleus. (b) The experimental giant dipole spectrum in the negative energy region for the two-stage processes shown of Fig. 10. The top corresponds to the solid line process while the bottom corresponds to the dashed line process of the previous figure.

be controlled. In Fig. 13a we plotted an indicator of the location of the nucleus inside the cigar as a function of Z_1, along with sketches of some possible orbits for the $m = 0$ case. This indicator is related to the ratio of the area of the orbit below the nucleus to that above the nucleus. The figure shows a remarkable property: for the fractional charges $Z_1 = 1/4$, $1/2$, and $3/4$ the nucleus is located at the center of the cigar (the atom has zero dipole moment). These fractional charges thus constitute lines across which the direction of the giant dipole reverses. In the first and third quarters the dipole is along the imposed field whereas in the other two quarters it is opposite to it. Given the size of the orbit (which can be calculated), one can use the above indicator to determine the magnitude of the giant dipole. Moreover, we have recipes to cook up giant dipoles of given Z_1 values. These features and others have been recently confirmed by our experiments. The giant dipole spectrum for $m = 0$ is shown in Fig. 13.

Figure 14a gives the indicator of the location of the nucleus as a function of Z_1 for the $|m| = 1$ case. The figure shows that only the fractions 0, 1/2 and 1 are of importance; they give two halves that classify the atoms. Fig. 14b gives the experimental spectrum confirming these results.

Examination of the spectrum indeed shows a variety of widths (lifetimes) that range from quite short to quite long. In fact there are giant dipoles that do not show up in our spectrum because they live longer than the time of measurement, which is 100 ns, or because they radiatively decay before they ionize and hence do not get detected. Again there are systematics to the ionization lifetime as a function of Z_1, and hence as a function of energy, that makes the selection of a giant dipole of given specification possible.

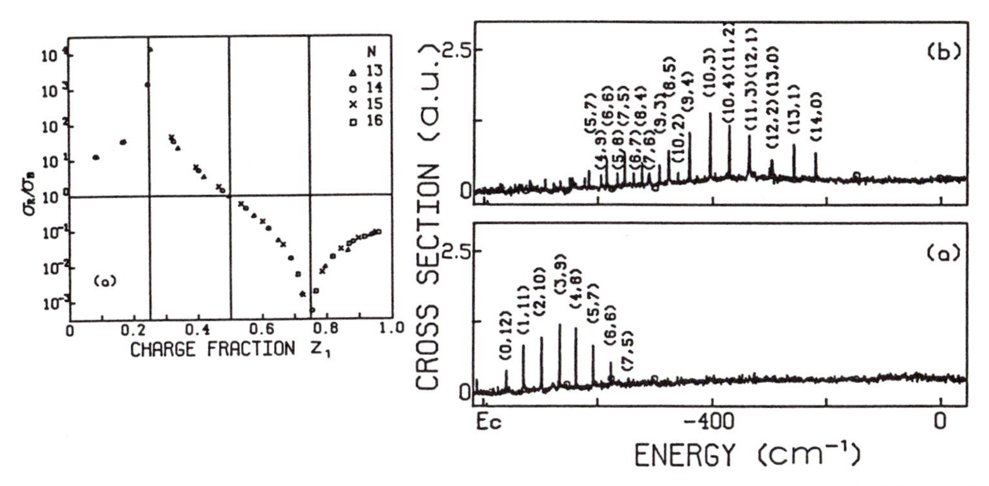

Fig. 14 (a) The indicator of the position of the nucleus, and (b) the hydrogen spectrum for |m| = 1 under similar conditions used in the studies given in Fig. 13.

4. Classical Chaos in Microwave Fields

There has been much interest in the question of the existence of chaotic behavior in quantum-mechanical systems whose classical analogues are known to be non-integrable and to exhibit chaotic behavior.[29-38] One system which received much attention theoretically and experimentally is highly excited atomic hydrogen.[29,34,38] We have theoretically analyzed stochasticity in highly excited atomic hydrogen in the presence of a microwave field and a dc electric field using a classical one-dimensional model similar to that of an electron system over a helium surface. We have previously determined in detail the effect of the dc field on the threshold of global stochasticity and on the number of states trapped in the nonlinear resonances.[38]

The basic procedure of our calculation which is based action angle variable has been published before, in general terms[39], for the case without a DC field[3], and for the case of a very high clamping DC field[38,40]. In the presence of strong dc fields, the action angle variables, however, cannot be determined analytically, and thus we did integrals numerically but did not do a numerical simulation. The nonlinear resonance overlap theory we used has proven to be a reliable estimate of the threshold of classical chaos in varying situations.[37,39] For example, the transition to global chaos happens within a factor of about 2 in AC field of the resonance overlap estimate in the problem with no DC field. In Fig. 15 we plot the classical chaos AC field threshold for an electron prepared in a state of given initial action for a number of different DC fields at an AC frequency of 30.5 GHz. For other frequencies all plots in the paper can be scaled; the scaling laws are given in the captions.

There are good theoretical reasons to believe that the number of quantum states trapped in the first nonlinear resonance determines whether the system is in the classical or purely quantum limit, at least in the low-frequency region (below the first resonance.)[41] The system is expected to behave classically when the number of states trapped is >>1, and non-classically when the number of states trapped is $<\approx 1$. With this in mind, we plot in Fig. 16a the number of states trapped in the first

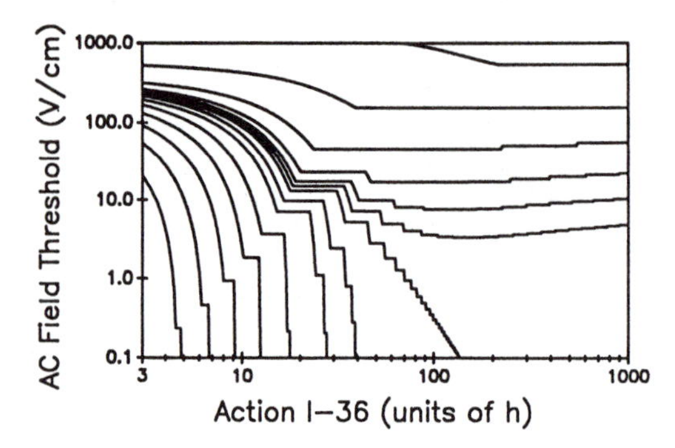

Fig. 15 AC field chaos threshold as a function of the initial action of the system for a number of DC field strengths. These from left to right are: -240, -160, -120, -80, -40, -20, 0, 10, 20, 40, 100, 300, and 1000 V/cm. The frequency of the AC field is 30.5 GHz.

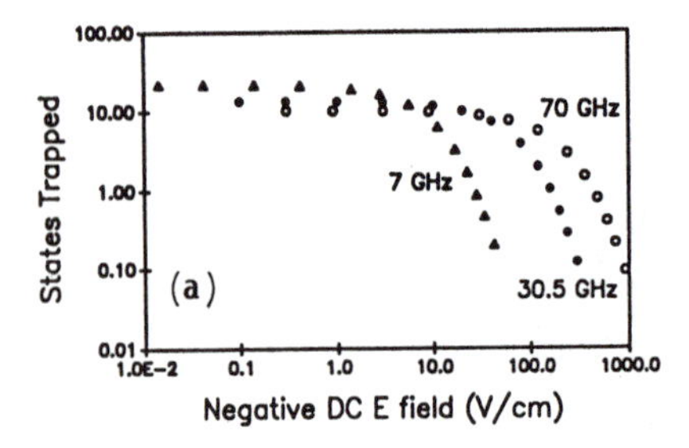

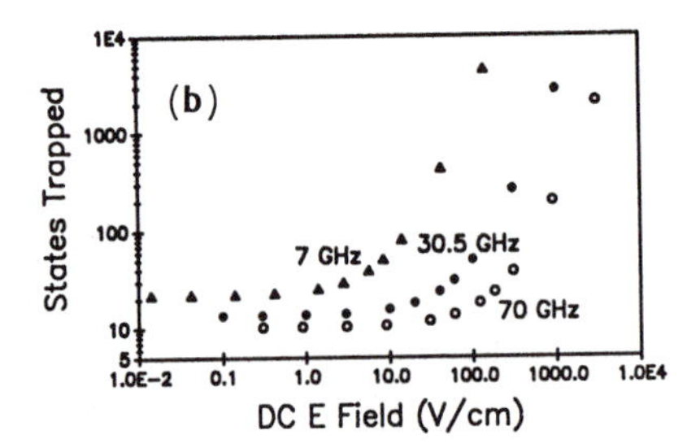

Fig. 16. Number of states trapped in the lowest nonlinear resonance at the threshold of chaos (a) as a function of positive DC E-field, and (b) as a function of negative DC E-field, for a number of frequencies of the AC field: 7GHz, 30.5GHz, and 70GHz.

resonance for a clamping DC field, and in Fig. 16b the same for an unclamping field. These quantities can be read off the Fig. 15 graph; they are simply the widths of the first nonlinear resonances for the different DC fields. The action measured in units of h is equal to the principal quantum number n in the Wilson-Sommerfeld semiclassical theory of quantum mechanics; it is approximately equal for large n in the WKB approximation. Thus, the width of a resonance is a good approximation for the number of states trapped. It is clear that the unclamping field provides a parameter that allows us to go from the classical limit to the uniquely quantum limit. One can see from the graphs that this limit can be reached with easily available microwave frequencies and DC fields. Another advantage of the unclamping DC field is that, by reducing the inherent frequency of the 1D system, it may allow one to more easily reach the regime in which the externally imposed microwave frequency is greater than the inherent system frequency. Uniquely quantum effects which do not occur in the low-frequency regime

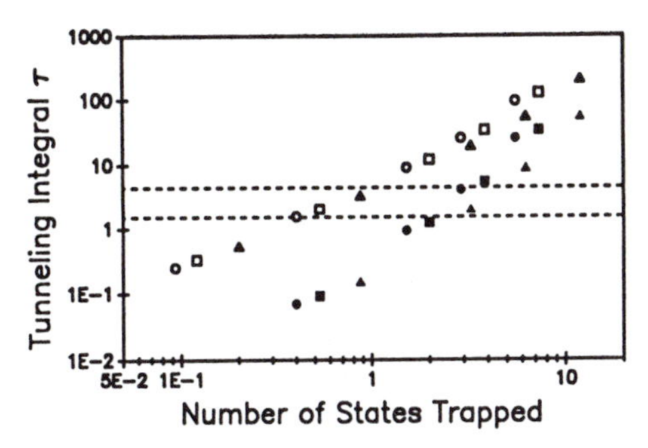

Fig. 17 Tunneling integral τ as a function of the number of state s trapped in the m=1 nonlinear resonance, from the bottom of the m=1 (open symbols) and m=2 (filled symbols) resonances at the threshold of chaos. The triangles are 7 GHz, the squares 30.5 GHz, and the circles 70 GhZ.

have been predicted for this high frequency regime by a quantum-mechanical numerical simulation of the system[42], including the quenching of classical chaos by the quantum system. It appears that the DC field allows one to quickly and easily examine the system for various values of the number of trapped states, down to the quantum regime, and helps one to reach a new regime in the frequency ratio as well. Note that if the hypothesis of the quantum quenching of chaos is true, then the application of the unclamping field, a field trying to rip the electron away from the atom or surface, might actually be a stabilizing factor because it brings the system into the quantum regime.

A question of interest is how few states trapped can be reached without tunneling becoming a problem. With that in mind, we plot in Fig. 17 the tunneling integral τ from states at the bottom and at the top of the m = 1 resonance as a function of the number of states trapped in that resonance at the threshold of chaos. Comparing with tunneling from the initial state, one clearly reaches the quantum regime at τ=1.5, with about 2 states trapped at 30.5 GHz. Note that the number of states trapped at τ=1.5 is fairly insensitive to frequency.

If one is trying to obtain a spectrum, instead of just looking at ionization by chaos, then there is also a maximum τ criterion. For a typical experimental setup (ref. 12), the atom must ionize by tunneling within a microsecond in order to be detected. This criterion gives a maximum τ of about 4.5 for the RIS experiment at 7 GHz (the τ criterion, is slightly larger for the other frequencies plotted). This maximum τ criterion, as well as the minimum, is plotted in Fig. 17. Although the band of observable spectra is fairly narrow in τ, it spans a wide range in the number of states trapped. If one looks at the spectrum low in the first resonance, one can go down to about 0.5 states trapped, and if one looks high in the first resonance, one can see the spectrum for as many as 5 states trapped. Since agreement with classical chaos predictions has been observed for as few as 20 states trapped[43], one might expect 5 states to show at least some classical chaos character. In any case, one could increase the number of states trapped by working at lower frequency. Thus, the spectrum measurement would span the region from the clearly quantum-mechanical to the classical.

5. Frequency Modulation of Hydrogen--Effect of Strong Field on the Dimensionality

The response of hydrogen to strong low frequency radiation has been extensively analyzed classically and quantum mechanically in recent years with the use of a one-dimensional model. In spite of the extensive use of this model theoretically and experimentally, there is still an uneasy feeling lingering about its validity and applicability in the presence of intense external fields.[42] Although there is some experimental evidence for the preservation of the one-dimensionality during the interaction time, this conclusion was drawn from experiments in which the field was well below the threshold for classical chaos.[34]

We recently examined the effect of the external field on the dimensionality of this system.[19] We attacked this problem by taking some hints from the interaction of polar molecules with intense radiation. Recently it was noted that the existence of a permanent electric dipole in a polar molecule allows absorption of a large number of photons in restricted two level system.[44] These ideas were also extended to the one-dimensional hydrogen atoms.[45] Here we reexamine the interaction and present a view in which the absorption proceeds via single photon absorption by sidebands that are created by the interaction of the intense field with the permanent dipole. In fact the idea of sidebands is not new since Townes reported their existence in his early work on masers. However, this idea has not surfaced in the literature with regard to the recent work of interaction of polar molecules with intense radiation.

Specifically, we analyze the interaction of an intense EM field of amplitude E_0 and frequency ω with a low lying (n in the range of 2 to 9) two level system of "one-dimensional" hydrogen (transition frequency: $\omega_0 \equiv (N+1)\omega$, and transition dipole moment μ).

Since in the presence of the dc field each n state splits into a number of one-dimensional states that are specified by the parabolic quantum numbers n_1, n_2 and m, then there are several transitions within a pair of n manifolds. Ladder transitions are defined by $\Delta n_1 = \Delta n$ for $n_1 \geq n_2$ or by $\Delta n_2 = \Delta n$ for $n_2 \geq n_1$. Off-ladder transitions, on the other hand, are defined by $\Delta n_1 = \Delta n$ for $n_1 \leq n_2$ or by $\Delta n_2 = \Delta n$ for $n_2 \leq n_1$. Finally all other possible transitions will be called "veryoffladder" transitions. Fig. 18 gives the permanent dipole moments for the various states in this range, and the transition dipole moments for some selected transitions. These were calculated using wavefunction of the parabolic states in the zero field limit. Exact numerical calculations indicate that the effect of a field of 3kV/cm on these is no more than a few percent.

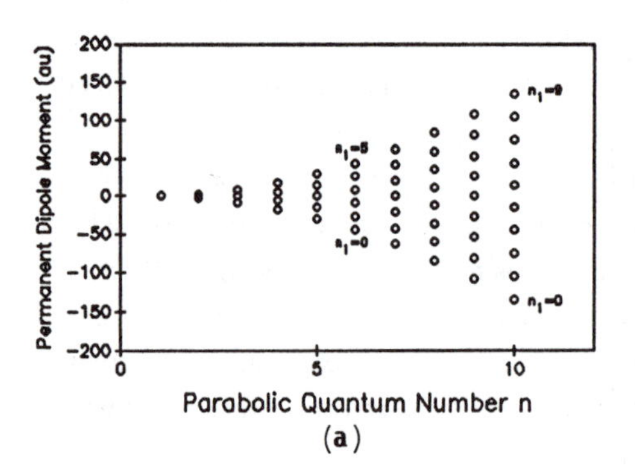

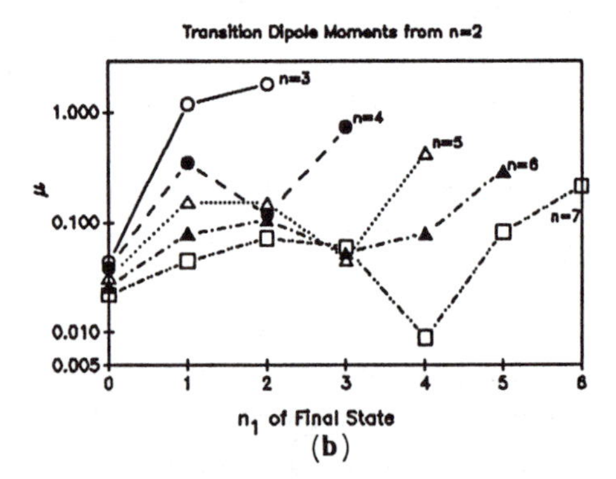

Fig. 18 (a) The permanent dipole moment of the Stark components in the n=2 to n=q manifold calculat ed in the zero field limit. (b) The transition dipole moment to some selected state in n=3 to n=9 manifold from the bluest component of n=2 (1,0,0).

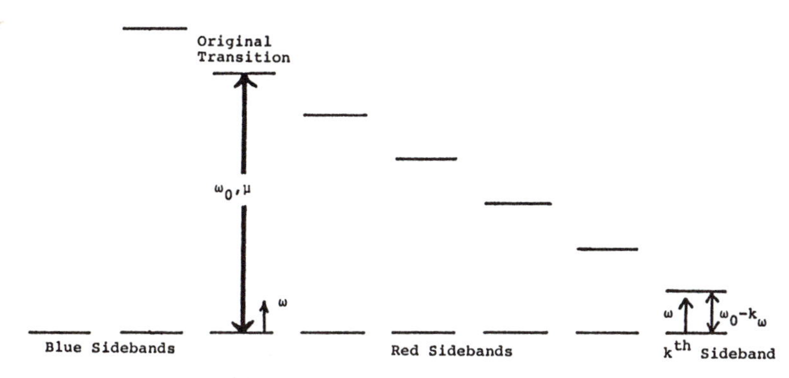

Fig. 19 A schematic of the energy diagram of the field induced sidebands of a given one-dimensional two level system, showing semi-infinite red and blue bands.

The interaction can be outlined as follows. The field couples strongly to the permanent dipole moments of the two levels (d_1 and d_2), frequently modulating the system and hence creating a large number of equally spaced sidebands (spacing equal to ω that share μ among themselves). This is shown schematically in Fig. 19 for illustration. The distribution of μ_n depends on $(d_2-d_1)E_0/\omega$. Figure 20 shows, as an example, the oscillator strength distribution for the transition between the two extreme blue states (ladder transition) of n=2 and n=3 manifolds, namely (1,0,0) and 2,0,0). The rest of the infinite number of the sidebands are not shown because their oscillator strength is very negligible and will not affect the effective band width of the modulation.

Because the Nth sideband where $(N+\omega)\omega \cong \omega_0$ is the only sideband which is in near single photon resonance, it will dominate the interaction especially at high enough intensities where its fraction of the transition dipole moment is sizeable. We determined the transition moment of the Nth sideband for a variety of ladder, offladder and veryoffladder systems. Our results, an example of which is shown in

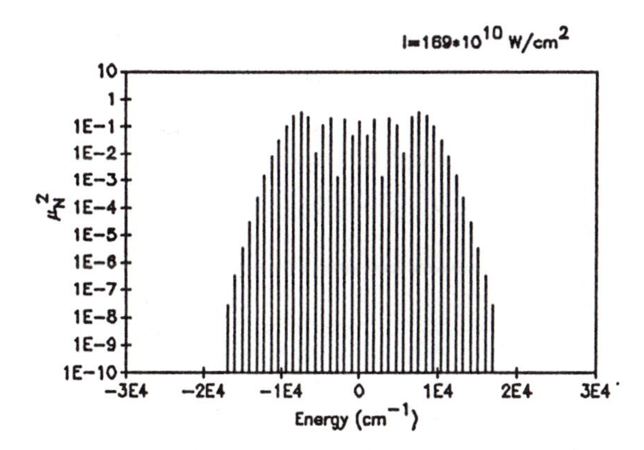

Fig. 20 The square of the induced oscillator strengths of the sidebands ot a two level system of the bluest components of n=2 and n=3 of hydrogen: 1,0,0 and 2,0,0 respectively. The weak bands are not shown and they will not affect the effective bandwidth of the system.

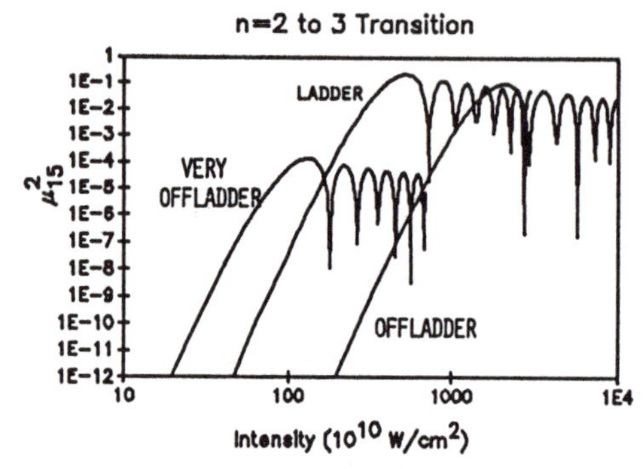

Fig. 21 The oscillator strength of the 15th side band of three transitions all originating from 1,0,0, the bluest component of n=2 as a function of the intensity of the 10.6μ radiation. The ladder transition is to (2,0,0), and off ladder is to (1,1,0), and the very off ladder is to (0,2,0) of the n=3 manifold. The local minima shown are actually zeroes. They are not plotted as such because only a finite number of points are plotted, and to improve the graph's readability.

Fig. 21, indicate that the veryoffladder transitions are very weak at all intensities. However at sufficiently high intensities, offladder sidebands may become as strong as ladder sidebands, thus breaking the one-dimensionality of the system. It is interesting to note that this loss of one-dimensionality occurs at powers close to those needed to induce global classical chaos of the system (this threshold marked on the figure). The onset of classical chaos was calculated using a one dimensional classical model similar to that of an electron near a helium surface as described in the previous section.

The calculations also show that the intensity I_b needed to break the dimensionality for transition in the n and n+1 manifold drops as n increases. These are shown in Fig. 22. It is interesting to plot, for

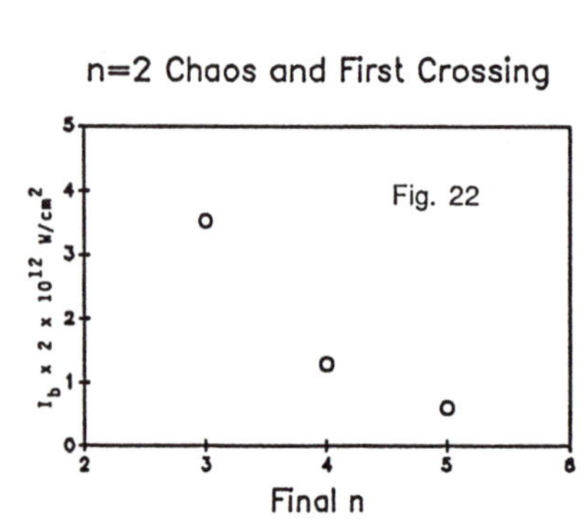

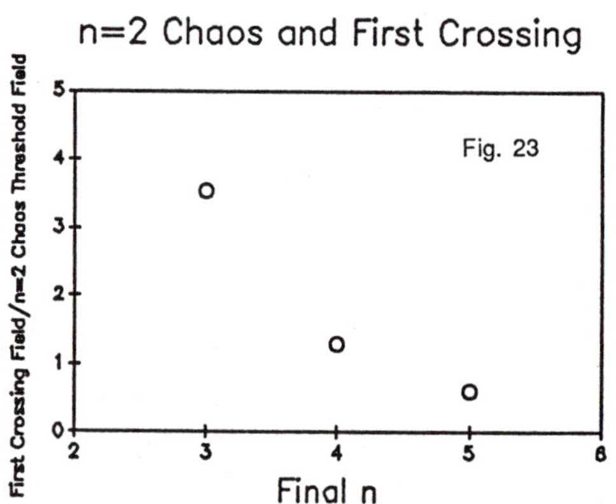

Fig. 22 The intensity of the ac field needed to break the one dimensionality of transition between n-1 and n manifolds, i.e. the intensity at which ladder and off ladder transitions become of equal strength.

Fig. 23 The one-dimensionality window, i.e. the ratio of the intensity needed to break the one-dimensionality of transition between n-1 and n where n=2, 3,. . . to the chaos threshold of n_0=2, as a function of n.

a given initial manifold, n_0, the ratio $R = I_b/I_c$ as a function of n where I_c is the intensity at which the initial state gets into the chaos regime. This ratio or what we call window is useful in the discussion of dimensionality of a chaotic system. As an example we plot in Fig. 23 the window for a system that is initially prepared in (1,0,0) of n=2 of hydrogen for which the classical chaos threshold at 10.6μ is $1.7 \times 10^{12} W/cm^2$. The figure shows that the window closes ($R \cong 1$) as the excitation reaches n = 5. It is to be noted that the interaction of 10.6μ radiation with the transition between n=6and n=7 manifold falls in the high frequency regime, i.e. $\omega \geq \omega_0$

We also studied the frequency dependence of these results. Using radiation of 21.2μ, we find that the intensity at which the offladder and ladder strengths become comparable does not change. This is reasonable since the order of the Bessel function as well as its argument depend on the frequency in manner that causes cancellation of the overall dependence. For instance $(\kappa + 1)$ is proportional to $\frac{1}{\omega}$ where κ is the order of the Bessel function, and the Bessel function of order κ peaks at $\Delta dE/\omega$ which is proportional to $(\kappa+1)$. Thus the magnitude of E at which the offladder transition becomes just as strong as the ladder transition is essentially frequency independent.

The chaos threshold, on the other hand, depends on the frequency. However, at sufficiently low frequencies such as the ones being used here, the chaos threshold is expected to be somewhat insensitive to frequency, since the conditions are close to the dc limit.[37] Thus one expects the window R to be nearly frequency independent. Numerical results for the case of 21.2μ are shown on Fig. 23 which also displays the 10.6μ case, showing the frequency independence.

Finally the condition of high/low frequency, on the other hand, depends strongly on the frequency, and therefore it is possible to conceive situations such that as the ionization proceeds, the system loses one-dimensionality before the high frequency regime is reached.

6. Conclusion

The last few years have seen a lot of progress in the deliberate manipulation of atomic structure by imposing strong external fields. These include other than electric fields such as dc magnetic fields which were shown to allow construction of two-dimensional atoms. For example, we can now design and construct atoms which are effectively one dimensional. The main directions we perceive for future work are: First, further development and understanding of the potentiality for and constraints on effective manipulation. Second, elucidation of the ways in which these new species of atoms interact with the physical and chemical environment. And third, we will see their utilization in scientific and technological applications, especially those which take advantage of the lower dimensionality which makes complex problems more amenable to theoretical treatment and interpretation. One example, mentioned above, is the study of chaotic dynamics in nonintegrable systems, a subject that is encountered in research areas as diverse as classical mechanics, chemical reaction, and quantum mechanics.

REFERENCES

[1] R. R. Freeman, N. P. Economou, G. C. Bjorklund, and K. T. Lu, Phys. Rev. Lett. 41, 1463 (1978).

[2] C. W. Clark, K. T. Lu, and A. F. Starce, in Progress in Atomic Spectroscopy, eds. H. Beyer and H. Kleinpoppen (Plenum, New York, 1984).

[3] See articles in Atomic Excitation and Recombination in External Fields, ed. M. H. Nayfeh and C. W. Clark (Gordon and Breach, New York 1985).

[4] U. Fano, Colloq. Int. CNRS 273, 127 (1977); Phys. Rev. A 24, 619 (1981).

[5] M. H. Nayfeh, K. Ng. and D. Yao in SEICOLS, Laser Spectroscopy, T. Hänsch and R. Shen, eds. (Springer-Verlag, Berlin, Heidelberg, 1985). p. 71.

[6] M. H. Nayfeh, D. Yao, Y. Ying, D. Humm, K. Ng and T. Sherlock: In Advances in Laser Science - 1, AIP Conference Proceedings 146, 370 (1986).; M. H. Nayfeh, in Resonance Ionization Spectrocopy, U. K. Institute of Physics, Number 84, G. S. Hurst and C. G. Morgan, eds., 21 (1986); M. H. Nayfeh in Lasers, Spectroscopy and New Ideas, W. Yen and M. Levenson, eds. (Springer-Verlag, Berlin, 1987), p. 141.

[7] T. S. Luk, L. DiMauro, T. Bergeman, and H. Metcalf, Phys. Rev. Lett. 47, 83 (1981).

[8] S. Feneuille, S. Liberman, E. Luc-Koenig, J. Pinard, and A. Taleb, Phys. Rev. A 25, 2853 (1982).

[9] W. Sandner, K. A. Safinya, and T. F. Gallagher, Phys. Rev. A 23, 2448 (1981).

[10] W. Glab, G. B. Hillard, and M. H. Nayfeh, Phys. Rev. A 28, 3682 (1983).

[11] K. H. Welge and H. Rottke, in Laser Techniques in the Extreme Ultraviolet-OSA, Boulder, Colorado, 1984, edited by S. E. Harris and T. B. Lucatorto, AIP Conf. Proc. No. 119 (AIP, New York, 1984), pp. 213-219.
H. Rottke and K. H. Welge, Phys. Rev. A 33, 301 (1986).

[12] W. L. Glab and M. H. Nayfeh, Phys. Rev. A 31, 530 (1985).

[13] E. Luc-Koenig and A. Bachelier, Phys. Rev. Lett. 43, 921 (1979).

[14] A.R.P. Rau and K. T. Lu, Phys. Rev. A 21, 1057 (1980).

[15] D. A. Harmin, Phys. Rev. A 24, 2491 (1981); D. A. Harmin, Phys. Rev. Lett. 49, 128 (1982); Phys. Rev. A 26, 2656 (1982).

[16] W. D. Kondratovich and V. N. Ostrovsky, Zh. Eksp. Teor. Fiz. 4, 1256 (1982).

[17] K. Ng, D. Yao and M. H. Nayfeh, Phys. Rev. A 35, 2508 (1987).

[18] D. Yao, K. Ng, M.H. Nayfeh, Phys. Rev. A. (Dec.) 1987.

[19] M. H. Nayfeh, Bulletin Am. Phys. Soc., March (1988).

[20] J. E. Bayfield, L. D. Gardner, and P. M. Koch: Phys. Rev. Lett. 39, 76 (1977);
P. M. Koch: Phys. Rev. Lett. 41, 99 (1978).

[21] H. C. Bryant et al. Phys. Rev. A 27, 2889, 2912 (1983); W. W. Smith et al. in Atomic Excitation and Recombination in External Fields, M. H. Nayfeh, and C. Clark eds., (Gordon and Breach, New York, 1985).

[22] G. C. Bjorklund, R. R. Freeman, and R. H. Storz, Opt. Comm. 31, 47 (1979).

[23] W. L. Glab and M. H. Nayfeh, Opt. Lett. 8, 30 (1983); W. Glab, Ph.D Thesis, University of Illinois, 1984 (unpublished).

[24] M. H. Nayfeh, K. Ng, and D. Yao in Atomic Excitation and Recombination in External Fields, M. H. Nayfeh and C. W. Clark eds., (Gordon and Breach, New York, 1985).
[25] T. W. Hänsch, S. A. Lee, R. Wallenstein, and C. Wieman, Phys. Rev. Lett., 34, 307 (1975).
[26] W. L. Glab, K. Ng, D. Yao, and M. H. Nayfeh, Phys. Rev. A 31, 3677 (1985).
[27] G. Luders, Ann. d. Phys. [6], 8, 301 (1951).
[28] Y. Ying and M. H. Nayfeh, Phys. Rev. A 35, 1945 (1987).
[29] G. Casati, B. V. Chirikov, F. M. Israelev, J. Ford: In Stochastic Behavior in Classical and Quantum Systems, ed. by G. Casati and J. Ford, **Lecture Notes in Physics**, Vol. 93, (Springer, New York, 1979).
[30] D. R. Grempel, S. Fishman, and R. E. Prange: Phys. Rev. Lett. 49, 833 (1982).
[31] M. V. Berry: Physica (Amsterdam) 10D, 369 (1984).
[32] S. J. Chang and K. J. Shi: Phys. Rev. Lett. 55, 269 (1985).
[33] E. V. Shuryak: Zh. Eksp. Teor. Fiz. 71, 2939 (1975); Sov. Phys. JETP 44, 1070 (1976); P. I. Belobrov, G. P. Berman, G. M. Zaslavski, and A. P. Slivinskii: ibid. 76, 1960 (1979); 49, 993 (1979).
[34] J. E. Bayfield and L. A. Pinnaduwage: Phys. Rev. Lett. 54, 313 (1985).
[35] P. M. Koch: J. Phys. (Paris), Colloq. 43, C2-187 (1982).
[36] D. Humm and M. H. Nayfeh: to be published.
[37] R. V. Jensen: Phys. Rev. A 30, 386 (1984); in Chaotic Behavior in Quantum Systems, ed. by G. Casati, (Plenum, London 1985).
[38] M. H. Nayfeh and D. Humm in Proceedings of the International Workshop on Photons and Continuum States, ed. by N. Rahman, (Cortona, Italy 1986), (Springer, Berlin, Heidelberg 1987).
[39] B. V. Chirikov, Phys. Rep. 52, 549 (1979); G. M. Zaslavsky and B. V. Chirikov, Usp. Fiz. Nauk. 105, 3(1971) [Sov. Phys. Usp. 14, 263(1971).
[40] G. P. Berman, G. M. Zaslavsky, and A. R. Kolovsky, Zh. Elesp. Teor. Fiz. 88, 1551(1985) [Sov. Phys. JETP G1, 925 (1985)].
[41] G. P. Berman, G. M. Zaslavsky, and A. R. Kolovsky, Phys. Lett. 87A, 152 (1982).
[42] G. Casati, B. Chirikov, D. Shepelyansky, and I. Guarneri, Phys. Rep. 154, 29 (1987).
[43] R. V. Jensen and S. M. Susskind, in Photons and Continuum States of Atoms and Molecules, eds. N. Rahman, C. Guidotti and M. Allegrini, 13 (Springer-Verlag 1987).
[44] W. J. Meath and E. A. Power, Canadian Molecular Phys. 51, 585 (1984).
[45] J. E. Bayfield, in Photons and Continuum States of Atoms and Molecules, eds. N. Rahman, C. Guidotti, and M. Allegrini, (Springer-Verlag) 8, 1987.

Hydrogen Atoms in Strong Magnetic Fields – in the Laboratory and in the Cosmos

G. Wunner, W. Schweizer, and H. Ruder

Lehrstuhl für Theoretische Astrophysik, Universität Tübingen,
D-7400 Tübingen, Fed. Rep. of Germany

1. Introduction

Recent years have seen tremendous progress in studies of the properties of hydrogen atoms in strong magnetic fields. Decisive stimulus came from the discovery of huge magnetic fields in astrophysical "laboratories", viz. field strengths of order $\sim 10^7 - 10^9$ T in neutron stars and of order $\sim 10^2 - 10^4$ T in white dwarf stars. At these field strengths the magnetic forces acting on an atomic electron outweigh the Coulomb binding forces even in low-lying states, and thus atomic structure is completely changed. On the other hand, the rapid advancement of high-resolution laser spectroscopy has made it possible to produce atoms in highly excited states, with principal quantum numbers ranging up to $n \cong 520$ (in Ba I) [1], and therefore Rydberg states can be used to investigate the effects of magnetic dominance on atomic structure also in terrestrial laboratories with magnetic fields of a few Tesla, or less.

A detailed description of the astrophysical scenarios which finally led to the conclusive determination of magnetic field strengths up to $5 \cdot 10^8$ T in neutron stars, and an account of theoretical results for atomic level schemes, wavefunctions, transition rates, etc. at these field strengths, where the system becomes highly nonseparable, has been presented elsewhere [2]. Here we present a selection of more recent results. We briefly review, in section 2, the basics of the theory and then concentrate, in section 3, on the wavelength spectrum of the hydrogen atom in magnetic fields of *arbitrary strength.* We then turn to atoms in terrestrial laboratories and discuss, in section 4, the progress that has been made in the theoretical description of magnetized hydrogen Rydberg states. We compare spectroscopic predictions of theory with results of actual experiments, and demonstrate, in section 5, that phenomena which have turned out characteristic of the onset of "quantum stochasticity" in investigations of model Hamiltonian systems are recovered in the quantal properties of highly excited hydrogen atoms (e.g. in the statistics of level sequences and transition strengths in the classically chaotic domain). This highlights the fact that the hydrogen atom in a strong magnetic field is an ideal example of a simple but real physical system displaying all the features which are currently causing so much excitement in the classical and quantum mechanical study of nonintegrable systems.

2. Hydrogenic Atoms in Magnetic Fields of Arbitrary Strength

The Hamiltonian which describes the motion of an electron under the combined influence of a fixed Coulomb potential and a uniform magnetic field $\mathbf{B} = B \cdot \mathbf{e}_z$ reads, (in atomic units, $\beta = B/B_o$ with $B_o = 2(\alpha m_e c)^2/(e\hbar) \cong 4.70 \cdot 10^5$ T),

$$H = -\Delta - \frac{2}{r} + 2\beta l_z + \beta^2(x^2 + y^2). \tag{1}$$

The reference magnetic field B_o is chosen in such a way that at B_o the cyclotron energy of the electron becomes equal to four times the Rydberg energy, or, equivalently, the Larmor radius equals the Bohr radius. For small or extremely large magnetic field strengths, where either the Coulomb forces dominate the Lorentz forces or vice versa, corresponding

The Hydrogen Atom Editors: G.F. Bassani · M. Inguscio · T.W. Hänsch

perturbational approaches to solving the Schrödinger equation for the Hamiltonian (1) are of course appropriate. In the *intermediate* (or *strong-) field regime*, the electron experiences electric and magnetic forces of comparable strength, and perturbative approaches fail. The mathematical reason for this lies in the fact that the spherical symmetry of the Coulomb potential, on the one hand, and the cylindrical symmetry of the magnetic field on the other, prevent a separation of variables so that closed-form analytical solutions are not possible. Thus the Hamiltonian (1) belongs to the class of nonintegrable Hamiltonians. The absolute sizes of the field strengths at which one lies in this regime depend on the state of excitation of the electron. By considering the equality of Coulomb and Lorentz forces for an electron in a circular Bohr orbit with principal quantum number n one obtains as a rough measure $B_n \cong B_o/(2n^3) \cong (8.7\,T)(30/n)^3$. Therefore for white dwarf and neutron star magnetic fields low-lying states are found to be subjected to a strong-field situation, while at laboratory field strengths studies of the strong-field regime are restricted to Rydberg states.

The nonintegrability of the Hamiltonian (1) implies that one is forced, in the complete quantum theoretical treatment of the problem, to resort to numerical methods. In our calculations we expand the wavefunctions either in terms of spherical harmonics and a complete set of radial functions, or in terms of Landau functions and a complete longitudinal basis. The different expansions are employed depending on whether one approaches the intermediate-field regime from the side of low or intense fields, and overlapping results are expected in the transitional region.

Except at very high fields ($\beta \geq 0.1$) [3] calculations using the Landau function expansion have confined themselves so far to low-lying states. Calculations using expansions based on the field-free symmetries have been performed for both low-lying and highly excited states, and in this case energies and wavefunctions were obtained by either direct integration [4], or diagonalization of the Hamiltonian matrix in large basis sets. In our own computations in this regime we expanded the wavefunctions in terms of spherical harmonics, $\psi_m = \sum h_\ell(r)\, Y_{\ell,m}(\theta,\phi)$ (m is the magnetic quantum number), and the radial functions in the complete, orthonormal set of functions

$$G_{nl}^{(\zeta)}(r) = \zeta^{3/2}\left[\frac{n!}{(n+2l+2)!}\right]^{1/2} \exp[-\zeta r/2]\,(\zeta r)^l\, L_n^{(2l+2)}(\zeta r)\,, \qquad (2)$$

where ζ denotes an inverse-length parameter, and the $L_n^{(2\ell+2)}$ are generalized Laguerre polynomials. Matrix elements with respect to this basis can be expressed in closed analytical form and give rise to a banded Hamiltonian matrix which can be diagonalized by efficient standard algorithms. In our calculations we used basis sizes of up to 220000 for determining both eigenvalues and eigenvectors of the first $\sim$ 700 states in given m-parity subspaces in the intermediate-field regime. Convergence was established by varying the size of the basis and the scale parameter ζ.

As a result of our efforts we are able to continuously trace the energy values and oscillator strengths of low-lying states ($n \leq 5$) from zero field up to neutron star magnetic fields (for accurate numerical values we refer the reader to Rösner et al. [4], Forster et al. [5], and Ruder et al. [6]), while we can calculate the energies and oscillator strengths of Rydberg states in laboratory fields of a few Tesla up to the field-free ionization threshold E = 0 . The following sections present examples of our results.

3. Low-Lying States in Magnetic Fields of Arbitrary Strength

3.1. Wavelength Spectrum of the Hydrogen Atom

The behaviour of the energy values and wavefunctions over the range from laboratory fields up to neutron star magnetic fields has been discussed in great detail elsewhere (Ruder et al. [2]). Here we concentrate on the behaviour of *wavelengths* and present, in Fig. 1, the

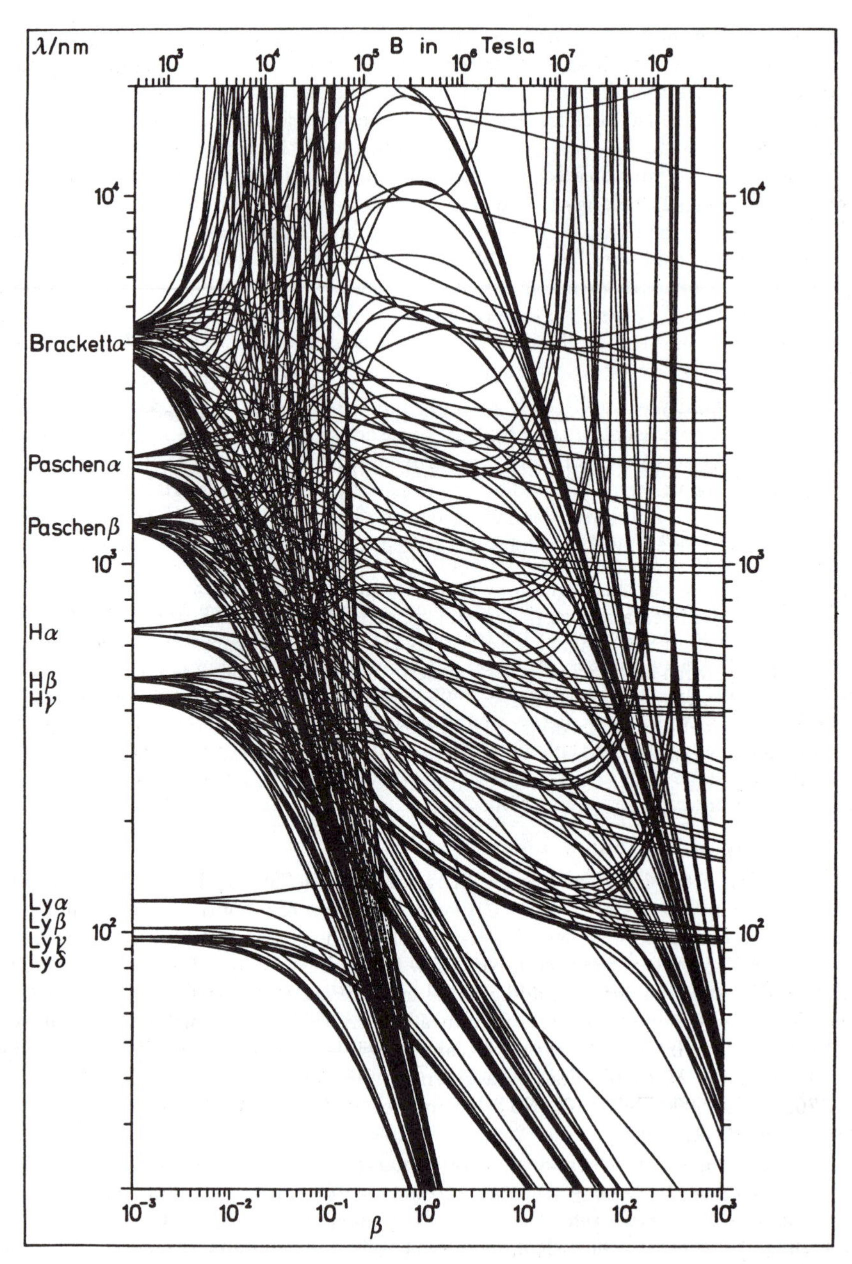

Fig. 1: The wavelength spectrum of the hydrogen atom from the soft X-ray range (20 nm) up to the far infrared (10000 nm) as a function of the magnetic field strength in the interval 470 T to $4.7 \cdot 10^8$ T. Effects of the finiteness of the proton mass are taken into account. The two rapidly declining bunches of lines correspond to cyclotron-like transitions of electrons (left-hand bunch) and protons (right-hand bunch), respectively. Note the occurrence of "stationary" lines in the intermediate region.

spectrum of the hydrogen atom between 20 nm and 10000 nm as a function of β in the range $10^{-3} \leq \beta \leq 10^3$ which we have computed from our energy values for a total of 364 transitions between states with $n \leq 5$. This is the most comprehensive compilation of the hydrogen spectrum to date.

One recognizes, at small field strengths, the splittings of the unperturbed lines into three equidistant Zeeman components. For larger β these components continue splitting by the quadratic Zeeman effect. The onset of the quadratic Zeeman effect is shifted to smaller β values with increasing wavelengths. Beyond this region ($\beta \approx 10^{-2}$), where the perturbation theory treatment breaks down, the lines are completely torn apart by the magnetic field within one β decade and the spectrum becomes totally distorted. Since the energy levels of states with different m and different z parity are allowed to cross, the wavelengths of corresponding transitions go to infinity at certain values of β.

Order reappears only in intense fields, indicative of the fact that in the limit $B \rightarrow \infty$ the level scheme approaches that of the one-dimensional Coulomb problem, which consists of tightly bound levels and levels whose energies equal those of the field-free H atom. As a consequence numerous lines tend to the wavelengths of the unperturbed hydrogen series at the right-hand side of Fig. 1.

Clearly, any attempt to observe, and resolve, a line spectrum of hydrogen at a given magnetic field strength in the intermediate regime is doomed to failure. An element of order is, however, brought in even in this domain by several transitions whose wavelengths go through minima and maxima in certain intervals of the magnetic field strength, that is they are less sensitive to variations of the magnetic field than the many fast running components. An inhomogeneous field with a variation of, say, a factor of two (as is the case for a dipolar field) around extrema of wavelengths will therefore filter out exactly these stationary components, thus opening the possibility of observing, in this instance, a clearly arranged spectrum with few well resolved features. Speculative as this may sound, nature has indeed provided us with a cosmic laboratory to test this hypothesis, and in fact this has opened a totally new era of stellar atomic spectroscopy, namely the "spectroscopy of stationary lines".

3.2 Spectroscopy of Stationary Lines

The spectrum of the white dwarf star known as $Grw + 70^o8247$ had been an unsolved puzzle for more than four decades. The circular polarization of its optical continuum [7] had given a clue to the existence of a strong magnetic field in the vicinity of this object, and Angel [8] first proposed a tentative identification of a few of the features in terms of stationary lines using variational energy values of Praddaude [9] available at that time. On the basis of our very accurate computations of the energy values and transition rates of the hydrogen atom in magnetic fields of arbitrary strength it became possible to positively identify all observed features as stationary hydrogen lines in a magnetic field whose value was pinned down to between 17000 and 35000 T [10-12]. No other previously known white dwarf had a magnetic field even one tenth this value. Fig. 2 demonstrates the excellent agreement between the wavelength positions of the extrema of stationary components of H_β, H_γ and absorption features in the blue part of the spectrum of $Grw + 70^o8247$. In particular, the sharp blue edges and "red-shaded" extensions of the features at 3650 Å and 4135 Å are well accounted for by the minimum character of the corresponding stationary components. Thus for the first time *low-lying states exposed to a strong-field situation* were actually observed in nature. In the meantime a second object has been found (PG 1031+234, Schmidt et al. [14]), in the spectrum of which stationary components of hydrogen lines, in a field ranging up to $\sim$ 50000 Tesla, have positively been identified. Here the observed phase modulations of the spectrum yield additional information from which the morphology of the magnetic field of this star can be extracted.

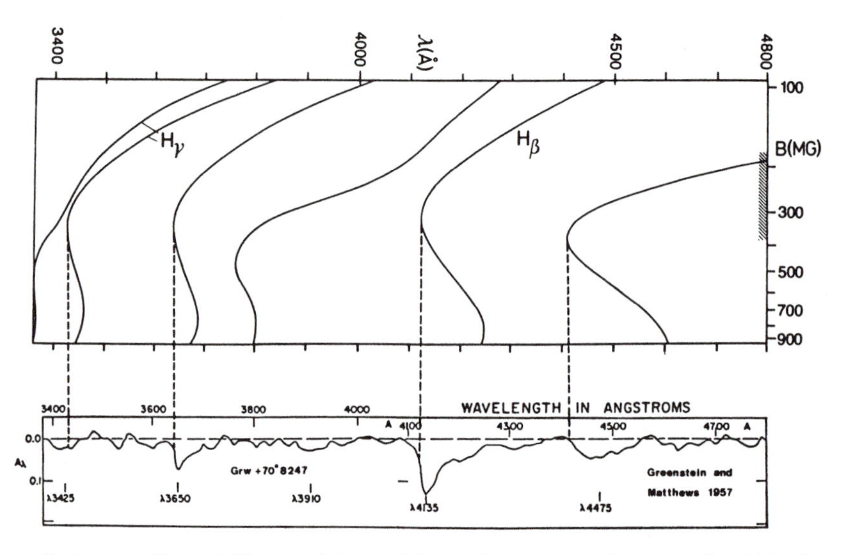

Fig. 2: Stationary H_β and H_γ transitions of the hydrogen atom in magnetic fields from 100 to 900 million Gauss in comparison with the short-wavelength part of the spectrum of Grw+70°8247 taken by Greenstein and Matthews [13]. All the features are explained in a consistent way in terms of stationary Balmer transitions of hydrogen in an extended magnetic field whose strength varies from $\sim$ 170 to $\sim$ 350 million Gauss (see the corresponding hatching along the B ordinate). Such a variation is present, e.g., in a dipolar magnetic dipole field.

4. Highly Excited States in Strong Laboratory Fields — Comparison between Experimental and Theoretical Spectra

For many years research of magnetized Rydberg atoms was characterized by the situation that experimental work concentrated on non-hydrogenic atoms (e.g. Ba, Na, Li) while most of the theoretical papers were devoted to hydrogenic systems. Semiclassical and quantal calculations were performed to account for the nature of the quasi-Landau-resonances observed experimentally around and above the field-free ionization threshold, but so far no detailed quantitative comparisons were possible for the rich and complicated line structure seen in experimental spectra of magnetized Rydberg atoms. Quite recently this situation has changed in that, on the one hand, actual experiments have been performed with highly excited *hydrogen* and *deuterium* atoms in 4 – 6 T fields by the Bielefeld group (Holle et al. [15]), and, on the other, the computer codes developed over the last years allow an almost routine calculation of highly accurate energy values and wavefunctions of one-electron systems in uniform laboratory magnetic fields up to shortly below the field-free ionization threshold. To illustrate the state of the art, Fig. 3 provides a comparison between the experimental and theoretical spectra in the range -24 cm^{-1} to -12 cm^{-1} at 5.96 T. The agreement between theory and experiment can be considered excellent; moreover, theory reveals where neighbouring lines were no longer resolved in the experiment. The broad experimental feature around -13 cm^{-1}, e.g., is found to be composed of $\sim$ 10 lines of different intensities. Obviously this calls for increased experimental efforts to refine resolution and actually check the theoretical spectra also in these ranges of energy.

All in all we have compared successfully more than 1000 lines so far in the π and σ spectra in the strong-field regime, and this certainly can be considered a hallmark of modern quantitative spectroscopy in magnetic fields.

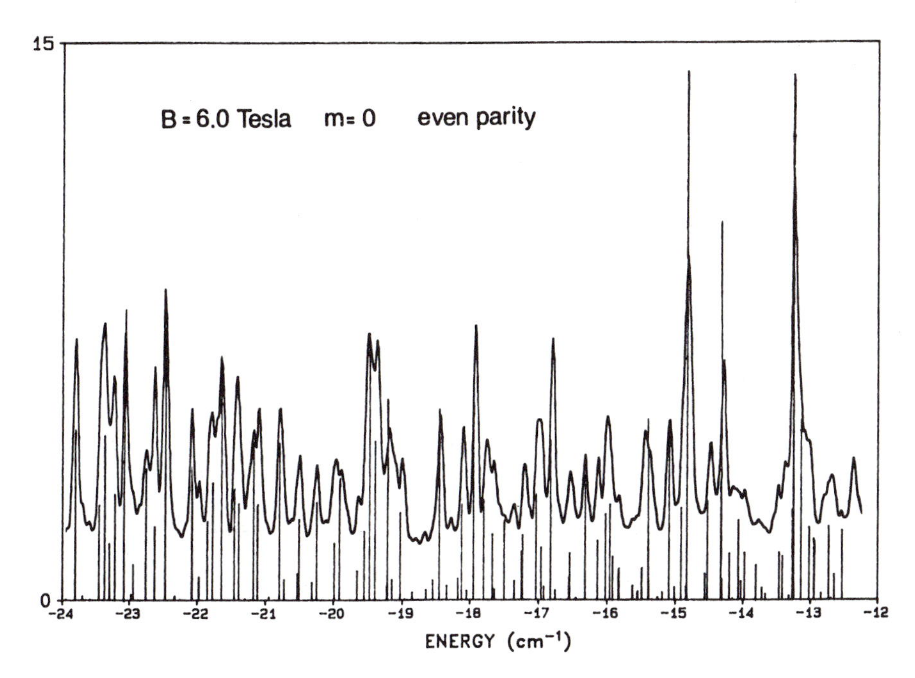

Fig. 3. Comparison between the experimental spectrum and the theoretical photoabsorption spectrum of $\Delta m = 0$ transitions from $2p_o$ to even parity final Rydberg states with energies between $-12\ cm^{-1}$ and $-24\ cm^{-1}$ in a magnetic field of 6 T. For the theoretical results (straight lines) the ordinate respresents the oscillator strength in units of 10^{-6} . The experimental results have kindly been provided by the Bielefeld group [15]. Note that the wealth of theoretical spectral structure can no longer be resolved by the experiment. The range of energy shown lies in the domain of chaotic motion of the corresponding classical system.

5. The Hydrogen Atom in Strong Magnetic Fields – an Object Lesson in "Quantum Chaology"

5.1 Quantum Physics in Classically Chaotic Regions

It has been realized, in recent years, to an increasing extent that in classical Hamiltonian systems the occurrence of "chaotic" behaviour – in the sense of exponential separation of initially infinitesimally neigbouring trajectories in phase space – is the rule rather than the exception. Furthermore, theoretical and experimental techniques have progressed to a point where it is now possible to study in detail quantal systems in the limit of large quantum numbers, in which the laws of quantum theory should reduce to the laws of classical physics. All this has combined in creating a new field of research, for which the term "quantum chaology" has been coined by Berry [16], who defines it as "the study of semiclassical – but still *non-classical* – behaviour characteristic of systems whose classical motion exhibits chaos". It should be emphasized that the term "quantum chaos", which has become rampant in literature, still lacks a rigorous definition. Although finally aiming at such a definition quantum chaology is taking, at present, a more empirical point of view: what one is looking for is archetypal quantum phenomena in a range of parameters where the classical analogs of the systems turn chaotic.

One of the most intriguing features of the spectroscopy of hydrogen atoms in magnetic fields, lies in the fact that magnetize hydrogen atoms are able to play a leading rôle in the search for manifestations of "quantum" chaos. On the one hand, the Kepler problem with Lorentz forces has been shown by a number of authors [17-19] to undergo a transition to chaos once the Lorentz forces acting on the highly excited electron become of the order of, or larger than, the Coulomb forces. On the other hand, high-resolution spectroscopic experiments on "real", quantal, magnetic Rydberg atoms are performed by a number of groups, and, as demonstrated in the foregoing section, large-scale computations have reached a point where it is possible to determine theoretically the energetic positions of, and oscillator strengths of transitions to, Rydberg states in Tesla fields up to almost the field-free ionisation limit. Therefore the hydrogen atom in a strong magnetic field indeed offers itself as a paradigm for studying, both theoretically and experimentally, the basic question of how quantum systems behave in a range of parameters where their classical counterparts exhibit phenomena of chaos.

A very convenient way of visualizing the appearance of chaotic orbits in a classical problem is to look at the Poincaré surfaces of section of the trajectories in phase space. To find the range of energy where, for the parameters used in the Bielefeld experiments, the transition from regularity to chaos occurs in the classical magnetic Kepler problem we computed trajectories as a function of energy for a sufficiently large number of random initial conditions, and determined the area fraction of regular orbits in the Poincaré surfaces of section at $z = 0$. In Fig. 4a the breakdown of regularity in the magnetic Kepler problem is

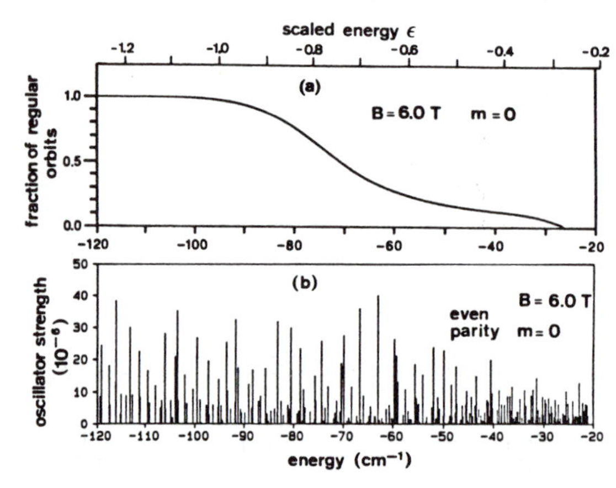

Fig. 4: The area fraction of the Poincaré surfaces of section of classical motion at z=0 which are filled by regular orbits as a function of orbital energy *(a)* compared with the oscillator strengths of $\Delta m = 0$ transitions from $2p_o$ to hydrogen Rydberg states as a function of the quantal energy of the Rydberg states *(b)*. The breakdown of regularity in classical motion is reflected by an increasing complexity of the quantal spectrum.

quantified by plotting the fraction of regular orbits as a function of energy for $\beta = 1.275 \cdot 10^{-5}$ ($B = 6.0$ T) and $m = 0$. Fig. 4a is in fact universal, and not restricted to this set of parameters. The reason for this is a scaling property [18] of the (classical) Hamiltonian (1): by the replacements $r \rightarrow \beta^{2/3} r$, $p \rightarrow \beta^{-1/3} p$ the Hamiltonian is brought into a form where it no longer depends on *two* parameters, viz. magnetic field strength and energy, but solely on one parameter, the "scaled" energy $\epsilon = E/(2\beta)^{2/3}$ (E in Rydbergs). Converting, in Fig. 4a, absolute to scaled energies we arrive at the ϵ scale shown at the top horizontal axis of Fig. 4a. For $B = 6\ T$ it is found that irregular orbits become noticeable around $\sim -104\ cm^{-1}$ ($\epsilon \approx -1.10$), and the regions filled with regular orbits virtually all vanish in the Poincaré surfaces of section at $E_c \approx -24.5\ cm^{-1}$ ($\epsilon_c \approx -0.25$).

Fig. 4b shows the quantal oscillator strength spectrum of $\Delta m = 0$ transitions from $2p_o$ to even-parity Rydberg states for the same magnetic field strength as in Fig. 4a. The comparison between classical chaos and the oscillator strength spectrum conveys a *qualitative* impression of the increasing complexity of the quantal spectrum as one penetrates into ranges of energy where classical motion becomes more and more chaotic.

To put the notion of increasing complexity of the quantal spectrum in the classically chaotic region on a more *quantitative* footing studies so far concentrated on properties of the energy level spectra, and, in particular, it was demonstrated in a number of systems [20] that nearest-neighbour level spacing distributions exhibit a universal structural change from a Poisson type to a Wigner type as classical motion becomes increasingly chaotic.

Because of space limitations it is impossible to touch on all the lines of research that are currently pursued in the interrelations between classical physics and quantum physics of the hydrogen atom in a strong magnetic field in the chaotic regime. On the classical side, computations of Liapunov exponents, in particular of unstable periodic orbits, in the chaotic regime have been performed to quantify the degree of irregularity of the trajectories [21-23]. On the quantum side, statistical analyses were performed of the distributions of the spacings of adjacent energy levels, and the change from a Poisson to a Wigner distribution was confirmed also in this system [24-26]. Much effort is presently placed on revealing the connection between unstable periodic orbits and long-range modulations in the quantum energy spectra [27, 28].

5.2 Statistical Analysis of Transition Strengths

Here we will report, as a new result, the first analysis of statistical properties of *transition strength spectra* of hydrogen atoms in strong magnetic fields in the chaotic regime, that is we go beyond investigating energy level sequences and look at quantities that are related to wavefunctions in a sensitive way.

Fig. 5 shows, for a magnetic field strength of 5.96 T, our computed oscillator strength spectrum of the $\Delta m = 0$ dipole transitions from $2p_o$ to even-parity Rydberg states with energies from -20 cm^{-1} up to the field-free ionization limit. Remember that at this field strength, classical motion becomes irregular around -25 cm^{-1}.

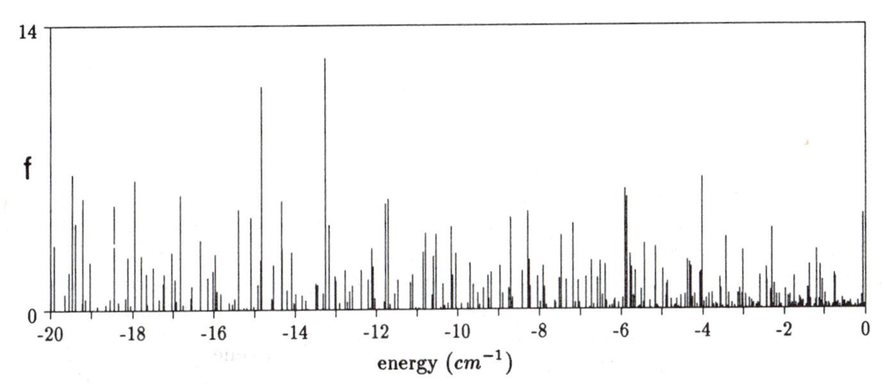

Fig. 5. Oscillator strength spectrum of $\Delta m = 0$ Balmer transition from $2p_o$ to even-parity Rydberg states with energies between $-20\ cm^{-1}$ and the field-free ionization limit at a field strength of 5.96 T (oscillator strengths in units of 10^{-6}). In our computations convergence of the f values is obtained to within a few percent. The range of energy shown lies in the domain of chaotic motion of the classical hydrogen atom.

In quest of further manifestations of chaos in quantal spectra we have analyzed the fluctuations of oscillator strengths of Balmer transitions to Rydberg states in the near-regular, near-irregular, and irregular regime. Following Alhassid and Levine [29] we consider the probability $P(y)dy$ of finding a value y of the strength of an allowed transition from a given to an arbitrary final state between y and y+dy. For a structureless spectrum the maximum-entropy principle, together with the completeness relation of the states and the normalization condition, implies the Porter-Thomas form for $P(y)$, viz. $P(y) \propto exp(-y/2\overline{y})/\sqrt{y}$. For situations where structures in the spectrum impose additional constraints we adopt the form

$P(y) \propto y^{(\nu-1)/2} exp(-\nu y/2\overline{y})/\sqrt{y}$. Here ν denotes a parameter characterizing the deviance of y from its averaged value $\overline{y}$ (see [29] for details). Fig. 6 shows three histograms of computed

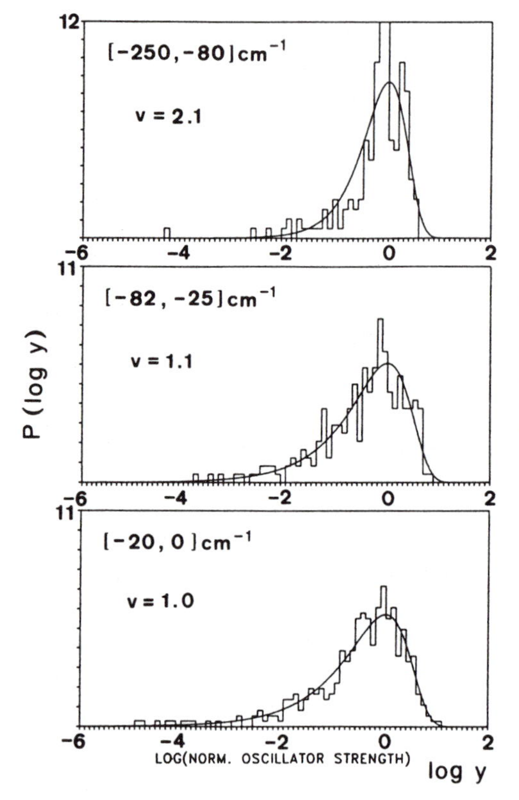

Fig. 6: Three histograms of computed Balmer transition strengths at 5.96 T and their fits (continuous curves) by the Alhassid-Levine form of P(y). Each panel gives the energy range of the final Rydberg states. In the classical system, motion changes from near-regular over near- irregular to irregular, as one passes through these intervals. In the quantum system we find a decrease of ν towards 1 (the Porter-Thomas limit). Oscillator strengths have been scaled by the secular variation prior to binning and normalized in such a way that $\overline{y} = 1$. On the vertical axis, $P(\log y)$ is given in units of 10^{-2}.

oscillator strengths of Balmer transitions to Rydberg states in a magnetic field of 5.96 T and their fits by this functional form. In the three energy ranges of the final states covered in Fig. 6 classical motion changes from almost completely regular to irregular. The results presented in Fig. 6 reveal an increase in the spreading of the fluctuations of the quantal oscillator strengths as this change takes place in the classical system. Correspondingly ν decreases from top to bottom in Fig. 6, and, in particular, we find that the fluctuations in the spectrum shown in Fig. 5 are best fitted by a value of $\nu = 1.0$, that is the Porter-Thomas limit for "chaotic" spectra. The characteristic differences in the distributions of transition strength fluctuations between classically regular and chaotic parameter ranges evident from Fig. 6 confirm findings [29] for transition strengths in a Hénon-Heiles-type model system. Our results thus point to another universal signature of classical chaos in the spectra of quantal systems.

5.3 What Do Hydrogen Atoms in Strong Magnetic Fields Look Like?

"Blessed are they that have not seen, and yet have believed" [30]. In the light of this word physicists can be reckoned among the group of blessed people since they believe in the reality of atoms, although no one has ever "seen" an atom. It is, however, a very human desire to really *see* the objects of one's occupation, and we have therefore represented pictorially hydrogen atoms in strong magnetic fields using the graphics display media available today. Starting point is of course the quantum mechanical interpretation of the squared modulus of the wavefunction as the spatial probability of presence of the electron. Imagining the equivalence of this scalar field to an optically thin self-radiating gaseous nebula, that is, assuming that the scalar field emits "light" with an intensity proportional to the local numerical value of the scalar field, and which propagates without absorption, the intensity a distant observer receives from such an object out of a specific direction is simply the integral along the line-of-sight of the intensities emitted towards the observer. By scanning the whole object with the line-of-sight moving across it in a fine grid, a raster image can be computed which is a good approximation to what the observer actually sees. The raster image is mapped onto the bit planes of a raster graphics display using an intensity or a colour-coding scheme to yield a true picture of the self-radiating object. In our computations we employed a raster of 575 × 575 lines-of-sight and 256 intensity levels.

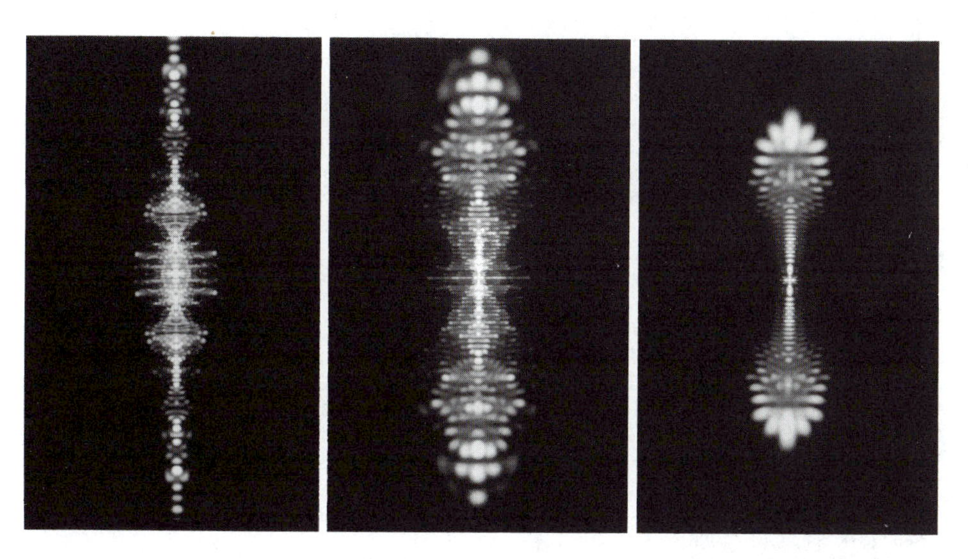

Fig. 7 Pictures of the probabilities of presence of the electron in different highly excited $m = 0$ states, with energies in the classically chaotic regime, of the hydrogen atom in a magnetic field of 5.96 T. The spatial extent of the bizarre atomic structures shown is on the order of 10^{-3} mm.

In Fig. 7 we present first examples of pictures of hydrogen Rydberg states with $m = 0$ in a magnetic field of 5.96 T. The energies of the states lie in the range between -30 and -10 cm^{-1}, and thus in the classically chaotic regime. The pictures demonstrate that the atoms have become extended objects along the direction of the magnetic field (linear dimensions $\sim$ 10000 Bohr radii), and are more or less delocalized, in the sense that they fill most of the configuration space available on account of the remaining conservation laws. Patterns in the shape of the atoms are caused by the nodal structures of the wavefunctions. We have not yet analyzed the pictures with respect to the presence or absence of "scars" [31] along classical periodic orbits, but evidently an analysis of this type is an obvious application of the optical visualization of the states.

The objective of this paper was to show that hydrogen atoms in strong magnetic fields exhibit features of both chaos and order in many respects. Fig. 7 illustrates that, in addition, they also contain a high element of beauty.

This work was supported by the Deutsche Forschungsgemeinschaft. We are grateful to our coworkers F. Geyer, G. Zeller, and R. Niemeier, who have been instrumental in obtaining the results presented in this paper.

References

1. J. Neukammer, H. Rinneberg, K. Vietzke, A. König, H. Hieronymus, M. Kohl, H.-J. Grabka, G. Wunner, *Phys. Rev. Lett.* **59**, 2947 (1987)
2. H. Ruder, H. Herold, W. Rösner, and G. Wunner, *Physica* **127B**, 11 (1984)
3. H. Friedrich and M. Chu , *Phys. Rev.* **A 28**, 1423 (1983)
4. W. Rösner, G. Wunner, H. Herold, and H. Ruder, *J. Phys. B: At. Mol. Phys.* **17**, 29 (1984)
5. H. Forster, W. Strupat, W. Rösner, G. Wunner, and H. Herold, *J. Phys. B: At. Mol. Phys.* **17**, 1301 (1984)
6. H. Ruder, F. Geyer, H. Herold, and G. Wunner, *Physics Reports*, in press (1987)
7. J. K. Daugherty and J. Ventura, *Astron. Astrophys.* **61**, 723 (1977)
8. J. R. P. Angel, *Ann. Rev. Astron. Astrophys.* **16**, 487 (1978)
9. H. C. Praddaude, *Phys. Rev.* **A 6**, 1321 (1972)
10. J. R. P. Angel, J. Liebert, and H. S. Stockman, *Astrophys. J.* **292**, 260 (1985)
11. J. L. Greenstein, R. J. W. Henry, and R. F. O'Connell, *Astrophys. J.* **289**, L47 (1985)
12. G. Wunner, W. Rösner, H. Herold, and H. Ruder, *Astron. Astrophys.* **149**, 102 (1985)
13. J. L. Greenstein, and M. S. Matthews, *Astrophys. J.* **126**, 14 (1957)
14. G. D. Schmidt, S. C. West, J. Liebert, R. F. Green, and H. S. Stockman, *Astrophys. J.* **309**, 218 (1986)
15. A. Holle, G. Wiebusch, J. Main, B. Hager, H. Rottke, and K. H. Welge, *Phys. Rev. Lett.* **56**, 2594 (1986); K. H. Welge, *these proceedings*
16. M.V. Berry, *"The Bakerian Lecture 1987"*, in *Proceedings of the the Royal Society*, (1987)
17. M. Robnik, *J. Phys. A: Math. Gen.* **14**, 3195 (1981)
18. A. Harada and H. Hasegawa, *J. Phys. A: Math. Gen.* **16**, L259 (1983)
19. J. B. Delos, S. K. Knudson, and D. W. Noid, *Phys. Rev.* **A 30**, 1208 (1984)
20. O. Bohigas, M. J. Giannono, and C. Schmit, *Phys. Rev. Lett.* **52**, 1, (1984)
21. D. Wintgen, *J. Phys. B: At. Mol. Phys.* **20**, L511 (1987)
22. W. Schweizer, R. Niemeier, H. Friedrich, G. Wunner, H. Ruder, *Phys. Rev. A* **38** 1724 (1988)
23. H. S. Taylor, J. Zakrzewski, *Phys. Rev. A* in press (1988)
24. G. Wunner, U. Woelk, I. Zech, G. Zeller, T. Ertl, F. Geyer, W. Schweizer, and H. Ruder, *Phys. Rev. Lett.* **57**, 3261 (1986)
25. D. Wintgen and H. Friedrich, *Phys. Rev. Lett.* **57**, 571 (1986)
26. D. Delande and J.C. Gay, *Phys. Rev. Lett.* **57**, 2006 (1986)
27. D. Wintgen, *Phys. Rev. Lett.* **58**, 1589 (1987)
28. M. L. Du and J. B. Delos, *Phys. Rev. Lett.* **58**, 1731 (1987)
29. Y. Alhassid, R. D. Levine, *Phys. Rev. Lett.* **57**, 2879 (1986)
30. *Holy Bible*, John 20: 29, Oxford University Press, London
31. E.J. Heller, *Phys. Rev. Lett.* **53**, 1515 (1984)

Theory of Chaos in the Atomic Hydrogen

G. Casati

Dipartimento di Fisica dell'Università, Via Celoria, 16
I-20133 Milano, Italy

1. Introduction

Investigations on the possibility of quantum chaotic behaviour have now grown into an autonomous field of research, usually referred to as 'Quantum Chaos' /1,2/. Arguments under current investigation range from issues of a mostly foundational flavour to others directly related to the phenomenology of microsystems. Indeed, some very actual problems in Quantum Chaos are of direct interest to molecular and atomic physics, to solid state physics and to quantum optics. Such problems are currently being investigated in close connection with experimental research and are of major applicative interest.

One central problem of Quantum Chaos is the applicability of classical nonlinear models to microphysical systems in the atomic and molecular domains. This is an important point, because classical models often exhibit quite peculiar dynamical properties, connected with the onset of chaos. For example, classical models for atoms in microwave field or molecules in laser fields predict that, for sufficiently high fields, a regime of chaotic motion is entered, characterized by a diffusion in phase space eventually leading to ionization. This diffusive phenomenology would be quite beyond the qualitative predictions of standard quantum mechanical treatment.

As is known, chaotic motion in classical mechanics is a fairly general occurrence when an integrable system is subject to some perturbation. The latter can be both time-independent and time-dependent; in the latter case, it is important to remark that the appearence of chaos does not require any particular disorder in the time dependence of the perturbation; even a periodic perturbation may be effective in this sense.

The onset of chaos has usually a threshold-like character in the perturbation strength and modern ergodic theory provides a clean

The Hydrogen Atom Editors: G.F. Bassani · M. Inguscio · T.W. Hänsch

mathematical framework for the mechanism of chaos generation; nevertheless, effective estimates of chaotic thresholds and quantitative descriptions of chaotic motion rely on at least partially heuristic criteria and/or on computer simulations.

The dynamical chaos is of absolute relevance, both for theory and applications. In the first place, the fact that chaotic motion is in many respects practically indistinguishable from a pure stochastic motion , raises important questions about the deterministic nature of classical mechanics. On the other hand, chaos involves a very peculiar phenomenology; e.g., in case that the perturbation is time-dependent, it may lead to indefinite diffusion in phase space with obviously observable consequences. Classical models for microphysical systems are usually of this type, i.e., they exhibit a stochastic transition. It is then a fundamental problem to understand what happens upon quantizing such models.

Two classes of problems should be distinguished:

1) Conservative systems described by time-independent Hamiltonians.

2) Systems under time-periodic or quasi-periodic perturbations.

The question of persistence of classical chaos in the first class of problem has been rather extensively studied and the general answer is known to be negative, simply because conservative bounded quantum systems have in all cases a pure point energy spectrum; this implies recurrent behaviour of any observable, in sharp contrast with the nonrecurrent classical dynamics, which is associated with a continuous spectrum. There is of course no violation of the correspondence principle involved here because, in the semiclassical regime, the quantum dynamics does indeed reproduce the classical behaviour, over some finite time scale.

However it is important to stress that the complexity of classical motion appears here in the statistical distribution of eigenvalues and eigenfunctions.

For the second class of problems, the present state of knowledge can be summarized as follows:

a) Quantum mechanics has in general an inibitory effect on classical chaos. This inibitory mechanism is closely connected to the localization phenomenon that plays a central role in solid state physics and which is typical for wave propagation in random media: wave packets propagating in an infinite (statistically) homogeneous random medium do neither travel to infinity as they would in free space, nor do they spread indefinitely. Instead, they remain localized in essentially finite regions of space. No dissipative mechanism is involved in this phenomenon, which is in fact produced by complicated interference of partial waves randomly scattered at various places.

b) The quantum inhibition of classical chaos may not be a complete one: a more or less marked memory of the classical chaotic motion may survive in the quantum domain. Indeed, the existence of a kind of quantum regime retaining some features of classical chaotic diffusion is the only possible explanation for some experimental results by Bayfield and Koch on microwave ionization of highly excited atoms /3/.

In the following we present a theoretical analysis which leads to the understanding of the mechanism of excitation and ionization of hydrogen atoms under microwave fields. One of the main predictions of this analysis, namely the localization phenomenon, has been recently observed in laboratory experiments /4,5/.

2. The Kicked Rotator

In order to understand the modifications that quantum mechanics imposes on the classical picture of chaotic motion it is convenient to start from a model which is sufficiently simple but which display the typical, very rich behaviour of classical systems: the δ-kicked rotator. This model is described by the Hamiltonian:

$$H= p^2/2 + \omega^2 \cos\theta \sum_n \delta (t-nT) \tag{1}$$

where p is the rotator momentum, θ is the angular coordinate, T the kick period and ω the perturbation strength.

In classical mechanics, the study of this model has provided deep insight into the general behaviour of dynamical systems, since it shares almost all their main features. Correspondingly, one may expect that the study of the quantum properties of the kicked rotator, especially in regions of parameters where the corresponding classical model is chaotic, will be of great relevance for understanding the qualitative features of the quantum motion.

Due to the presence of the δ-function, the classical equations of motions for system (1) can be integrated and reduced to the mapping:

$$P_{n+1} = P_n + K\sin\theta_n$$
$$\theta_{n+1} = \theta_n + P_{n+1} \qquad (2)$$

where $K = \omega^2 T$, n is time measured in number of kicks and P is the dimensionless angular momentum $P_n = p_n T$.

Mapping (2) is the well-known "standard map", extensively discussed in the literature and frequently used, at a tutorial level, to illustrate the great complexity of motion of simple dynamical systems. Indeed the motion of system (1) presents completely different qualitative features depending on whether K is less or larger than $K_c \sim 1$. More precisely for $K < K_c$ most orbits lie on smooth curves and, in particular, the kinetic energy remains bounded with a variation $\Delta P \sim K$. Instead, when K exceeds the critical value K_c, the mapping orbits become chaotic and the system mimics a random walk in momentum space leading to a diffusive growth of the kinetic energy:

$$\overline{P^2} \sim (K^2/2)\, n \qquad (3)$$

and to the angular momentum distribution of the Gaussian type

$$f(P,n) = (K^2 \pi n)^{-1/2} \exp(-P^2/K^2 n) \qquad (4)$$

In order to turn to the quantum description, we write the quantum Hamiltonian:

$$H = \hbar^2 \partial^2/\partial\theta^2 + \omega^2 \cos\theta \sum_n \delta(t - nT) \qquad (5)$$

Then letting

$$\psi(\theta,t) = \sum_s c_s e^{is\theta} \qquad (6)$$

we may write the solution of the Schroedinger equation as a mapping for the wave function over one period.

$$\psi_{n+1}(\theta) = S\ \psi_n(\theta) = \exp\,(i\ k\ \cos\theta)\ \sum_s c_s \exp\,[i\ (s\theta - 2\pi s^2\ \tau)] \qquad (7)$$

where $k = \omega^2/\hbar$, $\tau = \hbar T/4\pi$ and n is again time measured in number of kick periods.

The quantum mapping (7) can be numerically iterated and , starting from a given set of $\{c_s(o)\}$, one may compute the probability distribution $f(s)=|c_s|^2$, the quantum average kinetic energy $\sum s^2|c_s|^2$ and compare the time dependence of these two quantities with the classical one given by eqs. (4) and (3) respectively.

Surprisingly enough, it turned out /6,7/ that in typical situations the quantum excitation would follow the classical pattern only up to a finite time t_B, called break-time, after which the quantum rotator appears to enter a stationary regime, where the average kinetic energy oscillates around a maximum value. In this stationary regime, the quantum distribution over eigenstates of the free rotator is "frozen", the excitation decreasing exponentially away from the initially excited state /7,8/.
This "saturation" in the energy growth was qualitatively and quantitatively explained by the pure-point nature of the quasi-energy (q.e.) spectrum /9,10/ which is essentially the spectrum of operator S in (7). Indeed, in the quantum system, diffusive excitation can take place on a time scale so small that the wave packet evolution cannot "perceive" the finite separation of quasi-energy eigenvalues. Therefore, if ΔE is the average spacing of the q.e. eigenvalues represented in the wave packet, the break-time should be defined in order of magnitude by $t_B\ \Delta E \sim \hbar$.

Moreover, a formal connection between the rotator problem and the one-dimensional tight-binding model with a time-independent pseudorandom potential was found /8/, which led to the recognition that the quantum suppression of the chaotic excitation of the rotator is a sort of a dynamical version of Anderson localization. Therefore, the most important lesson learned from the kicked-rotator model was that the classical diffusive excitation, taking place above the chaotic threshold, is quantum mechanically suppressed by interference effects that lead to exponential localization of excitation in momentum space.

3. The Hydrogen Atom

The main question now is whether this phenomenon of quantum suppression of classically chaotic diffusion is peculiar of the kicked rotator model or a general occurrence in quantum mechanics.
A seemingly negative indication was given by the analysis of a completely different problem, namely, the microwave ionization of highly excited hydrogen atom. Indeed, in laboratory experiments /3/ a strong ionization was observed by making a beam of hydrogen atoms prepared in states with initial quantum number n_0=66 traverse a microwave cavity of frequency $\omega/2\pi \approx$ 10GHz and peak intensity $\epsilon \approx$ 10V/cm. In this situation, ionization would required the absorption of about 100 photons. Theoretical analysis and numerical simulations showed that a classical model satisfactorily accounts for the experimental results.

Moreover, recent laboratory experiments /11/ showed a good agreement with classical numerical computations at least for microwave frequencies such that $\omega_0=\omega n_0^3<1$, with the initially excited state n_0 in the range 40-90. Experiments with $\omega_0>1$ were not available until recently; however, it was considered quite unlikely that the localization phenomenon would appear by further increasing n_0.

The Hamiltonian of the system we are discussing can be written as

$$H= p^2/2 - 1/r + \epsilon z \cos \omega t \tag{8}$$

where atomic units are used: ϵ and ω are the field intensity and frequency, and the z coordinate is measured along the direction of the microwave field.

As it was recently shown /12,13/, the one-dimensional version of (8) describes with a very good approximation the main qualitative features of the excitation process of the real atom. For this simple 1-dimensional model, it was also shown /12,13/ that the classical motion, for $\omega_0=\omega n_0^3>>1$ can be approximately described by the area-preserving map

$$\begin{aligned} \bar{N} &= N + k \sin\phi \\ \bar{\phi} &= \phi + 2\pi\omega (-2\omega\bar{N})^{-3/2} \; ; \quad k = 0.0822\pi\, \epsilon\omega^{-5/3} \end{aligned} \tag{9}$$

which gives the change of $N=E/\omega$ (E the energy : $E = -1/2n^2$) and $\phi = \omega t$ between two consecutive transits of the electron at the aphelion.

The interesting fact is that upon linearizing the map (9) around the initial value $N_0 = -1/(2n_0^2\omega)$ one obtains again the well known standard map:

$$\bar{N} = N + k \sin \phi$$
$$\bar{\phi} = \phi + T\bar{N} \qquad (10)$$

with $T=6\pi n_0^5\omega^2$. It is then well known that a transition to chaotic motion occurs for (10) when $K=kT>1$. From this, defining rescaled quantities $\epsilon_0=\epsilon n_0^4$, $\omega_0=\omega n_0^3$ we get the threshold for the transition to chaotic motion in the hydrogen atom model (8)

$$\epsilon_{cr} \approx 1/(49\omega_0^{1/3}) \qquad (11)$$

Therefore, if $\epsilon_0 > \epsilon_{cr}$, strong excitation and ionization takes place in the classical model (8). Moreover, by exploiting the similarity between eq. (9) and the rotator model, it has been argued, and numerically demonstrated /12,13/, that the quantum motion of the same model is localized, i.e. that the quantum excitation process leads to a steady-state distribution given in average by

$$\bar{\pi}(N) \approx (1/2l)(1+2|N-N_0|/l) \exp(-2|N-N_0|/l) \qquad (12)$$

with a localization length

$$l \approx D \approx k^2/2 = 3.33\ \epsilon^2\omega^{-10/3} \qquad (13)$$

Notice that localization described by (12) is exponential in $N=E/\omega$, i.e., in the number of absorbed photons. This localization picture is valid, provided that $l \ll N_I = 1/(2n_0^2\omega)$, the number of photons required for ionization. Instead, if l given by (13) becomes comparable with N_I, localization breaks down and strong ionization takes place. From the condition $l \approx N_I$ we get a threshold

$$\epsilon_q = \omega_0^{7/6}/(6.6\ n_0)^{1/2} \qquad (14)$$

In order that strong ionization may occur in the quantum model, the condition $\epsilon_0 > \epsilon_q$ must be satisfied, in addition to the classical condition $\epsilon_0 > \epsilon_{cr}$.

As hinted above, formula (11) actually holds for $\omega_0 > 1$. (This is discussed in Ref. /14/). When $\omega_0 < 1$, Eq. (11) must be suitably corrected, and for $\omega_0 \to 0$, ϵ_{cr} must approach the static field value $\epsilon_0 \sim 0.13$. The classical border is displayed in fig. 1 (dashed curve).

To summarize, our theoretical analysis of the model (8) shows that the quantum interference effects, that in the rotator case caused a complete arrest of the classical chaotic diffusion, are still at work in the hydrogen atom model. Unlike the rotator case, however, these localizing effects of a quantum origin can be overcome and strong excitation can occur more or less along the classical lines by increasing the field ϵ_0 above the delocalization border (14).

In Fig. 1 we present a numerical check of this theory. The straight line gives the critical field value for 1% ionization, computed according to the above sketched theory. We define the ionization probability as the total probability above the level $\bar{n} = 1.5n_0$; therefore, the threshold for 1% ionization is obtained from the condition

$$0.01 = \int_{\bar{N}}^{\infty} \bar{f}(N)\, dN \tag{15}$$

with $\bar{N} = [1/n_0^2 - 1/(1.5n_0)^2]/(2\omega)$. The rhs of eq. (15) can be computed as a function of ϵ_0 by using eq.(12); upon solving the resulting equation for ϵ_0 we get

$$\epsilon_1 \approx (0.18\ \omega^{1/6})\ \omega_0 = 0.023\ \omega_0 \tag{16}$$

which gives the straight line in Fig. 1.

The numerically computed 1% thresholds are represented in Fig. 1 by full circles (quantum results) and empty circles (classical results). For each value of $\omega_0 = \omega n_0^3$ they were found by numerically integrating the equations of motion (quantum and classical) by increasing ϵ_0 until a 1% total

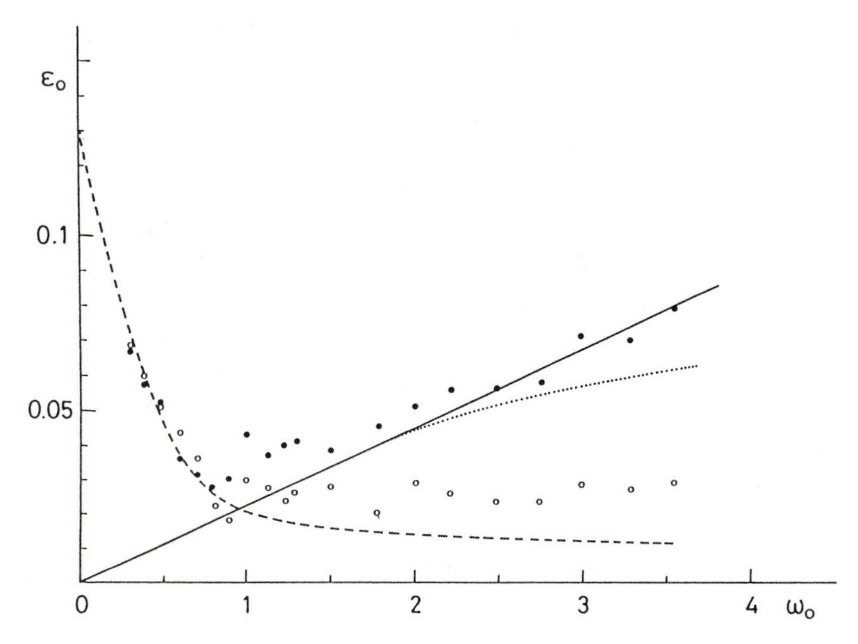

Fig. 1 - Classical (empty circles) and quantum (full circles) numerical threshold field values as a function of the rescaled frequency $\omega_0 = \omega n_0^3$, at fixed $\omega/2\pi$=26.41 GHz. The dotted curve gives the classical chaos border. The full straight line is the critical field value for 1% probability above level $\bar{n}$ = 1.5 n_0 (formula (16)). The dotted line gives the critical field for 1% probability above level n=120.

probability was reached on levels higher than 1.5 n_0. In the classical case, this probability was computed by integrating 350 orbits with given ϵ_0, ω_0, n_0 and homogeneously distributed phases in $(0,2\pi)$. As it is seen, for $\omega_0>1$, there is a quite good agreement between the quantum numerical results and our theoretical estimate (16).

It is also apparent from Fig. 1 that for $\omega_0 < 1$ the quantum excitation process closely follows the classical one /15/. This is due to the fact that for $\omega_0 < 1$ the quantum delocalization border falls below the classical chaotic threshold, so that there are no limitations of a quantum origin to the onset of chaotic excitation. The increase in the quantum numerical thresholds near $\omega_0 \approx 1$ is produced by the main resonance centered at $\omega_0=1$; see also Fig. 5 of Ref. /14/.

Of course, our theoretical analysis is a rather crude one, and should only be expected to account for the gross qualitative features of the quantum motion. In addition to quantum localization, other effects.are here at work, such as tunneling through KAM curves, resonance effects, etc., which are responsible for the fine structure of Fig. 1.

Fig. 1 also shows that the numerically computed classical thresholds are higher than the theoretical classical border. The reason is that, whereas the theoretical curve yields the threshold for the chaotic transition, the numerical data give the critical fields for 1% ionization after $\tau_{ex} \sim 400$ microwave periods. According to a theoretical estimate (formula (6) of Ref. /14/) the classical ionization time is $\tau_I \sim \omega_0^{7/3}/2\epsilon_0^2$ which, on the chaotic border $\epsilon_0 \sim 1/(49\omega_0^{1/3})$, and for $\omega_0 > 1$, takes the value $\tau_I \sim (10\omega_0)^3 > \tau_{ex}$. Therefore, due to the small interaction time, a stronger field than ϵ_{cr} is required in order to get 1% ionization.

We would like to add one final remark concerning the comparison with experimental data. In actual laboratory experiments it may be convenient to define the ionization as the total probability above some fixed level $\overline{n}$, (and not, as in our simulation, above $\overline{n} = 1.5n_0$ dependent on the initial level n_0). The corresponding 1% thresholds should not be very different provided that n_0 is kept sufficently smaller than $\overline{n}$. For sake of comparison, in Fig.1, we also plotted the theoretical curve for 1% probability above n=120.

4. Conclusions

The main qualitative feature of our theoretical and numerical data is the suppression of chaotic excitation produced by quantum mechanics in the region $\omega_0 > 1$, where the quantum thresholds rise significantly above the classical ones. Very recent laboratory experiments /4,5/ have shown the localization phenomenon for $\omega_0 > 1$ and the experimental data are in good agreement with the original theoretical predictions /16/. This is therefore the first experimental observation of the suppression of classically chaotic diffusion produced by quantum mechanics, a fact which may turn out to be of great relevance.

References

1. Proceedings of the NATO ARW on Chaotic Behaviour on Quantum Dynamics, Como, June 1983, edited by G. Casati, Plenum Press, NATO ASI Series B 120, (1985).

2. Quantum Chaos and Statistical Nuclear Physics, eds T. H. Seligman and H. Nishioka, Lectures Notes in Physics, 263, (Springer, Berlin 1986)

3. J.E. Bayfield and P.M. Koch, Phys. Rev. Lett. 33 (1974) 258.

4. J.E. Bayfield and D.W. Sokol: Excited Atoms in Strong Microwaves: Classical Resonances and Localization in Experimental Final State Distributions (Preprint).

5. J. E. Bayfield, G. Casati, I. Guarneri and D.V. Sokol: Localization of Classically Chaotic Diffusion for Hydrogen Atoms in Microwave Fields. Preprint.

6. G. Casati, B.V. Chirikov, F.M. Izrailev, J. Ford in Stochastic Behaviour in Classical and Quantum Hamiltonian Systems, Como, June 1977, Lectures Notes in Physics, Springer 93 (1979), 334.

7. B.V. Chirikov, F.M. Izrailev, D.L. Shepelyansky, Soviet Scientific Review 2C (1981), 209.

8. S. Fishman, D.R. Grempel, R.E. Prange, Phys. Rev. Lett. 49 (1982), 509; Phys. Rev. A. 29 (1984), 1639.

9. D.L. Shepelyansky, Physica, 8D (1983), 208.

10. G. Casati, J. Ford, I. Guarneri, F. Vivaldi, Phys. Rev. A 34 (1986) 1413.

11. K.A.H. Van Leeuven, G.V. Oppen, S. Renwick, J.B. Bowlin, P.M. Koch, R.V. Jensen, O. Rath, D. Richards and G. Leopold, Phys. Rev. Lett. Vol 55, (1985) 2231.

12. G. Casati, I. Guarneri and D.L. Shepelyansky, Phys. Rev. A 36 (1987) 3501.

13. G. Casati, I. Guarneri, D.L. Shepelyansky: Hydrogen Atom in Monochromatic Field: Chaos and Dynamical Photonic Localization, IEEE Journal of Quantum Electronics (July 1988).

14. G. Casati, B.V. Chirikov, I. Guarneri, D.L. Shepelyansky: Relevance of classical chaos in quantum mechanics: the hydrogen atom in a monochromatic field, Physics Reports 154 (1987) 77.

15. R. Blumel, V. Smilansky, Phys. Rev. Lett. 58 (1987) 2531.

16. G. Casati, B.V. Chirikov, D.L. Shepelyansky, Phys. Rev. Lett. 53 (1984) 2525.

Quantum Chaos and the Hydrogen Atom in Strong Magnetic Fields

D. Delande and J.-C. Gay

Laboratoire de Spectroscopie Hertzienne de l'ENS,
Tour 12, Etage 1, 4 Place Jussieu, F-75252 Paris Cedex 05, France

We study the classical and quantum dynamics of the hydrogen atom in a strong magnetic field. At low field, approximate dynamical symmetries exist, allowing a complete description of the system. As the magnetic field is increased, the classical dynamics smoothly evolves from regular to chaotic. However, the dynamical symmetries are not completely destroyed. There are "scars" of these symmetries which manifest in the energy spectrum and eigenstates. They also imply a partial phase-space localization of the quantum motion. In the experimental spectra, these scars are responsible for the modulations known as "Quasi-Landau" resonances.

1. THE HYDROGEN ATOM IN A MAGNETIC FIELD. EQUIVALENCE WITH A SYSTEM OF COUPLED OSCILLATORS

The hydrogen atom in a strong magnetic field is one of the simplest chaotic system. The classical and quantum dynamics can be studied (see sections 2,3,4), and stimulating experimental results have been obtained [1-4]. Conjectures on quantum chaos, established on model systems or observed on nuclear energy levels [5-8], can be tested on a "real" completely calculable system.

The hamiltonian of the hydrogen atom in a magnetic field is (in atomic units, neglecting relativistic effects and assuming an infinitely massive nucleus) :

$$H = \frac{\vec{p}^2}{2} - \frac{1}{r} + \frac{\gamma}{2} L_z + \frac{\gamma^2}{8} (x^2+y^2) \qquad (1)$$

where $\gamma = B/B_c$ is the magnetic field (along z-axis) measured in atomic units ($B_c = 2.35\ 10^5$ T). L_z (projection of the angular momentum on z-axis) and parity are constants of the motion.

In the following, we consider the $L_z = M = 0$ states (somewhat similar conclusions are obtained for the other low M series). The paramagnetic term $\gamma L_z/2$, constant over a given M series, can be disregarded.

As the hamiltonian (1) is time-independent, the total energy is constant. If another independent constant exists, the motion is regular (number of constants = number of degrees of freedom). Otherwise, the system is chaotic.

In the *strong-field regime,* the Coulomb potential (with spherical symmetry) and the diamagnetic potential $\gamma^2(x^2+y^2)/8$ (with cylindrical symmetry) are of the same order of magnitude. Their symmetries being not compatible, no constant exists and the system turns chaotic [9][10].

As the diamagnetic interaction actually couples almost any hydrogenic state with any other one, the study of (1) is not straightforward. A structure describing the hydrogenic spectrum in its whole is convenient : the dynamical group SO(2,2) (subgroup with fixed value of M of the full dynamical group SO(4,2)) [11-14].

It is easily built using the equivalence of the hydrogen atom with a pair of harmonic oscillators. The semi-parabolic coordinates are defined through :

$$\begin{cases} \mu = \sqrt{r+z} \\ \nu = \sqrt{r-z} \end{cases} . \tag{2}$$

In this system of coordinates, Schrödinger's equation is (E is the energy) :

$$\left\{ H(\mu) + H(\nu) + \frac{\gamma^2}{8}\mu^2\nu^2(\mu^2+\nu^2) \right\} |\Psi\rangle = 4\ |\Psi\rangle \tag{3}$$

with $$H(\mu) = -\frac{\partial^2}{\partial\mu^2} - \frac{1}{\mu}\frac{\partial}{\partial u} + \frac{M^2}{\mu^2} - 2E\mu^2 .$$

$H(\mu)$ is the radial part of the hamiltonian of a 2-dimensional harmonic oscillator with frequency $\sqrt{-2E}$ and angular momentum M. The dynamical group describing the radial motion of the 2-dimensional harmonic oscillator is SO(2,1), the hermitian generators of which are :

$$\begin{cases} S_1^{(\alpha)} = \frac{\alpha}{4}\left(\frac{\partial^2}{\partial\mu^2} + \frac{1}{\mu}\frac{\partial}{\partial\mu} - \frac{M^2}{\mu^2}\right) + \frac{1}{4\alpha}\mu^2 \\ S_2^{(\alpha)} = \frac{i}{2}\left(1 + \mu\frac{\partial}{\partial\mu}\right) \\ S_3^{(\alpha)} = -\frac{\alpha}{4}\left(\frac{\partial^2}{\partial\mu^2} + \frac{1}{\mu}\frac{\partial}{\partial\mu} - \frac{M^2}{\mu^2}\right) + \frac{1}{4\alpha}\mu^2 \end{cases} \tag{4}$$

with α a positive adjustable parameter.

They satisfy the commutation relations of a SO(2,1) Lie algebra :

$$\begin{cases} [S_1^{(\alpha)},S_2^{(\alpha)}] = -\ i\ S_3^{(\alpha)} \\ [S_2^{(\alpha)},S_3^{(\alpha)}] = \ i\ S_1^{(\alpha)} \\ [S_3^{(\alpha)},S_1^{(\alpha)}] = \ i\ S_2^{(\alpha)} \ . \end{cases} \tag{5}$$

Upon choosing $\alpha = 1/\sqrt{-2E}$, $S_3^{(\alpha)}$ is proportional to the hamiltonian $H(\mu)$. The eigenvalues of $S_3^{(\alpha)}$ are $n_1+(|M|+1)/2$, labelled with the non-negative integer n_1.

The three generators of the SO(2,1) group act like the three components of the angular momentum for the usual SO(3) rotation group. $S_\pm^{(\alpha)} = S_1^{(\alpha)} \pm i\ S_2^{(\alpha)}$ are ladder operators, rising n_1 by ± 1. The matrix elements are known from group representation theory and involve simple algebraic expressions [12][14].

The dynamical group of the hydrogen atom (or of the equivalent oscillator problem) is the direct product SO(2,1) $\otimes$ SO(2,1) = SO(2,2). The six generators are the three components of $\vec{S}^{(\alpha)}$ and those of $\vec{T}^{(\alpha)}$ (the latter is defined as $\vec{S}^{(\alpha)}$ but with the change $\mu \to \nu$). Using (2) and (3), Schrödinger's equation is cast under a pure algebraic form as a function of the generators :

$$\left\{ S_3^{(\alpha)}+T_3^{(\alpha)} - \alpha + \frac{\gamma^2\alpha^4}{2}(S_3^{(\alpha)}+S_1^{(\alpha)})(T_3^{(\alpha)}+T_1^{(\alpha)})(S_3^{(\alpha)}+S_1^{(\alpha)}+T_3^{(\alpha)}+T_1^{(\alpha)}) \right\} |\psi\rangle = 0 \tag{6}$$

with $\alpha = (-2E)^{-1/2}$.

2. CLASSICAL DYNAMICS

The hydrogen atom and the equivalent system of coupled oscillators have the same classical dynamics. The latter can be studied using the dynamical coordinates $(\vec{S}^{(\alpha)},\vec{T}^{(\alpha)})$. The hamiltonian of the system is the classical form of (6).

Another set of dynamical coordinates is the "energy-phase" coordinates for the oscillators along the μ and ν directions, defined as :

$$\begin{cases} \varphi_S = \tan^{-1}\left(S_2^{(\alpha)}/S_1^{(\alpha)}\right) = 2\tan^{-1}(p_\mu/\mu) \\ \varphi_T = \tan^{-1}\left(T_2^{(\alpha)}/T_1^{(\alpha)}\right) = 2\tan^{-1}(p_\nu/\nu) \ . \end{cases} \tag{7}$$

$(\varphi_S, \varphi_T, S_3^{(\alpha)}, T_3^{(\alpha)})$ is a set of canonical coordinates complying with the dynamical symmetries of the unperturbed system [14]. Indeed, in zero magnetic field, from (6), $S_3^{(\alpha)}$ and $T_3^{(\alpha)}$ are constant, and φ_S and φ_T are linear functions of time. This system of coordinates provides one with *the most simple description of the Kepler motion of the electron around the nucleus.*

There is a scaling law for the classical dynamics. It depends on the single parameter :

$$\beta = \frac{\gamma^2}{(-2E)^3} \tag{8}$$

which is the ratio of the diamagnetic force to the Coulomb force.

The various sets of trajectories associated with different values of γ and E, and same value of β, are simply homothetic. The strong-field regime where chaos takes place corresponds to values of β around unity.

The equations of the motion can be numerically solved. From the calculated trajectories in the 4-dimensional phase space, we deduce Poincaré surfaces of section. In order to preserve the $\vec{S}\leftrightarrow\vec{T}$ symmetry, we use a symmetrical $\varphi_S+\varphi_T=0$ surface of section and represent the section in the plane of canonically conjugate coordinates : $(X_3^{(\alpha)}=T_3^{(\alpha)}-S_3^{(\alpha)},\ \Delta\varphi=(\varphi_T-\varphi_S)/2)$. These simply represent the differences of the energies and phases of the two oscillators.

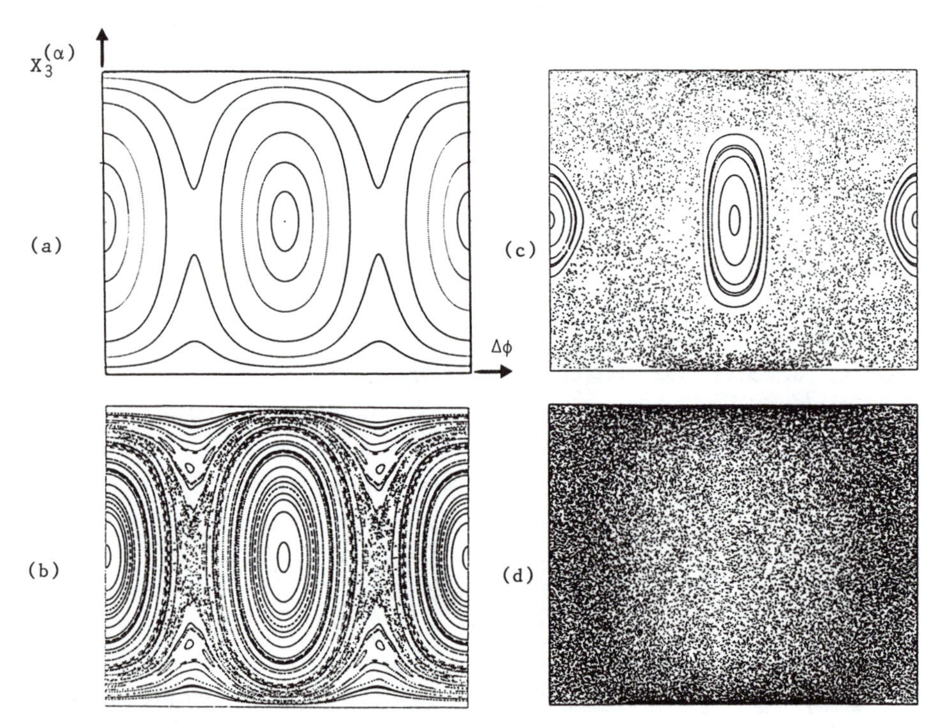

Figure 1 : Some Poincaré surfaces of section obtained in the $(-\alpha < X_3^{(\alpha)}=T_3^{(\alpha)}-S_3^{(\alpha)} < \alpha,\ -\pi < \Delta\varphi=(\varphi_T-\varphi_S)/2 < \pi)$ plane

(a) β=0.01 The motion is fully regular (see section 4)

(b) β=1 A small chaotic region appears

(c) β=5 Mixed regular-chaotic regime

(d) β=70 Fully chaotic regime

Figure 1 represents a set of Poincaré surfaces of section. At low field, the phase space trajectories are confined on 2-dimensional invariant tori, which intersect the surface of section along the 1-dimensional invariant curves of Fig. 1a. Near $\beta=1$, some invariant tori are destroyed (Fig. 1b) and chaos takes place. As the magnetic field is increased, the fraction of chaotic phase space smoothly increases (Fig. 1c). Above $\beta\simeq 60$ and up to the ionization limit ($\beta=\infty$), the whole phase space is chaotic, except for very small regions [14,15].

3. QUANTUM SPECTRUM AND EIGENSTATES

As the diamagnetic interaction mixes together all the hydrogenic states, including the continuum, understanding the quantum behaviour of the hydrogen atom in a strong magnetic field has been a challenging question for years.

An efficient way for calculating the energy spectrum and the eigenstates in the strong-field regime is to use the dynamical group SO(2,2) (see section 1) and (6). In an eigenbasis of the equivalent oscillator system (eigenstates of $S_3^{(\alpha)}$ and $T_3^{(\alpha)}$), the generators $(\vec{S}^{(\alpha)},\vec{T}^{(\alpha)})$ connect a given $|n_1\ n_2\ M\rangle$ state to its nearest neighbors $|n_1 \pm 1\ n_2 \pm 1\ M\rangle$ only. Hence, the hamiltonian matrix (6) has few non-zero elements, and leads to efficient numerical diagonalization [15-18].

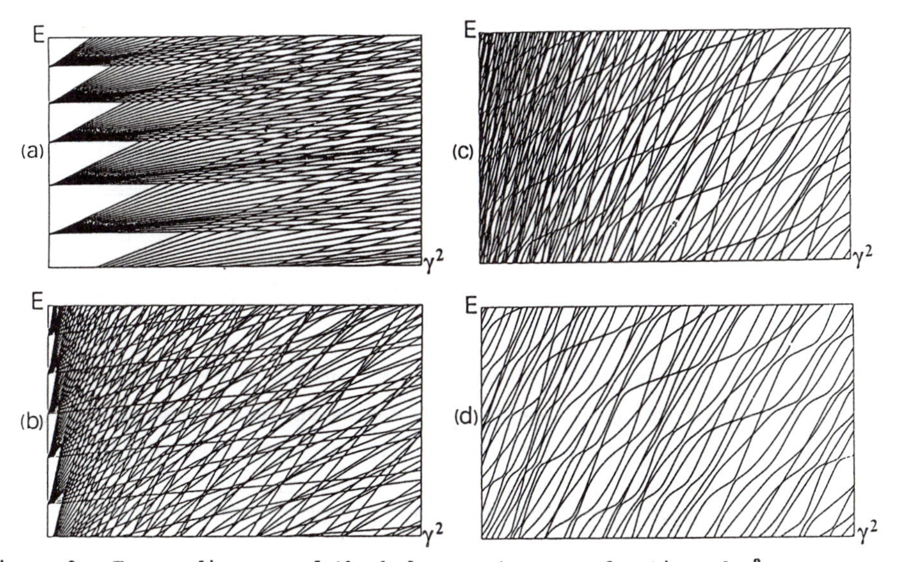

Figure 2 : Energy diagrams of the hydrogen atom as a function of γ^2 :

(a) Low field (regular) $-4\ 10^{-4} \leqslant E \leqslant -3\ 10^{-4}$; $0 \leqslant \gamma^2 \leqslant 1.4\ 10^{-10}$

(b) Moderate field (regular-chaotic) $-4\ 10^{-4} \leqslant E \leqslant -3\ 10^{-4}$; $0 \leqslant \gamma^2 \leqslant 1.4\ 10^{-9}$

(c) High field (regular-chaotic) $-4\ 10^{-4} \leqslant E \leqslant -3\ 10^{-4}$; $1\ 10^{-9} \leqslant \gamma^2 \leqslant 1.4\ 10^{-8}$

(d) High field (chaotic) $-2.5\ 10^{-4} \leqslant E \leqslant -2\ 10^{-4}$; $6\ 10^{-9} \leqslant \gamma^2 \leqslant 1.3\ 10^{-8}$

The convergency of the calculations is excellent, except very close to the ionization limit.

One of the key points for doing such calculations is to use a basis having the appropriate symmetry properties. The oscillator basis is such. Numerical diagonalization of the hamiltonian (1) in an hydrogenic basis would give very poorly converging results in the strong-field regime.

Figure 2 shows plots of the energy spectrum as a function of γ^2 (square of the magnetic field). The different plots correspond to increasing values of β, ranging from 0 (low-field limit) to about 100 (chaotic strong-field regime).

4. THE LOW-FIELD LIMIT

In the limit of vanishing magnetic field (the so-called inter-ℓ-mixing regime), the eigenstates are obtained by first order perturbation theory, that is, by diagonalizing the diamagnetic term inside a given n-manifold. The restriction of the diamagnetic perturbation to a n-manifold is then an adiabatic invariant, that is, a constant of the motion to first order in γ^2. It can be expressed as a function of the "energy-phase" coordinates (7) just by keeping in the diamagnetic term of equation (6), the part which does not change the $n = S_3^{(\alpha)} + T_3^{(\alpha)}$ value. The result is [12]:

$$1 + \Lambda = \frac{4}{n^2} S_3^{(\alpha)} T_3^{(\alpha)} \left(3 + 2 \cos(\varphi_S - \varphi_T) \right) \tag{9}$$

with $S_3^{(\alpha)} + T_3^{(\alpha)} = n$ (principal quantum number).

This result can be reexpressed in terms of the Runge-Lenz vector $\vec{A} = \vec{p} \times \vec{L} - \vec{r}/r$ as : $\Lambda = 4\vec{A}^2 - 5A_z^2$ [19,20].

The expression of constant of the motion (9) holds both in classical and quantum mechanics. It means that, beside E and L_z, another quantity is conserved during the classical motion. Hence, the motion is regular. Moreover, the invariant tori are known analytically. This can be checked in Fig. 1a where the invariant curves are in excellent agreement with the predictions of (9).

In the quantum formalism, the eigenstates are labelled with an integer K ranging from 0 (highest state) to $n-|M|-1$ (lowest state). The wavefunctions are usually very complicated in real space, showing lot of oscillations and, except for few of them, no clear spatial localization.

Performing a significant comparison between the classical and quantum dynamics requires to define a distribution function in phase space rather than in real space [21,22]. The most famous one is the Wigner density, but other definitions are possible, all of them sharing the same semi-classical behaviour. In the present case, the Q-density defined from the coherent states of the

SO(2,2) dynamical group is used [14]. For each eigenstate, the distribution function depends on the 4 dynamical variables $(\varphi_S, \varphi_T, S_3^{(\alpha)}, T_3^{(\alpha)})$. We verified that, as expected from general arguments [21], it is localized near the energy surface H = E with a gaussian dispersion. A further section in the plane $\varphi_S+\varphi_T=0$ allows us to plot it as a function of $(X_3^{(\alpha)}, \Delta\varphi)$, which is done in Fig 3 for the (n=89,K=40,M=0) state. As expected, <u>the phase-space distribution function is localized near a classical invariant curve</u> Λ = Cst (compare with Fig. 1a).

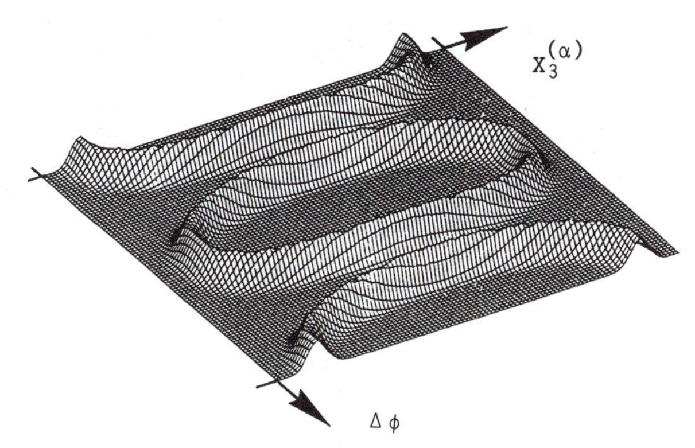

Figure 3 : Phase-space distribution function for the (n=89,K=40,M=0) state represented in the $(-\alpha < X_3^{(\alpha)} < \alpha,\ -\pi < \Delta\varphi < \pi)$ plane. It is localized near the invariant curve of the classical motion

From (9) and Fig. 1a, one easily shows that there exist two types of classical motion according to the sign of the constant Λ :

* $\Lambda > 0$ (inner invariant curves in Fig. 1a). The phase difference between the μ and ν oscillators is bounded. The motion along the invariant curve is around the central fixed point $(X_3^{(\alpha)}=0, \Delta\varphi=0)$, which corresponds to the motion of the atomic electron in the z=0 plane. The associated quantum eigenstates are the upper states of the diamagnetic manifolds (see Fig. 2a). Their symmetry is of an approximate "rotational" type [19] (However, the rotational character does not refer to a spherical property in real space [12,19]). The associated phase space distribution functions are localized near the central fixed point (see Fig. 4a).

* $\Lambda < 0$ (outer invariant curves in Fig. 1a). The phase difference between the two oscillators slowly and smoothly increases (or decreases) with time, which expresses that their correlation is weak. The invariant curves are localized near the $X_3^{(\alpha)} = \pm\alpha$ axis. The associated quantum eigenstates have an approximate "vibrational" symmetry [19]. They are the lower states of each diamagnetic manifold. The phase-space distribution functions are localized near the invariant curves (see Fig. 4b).

The $\Lambda = 0$ invariant curve is a separatrix between the two types of motion. Its quantum counterpart is visible in Fig. 2a : near the separatrix, the period of the classical motion tends to infinity ("critical slowing down"), and the spacing between consecutive diamagnetic components dramatically decreases.

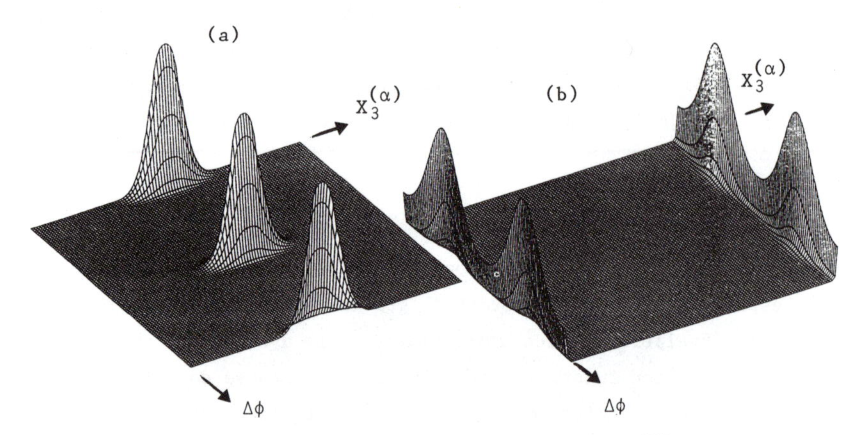

Figure 4 : Phase-space distribution functions $(-\alpha<X_3^{(\alpha)}<\alpha, -\pi<\Delta\varphi<\pi)$:
(a) Rotational (n=89,K=0,M=0) state
(b) Vibrational (n=89,K=88,M=0) state

5. STATISTICAL PROPERTIES OF THE ENERGY SPECTRUM

In the classically regular regime (Fig. 2a), there are only level crossings, or very small avoided crossings not visible at this scale, when inter-n-mixing takes place. This clearly follows from the fact that the different eigenstates are localized (see Fig. 1) near the invariant tori of the classical motion in phase space. Hence, two different eigenstates are localized in different regions and the energy levels cross.

When classical chaos appears, the energy levels anticross (right part of Fig. 2b and Fig. 2c). This is associated with the destruction of the phase-space invariant tori, leading to delocalization of the eigenstates and strong repulsion between them. Actually, the eigenstates associated with the motion close to the separatrix are the first ones to experience delocalization and strong anticrossings (see Fig. 2b).

As the magnetic field strength is increased, more and more energy levels anticross, which corresponds to the increasing size of chaotic regions (see Fig. 2c). Finally, above $\beta = 60$, all the energy levels do anticross (Fig. 2d).

The diagrams in Fig. 2 give a qualitative characterization of quantum chaos, that is, evolution from level quasi-crossing to level anticrossing when chaos develops. A quantitative measure is provided with the analysis of the statistical properties of the energy levels. It has been numerically shown that, in the regular regime, these statistical properties are well described by a model of non-interacting eigenstates while, in the chaotic regime, they obey a random matrix model (here, the Gaussian Orthogonal Ensemble, see [15,23,24]).

However, such a generic random matrix model cannot explain any property of an hamiltonian system. Indeed, non-generic properties of the system still exist, such as phase-space localization of the eigenstates.

6. PHASE SPACE LOCALIZATION - SCARS OF SYMMETRIES

In the chaotic regime, the wavefunctions do not present any clear localization. Actually, especially for highly excited states, they fail to properly exhibit the dynamical correlations which still underlie the dynamics of the system. The phase-space distribution functions directly highlight these correlations (Fig. 3 and 4). We calculate some of them for "chaotic" eigenstates of the system in the strong field conditions of Fig. 2d. We choose a typical rotational state and a typical vibrational state. The phase-space distributions are plotted in Fig. 5. They are normalized with respect to the classical phase-space volume, that is, they should be constant for an "ergodic" eigenstate. This is clearly not the case. Figure 5a is localized near the classical fixed point ($X_3^{(\alpha)} = 0$, $\Delta\varphi = 0$). Such a plot is not very different from the one for a "regular" rotational state (compare with Fig. 4a). This indicates that the phase-space distribution function is strongly scarred by the rotational symmetry. The same feature is obtained for the "chaotic" vibrational state (Fig. 5b to compare with Fig. 4b) which phase-space distribution function is localized near $X_3^{(\alpha)} = \pm\,\alpha$ and scarred by the vibrational symmetry. On the other hand, many "chaotic" phase-space distribution functions are not clearly localized.

Actually, the phase space localization manifests the existence of scars of the rotational or vibrational symmetries. By "scars", we mean that some localization still exists for the system in the chaotic region, in contradiction with Random Matrix Theories expectations [25]. In that sense, these scars are comparable to the scars found by Heller [26] in the wavefunctions of billiards, leading to weak and unexpected spatial localization effects.

These scars of symmetries have very important experimental consequences illustrated on the simulated spectra in Fig. 6 [27].

In Fig. 6a, we plot, for each eigenstate, a stick which height represents the amount of rotational symmetry (i.e. the projection of the eigenstate onto the subspace spanned by the pure hydrogenic rotational states). When the classical motion is regular (below -29cm^{-1}), the highest state of each hydrogenic manifold is strongly dominant, the other ones having nearly zero "rotational part".

When the classical motion turns chaotic, the rotational symmetry partly breaks down, leading to clusters of dominant states, approximately equally spaced, but with no regular structure inside the cluster. The cluster spacing at E=0 is 1.5 $\hbar\omega_c$, corresponding to the usual Quasi-Landau resonances [1].

Figure 6b is a simulation of a real spectrum using optical excitation from the (2p,M=-1) state (which are the experimental conditions of [4]). The clusters of rotational levels are still visible, though very weakened.

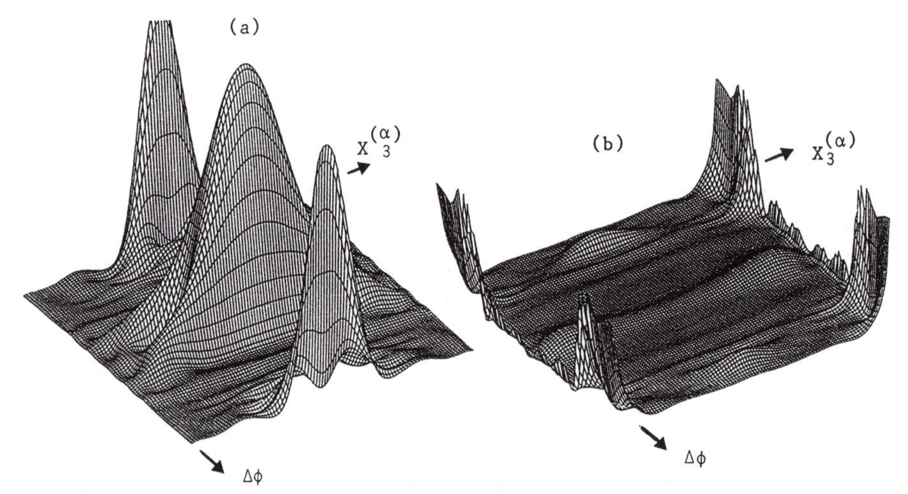

Figure 5 : Phase-space distribution functions for some "chaotic" eigenstates, represented in the $(-\alpha<X_3^{(\alpha)}<\alpha, -\pi<\Delta\varphi<\pi)$ plane.

(a) Rotational state

(b) Vibrational state

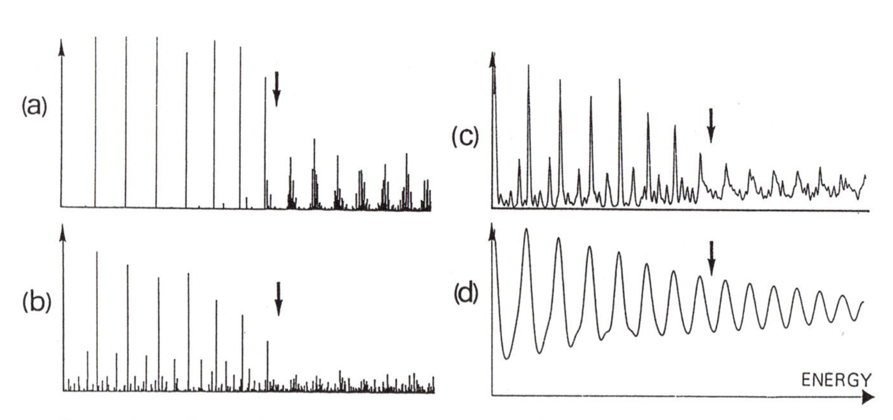

Figure 6 : Simulation of the M = 0, even parity spectrum of the hydrogen atom in a field of 8T between - 132 and 44 cm^{-1}

The arrow indicates the transition to complete chaos (-29 cm^{-1})

(a) Squared projections onto the rotational subspace

(b) Oscillator strengths from the (2p, M = -1) state

(c) Same than (b), with a resolution 0.5 cm^{-1}

(d) Same than (b), with a resolution 2 cm^{-1}

The symmetry of optical excitation is essentially spherical and does not comply with the internal symmetries of the magnetized atom : all the eigenstates are then efficiently optically excited. However, depending on the polarization being used, the intensities of the rotational or vibrational states may be favoured. With σ^+ polarization as used in Fig. 6b, the rotational states are preferably excited, which explains that the rotational clusters are visible. Figure 6c is a simulation of an experimental spectrum assuming finite resolution (modelled through convolution with a gaussian curve). The short-range fluctuations are partly smoothed out. At last, when a poor resolution is used (Fig. 6d), the spectrum looks "regular" continuously from the regular to the chaotic regime. The familiar aspect of the Quasi-Landau spectra with 1.5 $\hbar\omega_c$ spacing is rediscovered [1].

7. CONCLUSION

In summary, we have proved that the classically chaotic magnetized hydrogen atom has the following quantum characteristics :

* The short range fluctuations of the energy levels and eigenstates are essentially governed by a random matrix model.

* The long range behaviour traduces the essential part of the dynamics. At large scale, the spectrum is partly ordered : averaged over several states, the physical properties are regular and even present some periodicity. This is why low resolution experiments have shown regular features for a long time, and why high resolution experimental results are so difficult to interpret.

* The "fully chaotic" eigenstates of the quantum system are strongly scarred by the remnants of the low field symmetries, though the classical system is nearly completely chaotic and probably ergodic. These scars affect the phase-space distribution functions, expressing the internal dynamical correlations. Optical spectra reveal the existence of these scars, but the effects are weakened and entangled as optical excitation is a process not complying with the internal symmetries of the system.

REFERENCES

1. W.R.S.Garton and F.S.Tomkins, Astrophys.J. 158, 839 (1969)
2. J.C.Gay in "Atoms in unusual situations" Ed. J.P.Briand, Plenum (1986)
3. P.Cacciani, E.Luc-Koenig, J.Pinard, C.Thomas and S.Liberman, Phys.Rev.Lett. 56, 1124 (1986)
4. A.Holle, G.Wiesbuch, J.Main, H.Rottke, B.Hager and K.H.Welge, Phys.Rev.Lett. 56, 2594 (1986)

A.Holle, J.Main, G.Wiesbuch, H.Rottke and K.H.Welge, Phys.Rev.Lett. 61, 161 (1988)

5. O.Bohigas and M.J.Gianonni, in Lecture Notes in Physics 209 (1984) Springer-Verlag
6. T.H.Seligman, J.J.M.Verbaarschot and M.R.Zirnbauer, Phys.Rev.Lett. 53, 215 (1984)
7. T.Zimmerman, H.D.Meyer, H.Koppel, L.S.Cederbaum, Phys.Rev.A 33, 4334 (1986)
8. M.Gutzwiller, Phys.Rev.Lett., 45, 150 (1980)
9. If the (dynamical) symmetries are compatible, the system is regular whatever the field strength. A non-trivial exemple of such a system is the hydrogen atom in an electric field where the modified Runge-Lenz vector is a constant associated with the separability in parabolic coordinates
10. M.Henon in "Chaotic behaviour of deterministic systems", Les Houches Summer School (1983) North-Holland
11. M.J.Englefield, "Group theory and the Coulomb problem", Ed. Wiley, New-York (1972)
12. D.Delande and J.C.Gay, J.Phys.B Lett. 17, L335 (1984)
13. J.C.Gay and D.Delande in "Atomic excitation and recombination in external fields" Ed. M.H.Nayfeh and C.W.Clark, Gordon Breach (1985)
14. D.Delande, These de Doctorat d'Etat, Paris, unpublished (1988)
15. D.Delande and J.C.Gay, Phys.Rev.Lett., 57, 2006 (1986)
16. D.Delande and J.C.Gay, J.Phys.B Lett. 19, L173 (1986)
    D.Delande and J.C.Gay, Comm.At.Mol.Phys. 19, 35 (1986)
17. C.W.Clark and K.T.Taylor, J.Phys.B, 15, 1175 (1982)
18. D.Wintgen and H.Friedrich, J Phys.B 19, 991 (1986)
    D.Wintgen and H.Friedrich, J Phys.B 19, 1261 (1986)
19. D.R.Herrick, Phys.Rev.A 26, 323 (1982)
20. E.A.Solov'ev, JETP Lett., 34, 265 (1981)
21. M.V.Berry in "Chaotic behaviour of deterministic systems", Les Houches Summer School (1983) North-Holland
22. M.Hillery, R.F.O'Connell, M.O.Scully and E.P.Wigner, Physics Reports 106, 121 (1984)
23. D.Wintgen and H.Friedrich, Phys.Rev.Lett., 57, 571 (1986)
24. G.Wunner, U.Woelk, I.Zech, G.Zeller, T.Ertl, P.Geyer, W.Schweitzer and H.Ruder, Phys.Rev.Lett., 57, 3261 (1986)
25. D.Delande and J.C.Gay, in "Atomic excitation and recombination in external fields II", Ed. M.H.Nayfeh, C.W.Clark and K.T.Taylor, Plenum (1988)
26. E.J.Heller, Phys.Rev.Lett., 16, 1515 (1984)
27. D.Delande and J.C.Gay, Phys.Rev.Lett., 59, 1809 (1987)

Highly Excited Hydrogen in Strong Microwaves: Experimental Tests for Classically Chaotic Semiclassical Dynamics and for Quantum Localization

J.E. Bayfield and D.W. Sokol

Department of Physics and Astronomy, University of Pittsburgh, Pittsburgh, PA 15260, USA

In recent years the hydrogen atom strongly perturbed by external fields has become a major testing ground in the development of quantum dynamics, the study of general features for the time evolution of nonintegrable quantum systems. We discuss some reasons for this, using for illustration some recent results for the case of an external sinusoidally oscillating electric field.

1. Nonlinear dynamics and Quantum Mechanics

In classical Hamiltonian nonlinear dynamics the motion is deterministic and given by Hamilton's equations. On the other hand, quantum time evolution is unitary and controlled by Schrodinger's equation. Very little is known about the general properties of time evolution in quantum systems for which Schrodinger's equation cannot be solved either analytically or by means of perturbation theory. The path integral formulation of quantum mechanics introduced by Feynman is a conceptually useful starting point. When actions are large, a stationary phase situation can lead to a semiclassical regime where the time propagator for the system explicitly reflects the time evolution of the classical action function.

Classical evolution can exhibit interesting nonlinear dynamics. Changes in the character of the evolution can arise with changes in a control parameter, such as period doubling bifurcations and transitions from regular to chaotic

The Hydrogen Atom Editors: G.F. Bassani · M. Inguscio · T.W. Hänsch

simplicity of the one electron atom, both of these systems are amenable to accurate numerical calculations based on either quantum mechanics or on the mechanics of the system in the hard classical limit. Both systems are within the reach of modern laboratory techniques for direct experimental study. Both exhibit chaotic classical electron trajectories at high external field strengths. The magnetized HEHA is being studied using spectroscopy to ascertain the Wigner energy level separation distribution produced by the classical chaos, as is discussed by others at this Symposium. Again because of underlying chaos, at some microwave frequencies the microwave driven HEHA is predicted to exhibit a microwave energy absorption mechanism associated with a diffusion in electron classical action space that culminates in ionization of the atom [4]. In addition, at some microwave frequencies higher than the initial classical electron orbit frequency for the atom, a second predicted consequence of the chaos is a dynamical type of Anderson localization in quantum number or electron energy space that arises from destructive wave interference presumably produced by the degree of randomness present in the underlying chaos.

3. Developments in the Experimental Study of the Microwave Driven Weakly Bound Electron (Microwave Driven HEHA)

Since the initial experimental discovery of microwave ionization of HEHA requiring the absorption of some hundred microwave photons [5] and the initial discovery that classical models can predict the ionization probability [6], the experiments have improved through the development of new techniques. These have hinged upon various uses of applied static electric fields, where the separability of the quantum problem in parabolic coordinates n, n_1, m that is so special to the hydrogen atom plays a crucial role in maintaining the simplicity of the system. One development was an optical double resonance technique for the laser excitation of atoms into single Stark states, all parabolic quantum numbers being well defined [7]. Here a mixed state fast hydrogen atom beam was first passed into a region of strong static electric field, where a completely state selective m = 0 to m = 0 infrared CW laser transition to one n=10 Stark substate was made possible by large second order Stark energy level splittings. The atoms then passed into a weak static field, where a second infrared CW laser transition produced atoms in a desired final Stark state having the desired principal quantum number n_0 within the range 30 to 90. This ODR technique has been used to produce HEHA that are electrically polarized or "stretched" maximally along the static electric field direction. When a linearly polarized microwave electric field is also applied along the same direction, the resultant microwave driven HEHA constitutes a nearly one dimensional system, as has been verified by experiment [8]. The verification again crucially involved static electric fields that were used to selectively field ionize atoms with different quantum numbers, thereby separating the probabilities for different types of atoms being produced by the applied microwave field pulse. The measurement of final bound state probability distributions (in a one dimensional system) was made possible only by combining all the static field dependent developments mentioned above for the hydrogen atom [8,9].

3.1 Experimental Evidence of an Underlying Role for Classical Chaos in Microwave Driven HEHA

A schematic of the apparatus recently used for microwave driven HEHA studies is shown in Figure 1. A fast atom beam containing laser excited electrically polarized HEHA comes from the right, passes sideways through a WR-62 waveguide with holes electric discharge machined in its narrow sides, and goes to the left for both charge state analysis and final bound state distribution analysis via the differential state-selective static electric field ionization technique. The atom transit time between the planes of the waveguide holes is 7.5 nsec. A microwave source generates a traveling wave in the waveguide that is finally absorbed in a matched load. As the TE10 mode present in the waveguide has the sine dependence of the electric field shown in the figure, this is the pulse envelope seen by the atoms in their rest frames, ignoring the small loss of microwave power out of the holes which adds small long tails to the envelope. For a microwave frequency of 18 GHz, the nominal microwave total pulse length was 135 microwave periods with a total spread of 6 periods due to the velocity spread of the atom beam and a comparable additional spread expected from the range of different fringe fields seen by various atoms entering the waveguide at different points within the holes. The microwave electric field shown is along the single direction of all the static electric fields seen by the HEHA during their brief microsecond histories in the apparatus. (As spontaneous radiative decay times for HEHA increase with the cube of n, our HEHA are by themselves essentially stable atoms.) The primary source of microwaves is a frequency synthesizer. A measurement using a microwave spectrum analyzer placed an upper bound on the possible broadband noise r.m.s. power of 0.01% of typical sinewave output powers. The addition of a narrow pass filter tuned to the sinewave frequency that reduced the broadband noise bandwidth by a factor of 100 was found to not change experimental results. Before intersecting the atom beam, the microwave traveling wave passed into the vacuum system through a window; the remainder

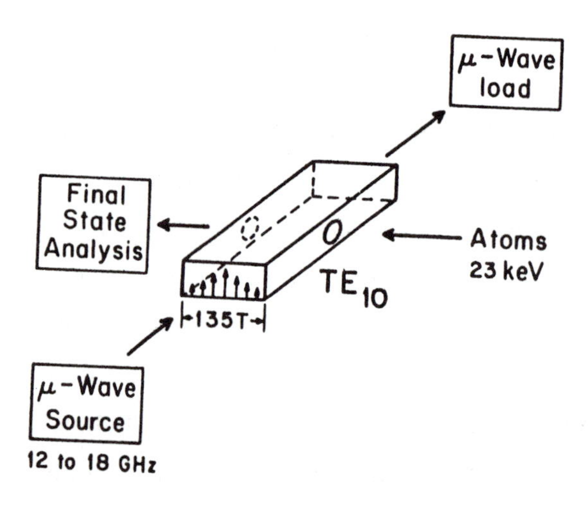

Fig. 1. Schematic of apparatus for HEHA in microwaves

of the microwave system was evacuated to eliminate the need for a second window, thus reducing the possible standing wave power seen by the atoms to less than 2% of the traveling wave power.

For microwave frequencies w below the initial classical electron orbit frequency, experimental "threshold" microwave field strengths for effective ionization are found to be close to classical threshold field strengths for chaos. This is true when the experimental threshold is set for 10% ionization probability, the experimental microwave pulse time is about 300 periods, and the initial quantum number value is well below a cutoff value of n=95 for bound state contributions included in the ionization probability. The classical threshold for chaos is defined by a 1% probability for an ensemble of atoms with initial action I_0 but uniform in action-angle variable to have chaotic classical trajectories while in the microwave field. For ionization and chaos thresholds so defined, Figure 2 shows the comparison [10,11], along with more recent ionization data obtained with the short pulse apparatus of Figure 1 [12]. For comparison purposes, the short pulse threshold field values have been multiplied by a factor of 0.70 in an attempt to account for the differences in pulse length. Although both sets of experimental data are not for the case of a pure beam of electrically polarized atoms, nevertheless the agreement with the predictions of the one dimensional classical model is excellent below $w_0=n^3w=1$. At higher microwave frequencies than this, the experimental thresholds are higher than the classical ones, suggesting a deviation due to quantum mechanics that might be the Anderson localization effect mentioned above. Such a deviation has been predicted to be observable for $w_0 = 1$ and above [13]. The short pulse data of Figure 2 do not definitely confirm this at present, as the normalization factor of 0.70 used for comparison has not been justified for the higher microwave frequencies.

The short pulse ionization data was obtained at a static electric field value of 0.87 V/cm within the microwave waveguide region that was produced by the atoms' motion in an externally applied uniform magnetic field. The dependence of the ionization threshold field on static field strength has been measured at constant bound state contribution cutoff quantum number value. As the static field is reduced from 6.25 V/cm to 0.87 V/cm, the ionization threshold rises in a way that depends on w_0. Thus the short pulse thresholds in Figure 2 should not be higher than that which would be obtained at the zero static field value assumed for the classical theoretical curve.

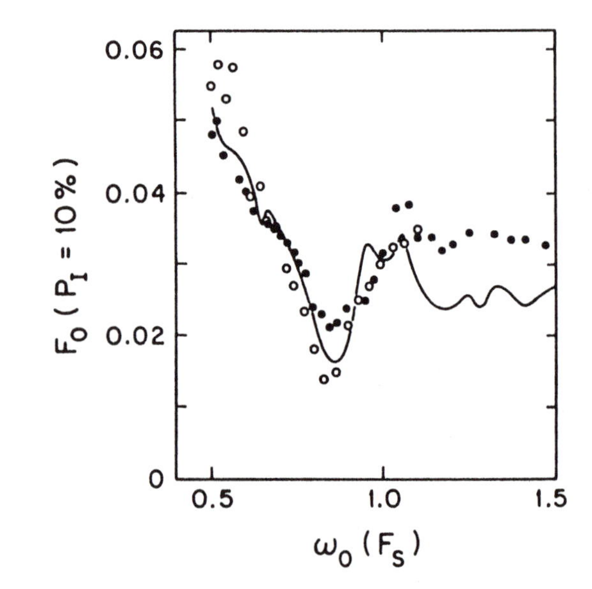

Fig. 2. Microwave field thresholds for classical chaos (solid line) and for observed 10% ionization probability ∘ [10-11], • [12]

Figure 3 shows a comparison of final bound state probability distributions for experiments with electrically polarized HEHA [14] and for purely classical predictions based on a one dimensional model [13]. The apparatus of Figure 1 was used for the experiments. A static field of 6.25 V/cm was included in both the theory and experiment. It is seen that the experimental final state distribution is near classical, having a large secondary maximum near n=66 away from the initial value of 72. For n above 72, the data drop on average exponentially with increasing n, a behavior predicted by the Fokker-Planck equation for the classical diffusion in action space associated with the random walk character of the chaotic electron trajectories [15].

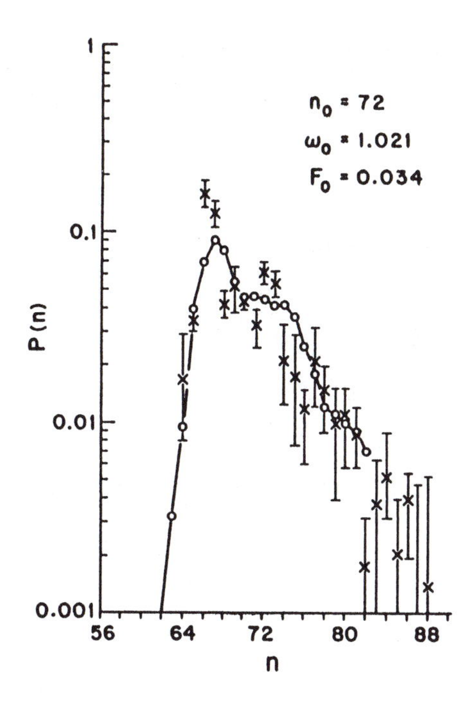

Fig. 3. Classical (circles [13]) and experimental (crosses, [14]) final bound state distributions

3.2 Experimental and Quantum Mechanical Localization Lengths in Action Space

The quantum theory of the dynamic Anderson localization predicts that as time evolves, the quantum distribution in action (quantum number) space first follows classical predictions, then differs from it after a quantum "break time", and finally stops changing after a quantum "freezing time" [4,15]. Let us characterize the exponential dependence of the final state distribution on increasing quantum number by the equation

$$P(n) = P_0 \exp\left[-2(n-n_0)/\ell\right], \quad n > n_0, \tag{1}$$

where ℓ (t) is the apparent localization length at the time t. Classically ℓ(t) grows indefinitely with time, while quantum mechanically it freezes to a constant value after a time that can be tens to hundreds of microwave periods. Figure 4 shows a comparison of experimental and quantum localization lengths obtained for microwave field strength, frequency and pulse time values theoretically expected to be in the quantum Anderson localization "freezing" regime [13, 16]. The experiments again used the apparatus of Figure 1 [14]. The agreement supports the predictions of numerical quantum mechanical calculations and suggests that dynamic Anderson localization may be a reality. However, definitive experimental verification of this requires a direct verification that the localization length freezes in time. New experiments at much longer microwave pulses times are needed to strongly discriminate between classical and quantum predictions for times much longer than the quantum freezing time.

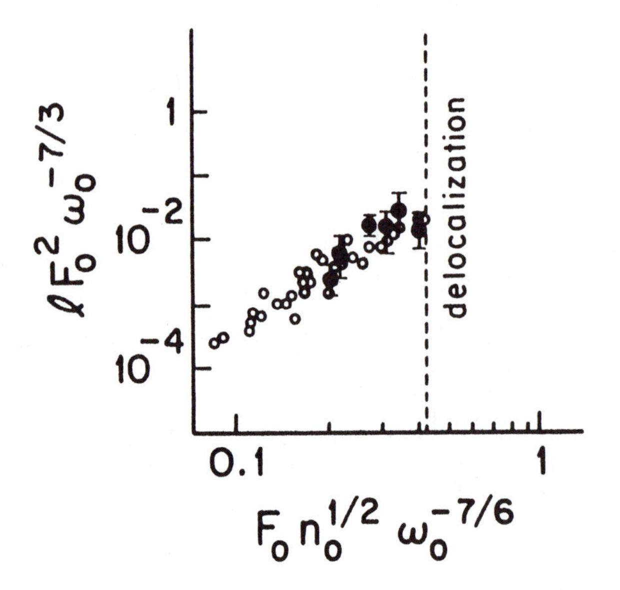

Fig. 4. Quantum mechanical [16] (open circles) and experimental values of scaled localization length [14]

We have seen that the hydrogen atom makes possible the study of three signatures of underlying classical chaos in semiclassical quantum systems, namely the Wigner nearest neighbor energy level separation distribution, the Fokker-Planck diffusive final state distribution and the dynamic Anderson localization effect. This progress promises major contributions to the development of the field of quantum dynamics.

1. G. Radons, R. E. Prange: "Wave Functions at the Critical Kolmogorov-Arnol'd-Moser Surface", Preprint, May 4, 1988, and references therein.
2. P. M. Koch, J. E. Bayfield: Phys. Rev. Lett. 34, 448 (1975)
3. J. E. Bayfield: Comments At. & Molec. Phys. 20, 245 (1987)
4. G. Casati, B. V. Chirikov, I. Guarneri, D. L. Shepelyansky: Phys. Reports 154, 77 (1987)
5. J. E. Bayfield, P. M. Koch: Phys. Rev. Lett. 33, 258 (1974)
6. J. G. Leopold, I. C. Percival: J. Phys. B 12, 709 (1979)
7. P. M. Koch, D. R. Mariani: Phys. Rev. Lett. 46, 1275 (1981)
8. J. E. Bayfield, L. A. Pinnaduwage: Phys. Rev. Lett. 54, 313 (1985)
9. J. E. Bayfield: in Quantum Measurement and Chaos, ed. by E. R. Pike and Sarben Sarkar (Plenum Press, 1987), page 1
10. R. V. Jensen: in Atomic Physics 10, ed. by H. Narumi and I. Shimamura (North-Holland, 1987), page 319
11. M. M. Sanders, R. V. Jensen, P. M. Koch, K. A. H. van Leeuwen: Nucl. Phys. B. (Proc. Suppl.) 2, 578 (1987)
12. D. W. Sokol, J. E. Bayfield: unpublished
13. G. Brivio, G. Casati, I. Guarneri, L. Perotti: Physica D, to be published
14. J. E. Bayfield, D. W. Sokol: submitted for publication
15. G. Casati, I. Guarneri, D. L. Shepelynasky: IEEE J. Quant. Electronics, to be published
16. G. Casati, B. V. Chirikov, D. L. Shepelyansky, Phys. Rev. Lett. 53, 2525 (1984)

Quasi-Landau Spectrum of the Chaotic Diamagnetic Hydrogen Atom

A. Holle, J. Main, G. Wiebusch, H. Rottke, and K.H. Welge

Fakultät für Physik, Universität Bielefeld,
D-4800 Bielefeld 1, Fed. Rep. of Germany

INTRODUCTION

The highly excited hydrogen atom in static magnetic fields has been in recent years a subject of intense experimental [1-4] and theoretical [4-9] studies which have led to substantial progress in the understanding of this previously unsolved elementary problem. Described by the Hamiltonian (in atomic units)

$$H = \frac{1}{2} p^2 + \frac{1}{2} \gamma L_z + \frac{1}{8} \gamma^2 \rho^2 - \frac{1}{r} \tag{1}$$

(cylindrical coordinates, $r = (\rho^2 + z^2)^{1/2}$; field parameter $\gamma = B/B_o$ with $B_o = 2.35 \times 10^5$ Tesla) the magnetized atom is of particular interest in the quasi-Landau regime of strong mixing of the Coulomb and diamagnetic interactions, i.e. where the two forces are of comparable strength. In this regime the motion of the Rydberg electron becomes classically chaotic [10]. It is this aspect which has recently attracted much attention, as the magnetized atom constitutes an ideal model case for detailed experimental studies of the quantum mechanics of a most simple atomic system in classical chaos.

Until recently it has been generally accepted that the physics of atoms in the quasi-Landau regime was essentially represented and determined by the quasi-Landau resonances discovered by Garton and Tomkins in the absorption spectrum of alkaline earth atoms [11] and explained first by Edmonds [12] as resulting from two-dimensional bound motion of the electron on closed classical orbits in the (z = 0)-plane perpendicular to the magnetic field axis. It was therefore with great surprise when new quasi-Landau resonances were unexpectedly discovered in first experiments with the hydrogen atom [1, 2] and that they are correlated to three-dimensional periodic orbits through the proton as origin [1, 2, 8]. In this paper we briefly present the state of the experimental investigation on the magnetized hydrogen atom.

The Hydrogen Atom Editors: G.F. Bassani · M. Inguscio · T.W. Hänsch

EXPERIMENTAL

Hydrogen atoms are excited by independently tunable pulsed laser light in two steps,

$$H(1s) + h\nu_1(vuv) \rightarrow H(2p) + h\nu_2(uv) \rightarrow H^* \tag{2}$$

in a crossed atom-laser beam arrangement (Fig. 1) at the center of the magnetic field, with the atomic beam directed parallel to the field axis. In the first step single sublevels of the Paschen-Back manifold ($m = 0, \pm 1$) of the 2p-state are selectivly prepared by linearly polarized vuv-laser light tunable in the region of the Lyman-α wavelength (121.6 nm). From there final states are excited with even parity and quantum numbers $|m| = 0$, 1, or 2 around the ionization limit by scanning the uv-laser, also linearly polarized. Further experimental details can be taken from previous papers [1-3].

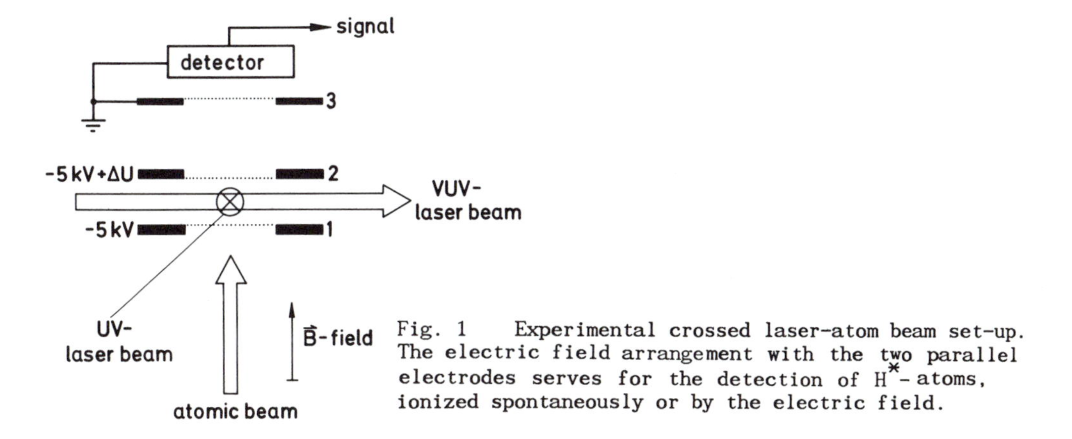

Fig. 1 Experimental crossed laser-atom beam set-up. The electric field arrangement with the two parallel electrodes serves for the detection of H^*- atoms, ionized spontaneously or by the electric field.

RESULTS AND DISCUSSION

Fig. 2 shows a spectrum taken at a magnetic field strength of B = 5.96 T and excited to final even parity states with magnetic quantum number m = 0. Also shown in Fig. 2 is a theoretical stick spectrum obtained by quantum mechanical calculation via numerical solution of the Schroedinger equation [4, 5]. Within the precision and resolution limits (1.5 GHz) of the experiment the theoretical and experimental spectra are in good agreement. More recently theoretical spectra have been calculated up to close to the ionization threshold [13].

Fig. 3a shows spectra with magnetic quantum numbers m = 0 and m = -1 in the chaotic regime from $-30\ cm^{-1}$ up to $+ 30\ cm^{-1}$ in the continuum region. At this resolution the energy quantum spectra have seemingly lost the oscillatory structures of quasi-Landau resonances discovered at lower resolution [1]. Periodic modulations

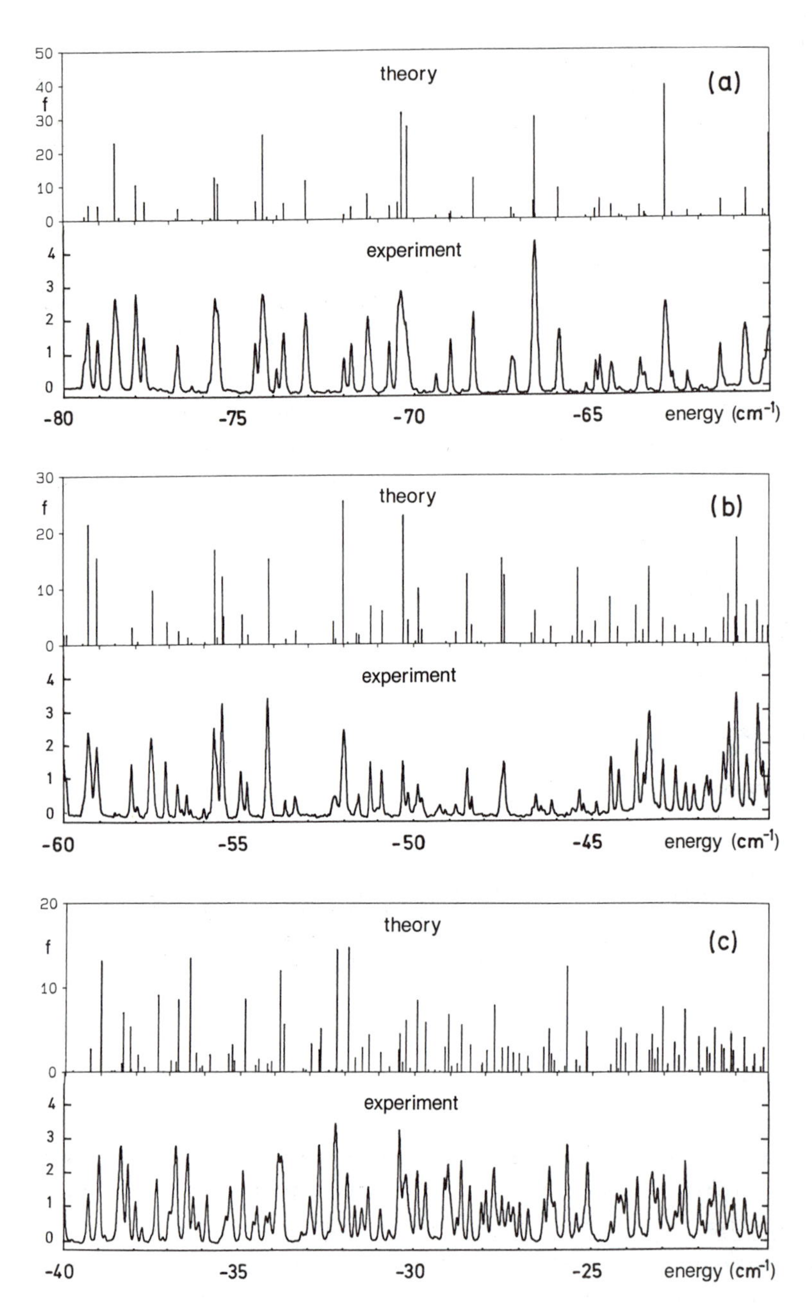

Fig. 2 Rydberg atoms in a magnetic field of 5.96 T: comparison between the theoretical oscillator strength spectrum and the experimental photoabsorption spectrum for $\Delta m = 0$ Balmer transitions to $m = 0$, even parity final states. Oscillator strengths are given in units of 10^{-6}, the experimental intensity scale is in arbitrary units.

(in line density and/or oscillator strength) in high resolution energy spectra can, however, be recovered by Fourier transformation in time-domain. Following Gutzwiller [14] such energy spectrum oscillations can be rationalized by long-range periodic orbits of the electron motion, also in the classically chaotic regime.

Fourier transformation of the spectra in Fig. 3a yields time-domain spectra (Fig. 3b) clearly exhibiting a number of resonances at times T_i, related to the energy modulation spacing $\Delta\epsilon_i$ by $\Delta\epsilon_i T_i = 2\pi\hbar$. The spectra in Fig. 3b exhibit a multitude of new resonances, in addition to the known Garton-Tomkins one. Each of them can be correlated to a closed classical three-dimensional orbit of the Rydberg electron through the proton as origin, with T_i the recurrence time of the respective orbit. The orbits shown in Fig. 3, plotted in (ρ,z)-projection, are members of a series of resonances ("fundamental" series) discussed and analysed in detail elsewhere [2, 8]. The Fourier spectra in Fig. 3b have recently also been treated theoreti-cally by Delos [9] on the basis of the periodic-orbit theory by Gutzwiller [14].

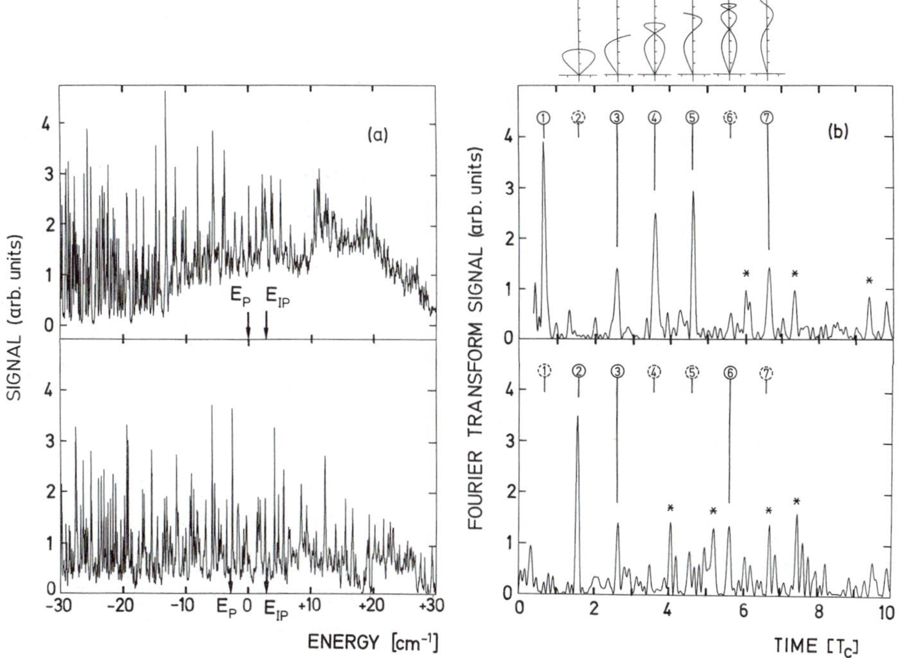

Fig. 3 (a) Excitation ionization spectra of H-atom even parity Balmer series around the ionization limit in a static homogenous magnetic field of B = 5.96 Tesla. Above: magnetic quantum number m = 0; below: m = - 1. (b) Fourier-transformed spectra of (a). Plotted is the absolute value squared. Abscissa with time scale normalized to cyclotron period T_c. Corresponding to resonances ② to ⑦ calculated closed classical orbits of electron motion are shown in (ρ,z)-projection.

The discovery of a multitude of new quasi-Landau resonances raises the basic problem as to the "entire" manifold of resonances arising as a function of the energy (E) and field strength (B) in the excitation of final states with given parity and magnetic quantum number. Spectra like in Fig. 3a taken at constant B as function of E can deliver resonances T_i in the Fourier transform only when the oscillation spacing $\Delta\epsilon_i$ in the energy quantum spectrum is not, or at most weakly, energy dependent within the integration intervall. We have solved this problem by employing "scaled-energy spectroscopy", which allows to observe also strongly energy dependent quasi-Landau resonances. Following theory [10], the technique is based on the scaling property of the Hamiltonian with respect to E and B. With the scaling relations

$$\tilde{\vec{p}} = \gamma^{-1/3}\vec{p},\ \tilde{\vec{r}} = \gamma^{2/3}\vec{r} \tag{3}$$

the Hamiltonian (eqn. 1) transforms to

$$\tilde{H} = \frac{1}{2}\tilde{\vec{p}}^{2} + \frac{1}{2}\tilde{L}_z + \frac{1}{8}\tilde{\rho}^2 - (\tilde{\rho}^2 + \tilde{z}^2)^{-1/2} \tag{4}$$

which is no more dependent on B and E independently, but on the scaled energy

$$\tilde{E} = E\gamma^{-2/3} \tag{5}$$

only. Transformation of the semi-classical quantization condition with the two non-separable coordinates ρ, z to scaled variables yields

$$\oint_i (\tilde{p}_z d\tilde{z} + \tilde{p}_\rho d\tilde{\rho}) \equiv C_i = n\gamma^{1/3} \tag{6}$$

where i denotes a given closed orbit. For $\tilde{E}$ = constant the action integral is constant, so that a spectrum taken on a scale linear in $\gamma^{-1/3}$ will consist of equidistant lines for each given i. Fourier transformation of the $\gamma^{-1/3}$-spectrum to conjugate coordinates $n\gamma^{1/3}$ thus results in one action resonance C_i only for each i. Fig. 4 shows, as an example, a scaled-energy spectrum taken at $\tilde{E} = -0.45$, and plotted as function of $\gamma^{-1/3}$. In taking this spectrum E and B have been varied simultaneously such that the condition $\tilde{E} = E\gamma^{-2/3}$ = constant was obeyed. The right part of the figure shows the action resonance spectrum obtained by the Fourier transformation of the left spectrum. Also shown are calculated closed orbits correlated to the respective C_i resonances. Such scaled-energy spectra have been measured over a range of $\tilde{E}$ from $\tilde{E} = -0.5$ to $\tilde{E} = +0.2$.

Fig. 5a shows in concise overlay fashion a set of action spectra taken at different scaled energies in steps $\Delta\tilde{E} = 0.01$, representing the evolution of the entire quasi-Landau spectrum as function of $\tilde{E}$ and C from the regular into the chaotic

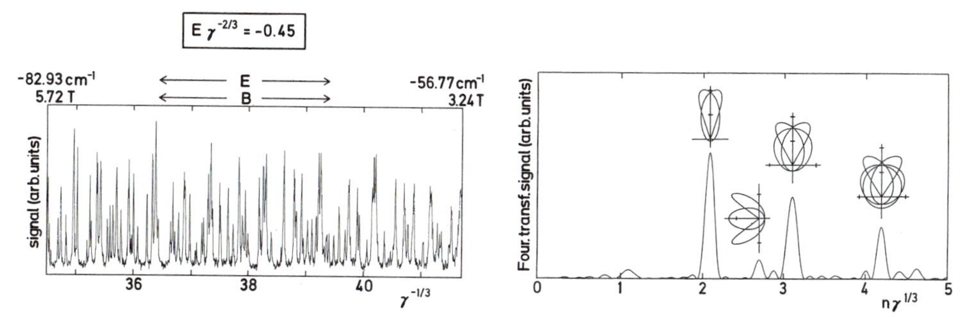

Fig. 4 (a) Scaled-energy spectrum at $\tilde{E} = -0.45$ as a linear function of $\gamma^{-1/3}$. (b) Fourier-transformed action spectrum of (a); closed orbits correlated to respective resonances in (ρ, z)-projection; z-coordinate vertically.

regime. Seen here for the first time it evolves as a remarkably well-structured, ordered system of branches and clusters of resonances.

To understand the experimental spectrum we have calculated the complete semiclassical action spectrum, that is the position of resonances in the $(\tilde{E}, C)$-plane correlated to periodic orbits with the proton as origin by numerical solution of the scaled quantization integral (6). The result is shown in Fig. 5b where each point represents a calculated resonance and thus a closed orbit.

We summarize the main results and observations derived from the experimental and theoretical quasi-Landau spectra in Fig. 5 and the corresponding trajectory calculation. A detailed discussion is given elsewhere [3]: 1) The experimental and theoretical spectra resemble each other in the overall $(\tilde{E}, C)$-dependence. They evolve from rootes of few "basic" resonances at low $\tilde{E}$ by bifurcation into branches which mix as $\tilde{E}$ increases. In the theoretical spectrum the density of resonances grows rapidly making identification of individual resonances in the experimental spectrum impossible in regions of high $\tilde{E}$ and C. On the other hand, however, the experimental spectrum remains distinctly structured and this despite the comparatively large width of the resonances. This fact shows that even in the classically fully chaotic regime only rather few resonances are preferentially excited or, in other words, correspondingly few periodic orbits dominate there the dynamics of the atom. 2) Three types of "basic" resonances with corresponding types of period orbits are identified: (a) "vibrators", one-dimensional orbits along the $(z = 0)$-plane, (b) "rotators", two-dimensional orbits in the $(z = 0)$-plane; and (c) "exotics", genuine three-dimensional orbits. They are labelled in Fig. 5, respectively, by V_μ, R_μ, X_μ $(\mu = 1, 2 \ldots)$. The basic vibrators and rotators constitute sets of resonances harmonic in the action with V_1 and R_1 being the fundamentals. From the V_μ's and R_μ's evolves the spectrum by bifurcation into higher generations of resonances with three-dimensional periodic orbits.

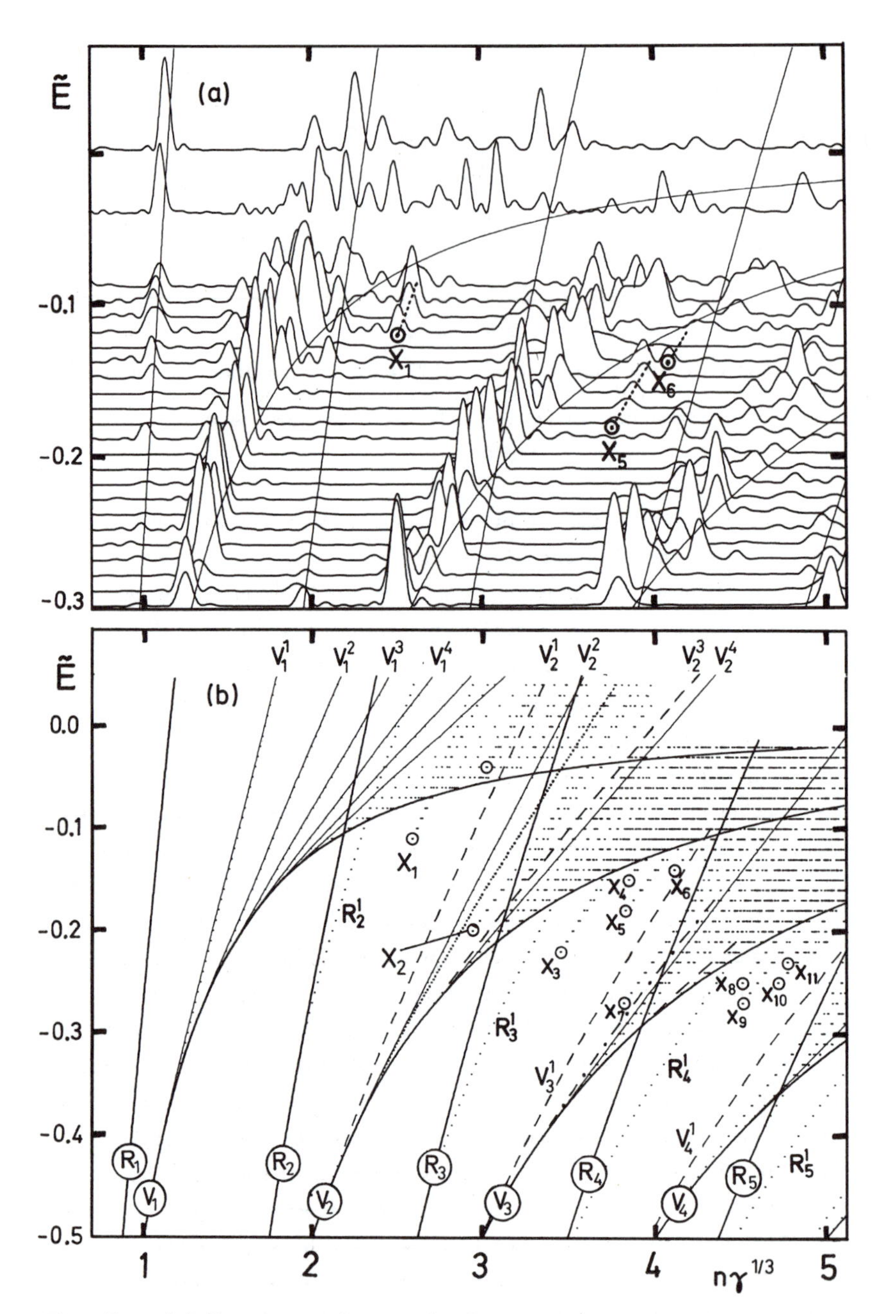

Fig. 5 (a) Experimental quasi-Landau resonance action spectrum as a function of scaled energy $\tilde{E}$ in overlay form. Even-parity, magnetic m = 0 final state. (b) Semiclassically calculated ($\tilde{E}$, C)-spectrum of quasi-Landau resonances correlated to closed classical orbits through origin.

The existence of "exotics" is experimentally discovered by observation of resonances (e.g. X_1, X_5, X_6) in regions of the ($\tilde{E}$,C)-plane not connected to vibrators or rotators. Main features and properties of them are: a) they occur apparently at random in the chaotic regime of the ($\tilde{E}$,C)-plane, increasing in density with $\tilde{E}$ and C, b) they are born at singular points and bifurcate right at origin, c) the correlated periodic orbits do not show any systematics or symmetry in their topology, which vibrators and rotators do [3], d) the X_1 exotic is peculiar in that it resides in a microscopically regular regime embedded in the otherwise fully chaotic regime.

In summary, studies during the last few years have revealed a much advanced picture of the physics of highly excited, magnetized atoms. Essentially new insight has been gained in the structure and dynamics of these systems under classically chaotic conditions by the discovery of a multitude of new quasi-Landau resonances with correlated three-dimensional periodic orbits and their evolution from the regular into the chaotic regime.

REFERENCES

[1] A. Holle, G. Wiebusch, J. Main, B. Hager, H. Rottke, and K.H. Welge, Phys. Rev. Lett. 56, 2594 (1986)

[2] J. Main, G. Wiebusch, A. Holle, and K.H. Welge, Phys. Rev. Lett. 57, 2789 (1986);
J. Main, A. Holle, G. Wiebusch, and K.H. Welge, Z. Phys. D6, 295 (1987)

[3] A. Holle, J. Main, G. Wiebusch, H. Rottke, and K.H. Welge, Phys. Rev. Lett. 61, 161 (1988)

[4] A. Holle, G. Wiebusch, J. Main, K.H. Welge, G. Zeller, G. Wunner, T. Erd, and H. Ruder, Z. Phys. D5, 279 (1987)

[5] G. Wunner, U. Woelk, I. Zech, G. Zeller, T. Ertl, F. Geyer, W. Schweizer, and H. Ruder, Phys. Rev. Lett. 57, 3261 (1986)

[6] D. Delande and J.C. Gay, Phys. Rev. Lett. 57, 2006 (1986), and 59, 1809 (1987)

[7] D. Wintgen and H. Friedrich, Phys. Rev. Lett. 57, 571 (1986) and Phys. Rev. A 36, 131 (1987)

[8] M.A. Al-Laithy, P.F. O'Mahony, and K.T. Taylor, J. Phys. B 19, L773 (1986)

[9] M.L. Du and J.B. Delos, Phys. Rev. Lett. 58, 1731 (1987)

[10] M. Robnik, J. Phys. A 14, 3195 (1981);
A. Harada and H. Hasegawa, J. Phys. A 16, L259 (1983)

[11] W. R. S. Garton and F. S. Tomkins, Astrophys. J. 158, 839 (1969)

[12] A.R. Edmonds, J. Phys. Pais 31 Colloque C 4, 71 (1970)

[13] G. Wunner and G. Zeller, private communication

[14] M.C. Gutzwiller, J. Math. Phys. 8, 1979 (1967) and 10, 1004 (1969), and 11, 1791 (1970), and 12, 343 (1971);
D. Wintgen, Phys. Rev. Lett. 58, 1589 (1987) and 61, 1803 (1988)

Index of Contributors